Graduate Texts in Mathematics
70

William S. Massey

Singular
Homology Theory

Springer-Verlag
New York Heidelberg Berlin

William S. Massey
Department of Mathematics
Yale University
New Haven, Connecticut 06520
USA

Editorial Board

AMS Subject Classifications (1980): 55-01, 55N10

With 13 Figures.

Library of Congress Cataloging in Publication Data

Massey, William S
    Singular homology theory.
    (Graduate texts in mathematics; 70)
    Bibliography: p.
    Includes index.
    1. Homology theory. I. Title. II. Series.
QA612.3.M36   514′.23   79-23309

9 8 7 6 5 4 3 2 1

ISBN 978-0-387-97430-9

Preface

The main purpose of this book is to give a systematic treatment of singular homology and cohomology theory. It is in some sense a sequel to the author's previous book in this Springer-Verlag series entitled *Algebraic Topology: An Introduction*. This earlier book is definitely not a logical prerequisite for the present volume. However, it would certainly be advantageous for a prospective reader to have an acquaintance with some of the topics treated in that earlier volume, such as 2-dimensional manifolds and the fundamental group.

Singular homology and cohomology theory has been the subject of a number of textbooks in the last couple of decades, so the basic outline of the theory is fairly well established. Therefore, from the point of view of the mathematics involved, there can be little that is new or original in a book such as this. On the other hand, there is still room for a great deal of variety and originality in the details of the exposition.

In this volume the author has tried to give a straightforward treatment of the subject matter, stripped of all unnecessary definitions, terminology, and technical machinery. He has also tried, wherever feasible, to emphasize the geometric motivation behind the various concepts.

In line with these principles, the author has systematically used singular cubes rather than singular simplexes throughout this book. This has several advantages. To begin with, it is easier to describe an n-dimensional cube than it is an n-dimensional simplex. Then since the product of a cube with the unit interval is again a cube, the proof of the invariance of the induced homomorphism under homotopies is very easy. Next, the subdivision of an n-dimensional cube is very easy to describe explicitly, hence the proof of the excision property is easier to motivate and explain than would be the case using singular simplices. Of course, it is absolutely necessary to factor out

the degenerate singular cubes. However, even this is an advantage: it means
that certain singular cubes can be ignored or neglected in our calculations.

Chapter I is not logically necessary in order to understand the rest of
the book. It contains a summary of some of the basic properties of homology
theory, and a survey of some problems which originally motivated the
development of homology theory in the nineteenth century. Reading it
should help the student understand the background and motivation for
algebraic topology.

Chapters II, III, and IV are concerned solely with singular homology with
integral coefficients, perhaps the most basic aspect of the whole subject.
Chapter II is concerned with the development of the fundamental properties,
Chapter III gives various examples and applications, and Chapter IV ex-
plains a systematic method of determining the integral homology groups of
certain spaces, namely, regular CW-complexes. Chapters II and III could
very well serve as the basis for a brief one term or one semester course in
algebraic topology.

In Chapter V, the homology theory of these early chapters is generalized
to homology with an arbitrary coefficient group. This generalization is
carried out by a systematic use of tensor products. Tensor products also play
a significant role in Chapter VI, which is about the homology of product
spaces, i.e., the Künneth theorem and the Eilenberg–Zilber theorem.

Cohomology theory makes its first appearance in Chapter VII. Much of
this chapter of necessity depends on a systematic use of the Hom functor.
However, there is also a discussion of the geometric interpretation of
cochains and cocycles. Then Chapter VIII gives a systematic treatment of
the various products which occur in this subject: cup, cap, cross, and slant
products. The cap product is used in Chapter IX for the statement and proof
of the Poincaré duality theorem for manifolds. Because of the relations
between cup and cap products, the Poincaré duality theorem imposes certain
conditions on the cup products in a manifold. These conditions are used in
Chapter X to actually determine cup products in real, complex, and quaternionic projective spaces. The knowledge of these cup products in projective
spaces is then applied to prove some classical theorems.

The book ends with an appendix devoted to a proof of De Rham's
theorem. It seemed appropriate to include it, because the methods used are
similar to those of Chapter IX.

Prerequisites. For most of the first four chapters, the only necessary
prerequisites are a basic knowledge of point set topology and the theory of
abelian groups. However, as mentioned earlier, it would be advantageous
to also know something about 2-dimensional manifolds and the theory of
the fundamental group as contained, for example, in the author's earlier
book in this Springer-Verlag series. Then, starting in Chapter V, it is assumed
that the reader has a knowledge of tensor products. At this stage we also
begin using some of the language of category theory, mainly for the sake of
convenience. We do not use any of the results or theorems of category theory,

however. In order to state and prove the so-called universal coefficient theorem for homology we give a brief introduction to the Tor functor, and references for further reading about it. Similarly, starting in Chapter VII it is assumed that the reader is familiar with the Hom functor. For the purposes of the universal coefficient theorem for cohomology we give a brief introduction to the Ext functor, and references for additional information about it. In order to be able to understand the appendix, the reader must be familiar with differential forms and differentiable manifolds.

Notation and Terminology. We will follow the conventions regarding terminology and notation that were outlined in the author's earlier volume in this Springer-Verlag series. Since most of these conventions are rather standard nowadays, it is probably not necessary to repeat all of them again.

The symbols $\mathbf{Z}$, $\mathbf{Q}$, $\mathbf{R}$, and $\mathbf{C}$ will be reserved for the set of all integers, rational numbers, real numbers, and complex numbers respectively. $\mathbf{R}^n$ and $\mathbf{C}^n$ will denote the space of all n-tuples of real and complex numbers respectively, with their usual topology. The symbols RP^n, CP^n, and QP^n are introduced in Chapter IV to denote n-dimensional real, complex, and quaternionic projective space respectively.

A homomorphism from one group to another is called an *epimorphism* if it is onto, a *monomorphism* if it is one-to-one, and an *isomorphism* if it is both one-to-one and onto. A sequence of groups and homomorphisms such as

$$\ldots \to A_{n-1} \xrightarrow{h_{n-1}} A_n \xrightarrow{h_n} A_{n+1} \to \ldots$$

is called *exact* if the kernel of each homomorphism is precisely the same as the image of the preceding homomorphism. Such exact sequences play a big role in this book.

A reference to Theorem or Lemma III.8.4 indicates Theorem or Lemma 4 in Section 8 of Chapter III; if the reference is simply to Theorem 8.4, then the theorem is in Section 8 of the same chapter in which the reference occurs. At the end of each chapter is a brief bibliography; numbers in square brackets in the text refer to items in the bibliography. The author's previous text, *Algebraic Topology: An Introduction* is often referred to by title above.

Acknowledgments. Most of this text has gone through several versions. The earlier versions were in the form of mimeographed or dittoed notes. The author is grateful to the secretarial staff of the Yale mathematics department for the careful typing of these various versions, and to the students who read and studied them—their reactions and suggestions have been very helpful. He is also grateful to his colleagues on the Yale faculty for many helpful discussions about various points in the book. Finally, thanks are due to the editor and staff of Springer-Verlag New York for their care and assistance in the production of this and the author's previous volume in this series.

New Haven, Connecticut   WILLIAM S. MASSEY
February, 1980

Contents

Chapter I

Background and Motivation for Homology Theory   1

§1. Introduction   1
§2. Summary of Some of the Basic Properties of Homology Theory   1
§3. Some Examples of Problems Which Motivated the Developement   3
of Homology Theory in the Nineteenth Century
§4. References to Further Articles on the Background and Motivation   10
for Homology Theory
Bibliography for Chapter I   10

Chapter II

Definitions and Basic Properties of Homology Theory   11

§1. Introduction   11
§2. Definition of Cubical Singular Homology Groups   11
§3. The Homomorphism Induced by a Continuous Map   16
§4. The Homotopy Property of the Induced Homomorphisms   19
§5. The Exact Homology Sequence of a Pair   22
§6. The Main Properties of Relative Homology Groups   26
§7. The Subdivision of Singular Cubes and the Proof of Theorem 6.3   31

Chapter III

Determination of the Homology Groups of Certain Spaces:
Applications and Further Properties of Homology Theory   38

§1. Introduction   38
§2. Homology Groups of Cells and Spheres—Application   38
§3. Homology of Finite Graphs   43

§4.   Homology of Compact Surfaces   53
§5.   The Mayer–Vietoris Exact Sequence   58
§6.   The Jordan–Brouwer Separation Theorem and   62
      Invariance of Domain
§7.   The Relation between the Fundamental Group and the First   69
      Homology Group
      Bibliography for Chapter III   75

Chapter IV
Homology of CW-complexes   76

§1.   Introduction   76
§2.   Adjoining Cells to a Space   76
§3.   CW-complexes   79
§4.   The Homology Groups of a CW-complex   84
§5.   Incidence Numbers and Orientations of Cells   89
§6.   Regular CW-complexes   94
§7.   Determination of Incidence Numbers for a Regular Cell Complex   95
§8.   Homology Groups of a Pseudomanifold   100
      Bibliography for Chapter IV   103

Chapter V
Homology with Arbitrary Coefficient Groups   105

§1.   Introduction   105
§2.   Chain Complexes   105
§3.   Definition and Basic Properties of Homology with   112
      Arbitrary Coefficients
§4.   Intuitive Geometric Picture of a Cycle with Coefficients in G 117
§5.   Coefficient Homomorphisms and Coefficient Exact Sequences   117
§6.   The Universal Coefficient Theorem   119
§7.   Further Properties of Homology with Arbitrary Coefficients   125
      Bibliography for Chapter V   128

Chapter VI
The Homology of Product Spaces   129

§1.   Introduction   129
§2.   The Product of CW-complexes and the Tensor Product of   130
      Chain Complexes
§3.   The Singular Chain Complex of a Product Space   132
§4.   The Homology of the Tensor Product of Chain Complexes   134
      (The Künneth Theorem)
§5.   Proof of the Eilenberg–Zilber Theorem   136
§6.   Formulas for the Homology Groups of Product Spaces   149
      Bibliography for Chapter VI   153

Chapter VII

Cohomology Theory   154

§1. Introduction   154
§2. Definition of Cohomology Groups—Proofs of the Basic Properties   155
§3. Coefficient Homomorphisms and the Bockstein Operator   158
     in Cohomology
§4. The Universal Coefficient Theorem for Cohomology Groups   159
§5. Geometric Interpretation of Cochains, Cocycles, etc.   165
§6. Proof of the Excision Property; the Mayer–Vietoris Sequence   168
     Bibliography for Chapter VII   171

Chapter VIII

Products in Homology and Cohomology   172

§1. Introduction   172
§2. The Inner Product   173
§3. An Overall View of the Various Products   173
§4. Extension of the Definition of the Various Products to   178
     Relative Homology and Cohomology Groups
§5. Associativity, Commutativity, and Existence of a   182
     Unit for the Various Products
§6. Digression: The Exact Sequence of a Triple or a Triad   185
§7. Behavior of Products with Respect to the Boundary and   187
     Coboundary Operator of a Pair
§8. Relations Involving the Inner Product   190
§9. Cup and Cap Products in a Product Space   191
§10. Remarks on the Coefficients for the Various
     Products—The Cohomology Ring   192
§11. The Cohomology of Product Spaces (The Künneth Theorem   193
     for Cohomology)
     Bibliography for Chapter VIII   198

Chapter IX

Duality Theorems for the Homology of Manifolds   199

§1. Introduction   199
§2. Orientability and the Existence of Orientations for Manifolds   200
§3. Cohomology with Compact Supports   206
§4. Statement and Proof of the Poincaré Duality Theorem   207
§5. Applications of the Poincaré Duality Theorem to Compact Manifolds   214
§6. The Alexander Duality Theorem   218
§7. Duality Theorems for Manifolds with Boundary   224
§8. Appendix: Proof of Two Lemmas about Cap Products   228
     Bibliography for Chapter IX   238

Chapter X

Cup Products in Projective Spaces and Applications of Cup Products   239

§1. Introduction   239
§2. The Projective Spaces   239
§3. The Mapping Cylinder and Mapping Cone   244
§4. The Hopf Invariant   247
     Bibliography for Chapter X   250

Appendix

A Proof of De Rham's Theorem   251

§1. Introduction   251
§2. Differentiable Singular Chains   252
§3. Statement and Proof of De Rham's Theorem   256
     Bibliography for the Appendix   261

Index   263

CHAPTER I

Background and Motivation for Homology Theory

§1. Introduction

Homology theory is a subject whose development requires a long chain of definitions, lemmas, and theorems before it arrives at any interesting results or applications. A newcomer to the subject who plunges into a formal, logical presentation of its ideas is likely to be somewhat puzzled because he will probably have difficulty seeing any motivation for the various definitions and theorems. It is the purpose of this chapter to present some explanation, which will help the reader to overcome this difficulty. We offer two different kinds of material for background and motivation. First, there is a summary of some of the most easily understood properties of homology theory, and a hint at how it can be applied to specific problems. Secondly, there is a brief outline of some of the problems and ideas which lead certain mathematicians of the nineteenth century to develop homology theory.

It should be emphasized that the reading of this chapter is *not* a logical prerequisite to the understanding of anything in later chapters of this book.

§2. Summary of Some of the Basic Properties of Homology Theory

Homology theory assigns to any topological space X a sequence of abelian groups $H_0(X)$, $H_1(X)$, $H_2(X)$, ..., and to any continuous map $f: X \to Y$ a sequence of homomorphisms

$$f_*: H_n(X) \to H_n(Y), \qquad n = 0, 1, 2, \ldots.$$

1

$H_n(X)$ is called the *n-dimensional homology group of X*, and f_* is called *the homomorphism induced by f*. We will list in more or less random order some of the principal properties of these groups and homomorphisms.

(a) If $f : X \to Y$ is a homeomorphism of X onto Y, then the induced homomorphism $f_* : H_n(X) \to H_n(Y)$ is an isomorphism for all n. Thus the algebraic structure of the groups $H_n(X)$, $n = 0, 1, 2, \ldots$, depends only on the topological type of X. In fact, an even stronger statement holds: if f is a homotopy equivalence[1], then f_* is an isomorphism. Thus the structure of $H_n(X)$ only depends on the homotopy type of X. Two spaces of the same homotopy type have isomorphic homology groups (for the definition of these terms, the reader is referred to *Algebraic Topology: An Introduction*, Chapter 2, §4 and §8).

(b) If two maps f_0, $f_1 : X \to Y$ are homotopic[2], then the induced homomorphisms f_{0*} and $f_{1*} : H_n(X) \to H_n(Y)$ are the same for all n. Thus the induced homomorphism f_* only depends on the homotopy class of f. By its use, we can sometimes prove that certain maps are *not* homotopic.

(c) For any space X, the group $H_0(X)$ is free abelian, and its rank is equal to the number of arcwise connected components of X. In other words, $H_0(X)$ has a basis in 1-1 correspondence with the set of arc-components of X. Thus the structure of $H_0(X)$ has to do with the arcwise connectedness of X. By analogy, the groups $H_1(X)$, $H_2(X)$, ... have something to do with some kind of higher connectivity of X. In fact, one can look on this as one of the principal purposes for the introduction of the homology groups: to express what may be called the higher connectivity properties of X.

(d) If X is an arcwise connected space, the 1-dimensional homology group, $H_1(X)$, is the abelianized fundamental group. In other words, $H_1(X)$ is isomorphic to $\pi(X)$ modulo its commutator subgroup.

(e) If X is a compact, connected, orientable n-dimensional manifold, then $H_n(X)$ is infinite cyclic, and $H_q(X) = \{0\}$ for all $q > n$. In some vague sense, such a manifold is a prototype or model for nonzero n-dimensional homology groups.

(f) If X is an open subset of Euclidean n-space, then $H_q(X) = \{0\}$ for all $q \geq n$.

We have already alluded to the fact that sometimes it is possible to use homology theory to prove that two continuous maps are not homotopic. Analogously, homology groups can sometimes be used to prove that two spaces are not homeomorphic, or not even of the same homotopy type. These are rather obvious applications. In other cases, homology theory is used in less obvious ways to prove theorems. A nice example of this is the proof of the Brouwer fixed point theorem in Chapter III, §2. More subtle examples are the Borsuk–Ulam theorem in Chapter X, §2 and the Jordan–Brouwer separation theorem in Chapter III, §6.

---

[1] This term is defined in Chapter II, §4.

[2] For the definition, see Chapter II, §4.

§3. Some Examples of Problems which Motivated the Development of Homology Theory in the Nineteenth Century

The problems we are going to consider all have to do with line integrals, surface integrals, etc., and theorems relating these integrals, such as the well-known theorems of Green, Stokes, and Gauss. We assume the reader is familiar with these topics.

As a first example, consider the following problem which is discussed in most advanced calculus books. Let U be an open, connected set in the plane, and let $\mathbf{V}$ be a vector field in U (it is assumed that the components of $\mathbf{V}$ have continuous partial derivatives in U). Under what conditions does there exist a "potential function" for $\mathbf{V}$, i.e., a differentiable function $F(x,y)$ such that $\mathbf{V}$ is the gradient of F? Denote the x and y components of $\mathbf{V}$ by $P(x,y)$ and $Q(x,y)$ respectively; then an obvious necessary condition is that

$$\frac{\partial P}{\partial y} = \frac{\partial Q}{\partial x}$$

at every point of U. If the set U is convex, then this necessary condition is also sufficient. The standard proof of sufficiency is based on the use of Green's theorem, which asserts that

$$\oint_C P\,dx + Q\,dy = \iint_D \left(\frac{\partial Q}{\partial x} - \frac{\partial P}{\partial y} \right) dx\,dy.$$

Here D is a domain with piece-wise smooth boundary C (which may have several components) such that D and C are both contained in U. By using Green's theorem, one can prove that the line integral on the left-hand side vanishes if C is any closed curve in U. This implies that if (x_0,y_0) and (x,y) are any two points of U, and L is any piece-wise smooth path in U joining (x_0,y_0) and (x,y), then the line integral

$$\int_L P\,dx + Q\,dy$$

is independent of the choice of L; it only depends on the end points (x_0,y_0) and (x,y). If we hold (x_0,y_0) fixed, and define $F(x,y)$ to be the value of this line integral for any point (x,y) in U, then $F(x,y)$ is the desired potential function.

On the other hand, if the open set U is more complicated, the necessary condition $\partial P/\partial y = \partial Q/\partial x$ may not be sufficient. Perhaps the simplest example to illustrate this point is the following: Let U denote the plane with the origin deleted,

$$P = -\frac{y}{x^2 + y^2} \quad \text{and} \quad Q = \frac{x}{x^2 + y^2}.$$

Then the condition $\partial Q/\partial x = \partial P/\partial y$ is satisfied at each point of U. However, if we compute the line integral

$$\int_C P\,dx + Q\,dy, \tag{1}$$

where C is a circle with center at the origin, we obtain the value 2π. Since $2\pi \neq 0$, there cannot be any potential function for the vector field $\mathbf{V} = (P,Q)$ in the open set U. It is clear where the preceding proof breaks down in this case: the circle C (with center at the origin) does not bound any domain D such that $D \subset U$.

Since the line integral (1) may be nonzero in this case, we may ask, What are all possible values of this line integral as C ranges over all piece-wise smooth closed curves in U? The answer is $2n\pi$, where n ranges over all integers, positive or negative. Indeed, any of these values may be obtained by integrating around the unit circle with center at the origin an appropriate number of times in the clockwise or counter-clockwise direction; and an informal argument using Green's theorem should convince the reader that these are the only possible values.

We can ask the same question for any open, connected set U in the plane, and any continuously differentiable vector field $\mathbf{V} = (P,Q)$ in U satisfying the condition $\partial P/\partial y = \partial Q/\partial x$: What are all possible values of the line integral (1) as C ranges over all piece-wise smooth closed curves in U? Anybody who studies this problem will quickly come to the conclusion that the answer depends on the number of "holes" in the set U. Let us associate with each hole the value of the integral (1) in the case where C is a closed path which goes around the given hole exactly once, and does not encircle any other hole (assuming such a path exists). By analogy with complex function theory, we will call this number the *residue* associated with the given hole. The answer to our problem then is that the value of the integral (1) is some finite, integral linear combination of these residues, and any such finite integral linear combination actually occurs as a value.

Next, let us consider the analogous problem in 3-space: we now assume that U is an open, connected set in 3-space, and $\mathbf{V}$ is a vector field in U with components $P(x,y,z)$, $Q(x,y,z)$, and $R(x,y,z)$ (which are assumed to be continuously differentiable in U). Furthermore, we assume that curl $\mathbf{V} = 0$. In terms of the components, this means that the equations

$$\frac{\partial R}{\partial y} = \frac{\partial Q}{\partial z}, \qquad \frac{\partial P}{\partial z} = \frac{\partial R}{\partial x}, \quad \text{and} \quad \frac{\partial Q}{\partial x} = \frac{\partial P}{\partial y}$$

hold at each point of U. Once again it can be shown that if U is convex, then there exists a function $F(x,y,z)$ such that $\mathbf{V}$ is the gradient of F. The proof is much the same as the previous case, except that now one must use Stokes's theorem rather than Green's theorem to show that the line integral

$$\int P\,dx + Q\,dy + R\,dz$$

is independent of the path.

In case the domain U is not convex, this proof may break down, and it can actually happen that the line integral

$$\oint_C P\,dx + Q\,dy + R\,dz \tag{2}$$

is nonzero for some closed path C in U. Once again we can ask: What are all possible values of the line integral (2) for all possible closed paths in U? The "holes" in U are again what makes the problem interesting; however, in this case there seem to be different kinds of holes. Let us consider some examples:

(a) Let $U = \{(x,y,z)\,|\,x^2 + y^2 > 0\}$, i.e., U is the complement of the z-axis. This example is similar to the 2-dimensional case treated earlier. If C denotes a circle in the xy-plane with center at the origin, we could call the value of the integral (2) with this choice of C the residue corresponding to the hole in U. Then the value of the integral (2) for any other choice of C in U would be some integral multiple of this residue; the reader should be able to convince himself of this in any particular case by using Stokes's theorem.

(b) Let U be the complement of the origin in $\mathbf{R}^3$. If Σ is any piece-wise smooth orientable surface in U with boundary C consisting of one or more piece-wise smooth curves, then according to Stokes's theorem,

$$\oint_C P\,dx + Q\,dy + R\,dz = \iint_\Sigma \left(\frac{\partial R}{\partial y} - \frac{\partial Q}{\partial z}\right)dy\,dz$$
$$+ \left(\frac{\partial P}{\partial z} - \frac{\partial R}{\partial x}\right)dz\,dx + \left(\frac{\partial Q}{\partial x} - \frac{\partial P}{\partial y}\right)dx\,dy.$$

We leave it to the reader to convince himself that any piece-wise smooth closed curve c in U is the boundary of such a surface Σ, hence by Stokes's theorem, the integral around such a curve is zero (the integral on the right-hand side is identically zero). Thus the same argument applies as in the case where U is convex to show that any vector field $\mathbf{V}$ in U such that curl $\mathbf{V} = 0$ in U is of the form $\mathbf{V} = \text{grad } F$ for some function F. The existence of the hole in U does not matter in this case.

(c) It is easy to give other examples of domains in 3-space with holes in them such that the hole does not matter. The following are such examples: let $U_1 = \{(x,y,z)\,|\,x^2 + y^2 + z^2 > 1\}$; let U_2 be the complement of the upper half ($z \geq 0$) of the z-axis; and let U_3 be the complement of a finite set of points in 3-space. In each case, if $\mathbf{V}$ is a vector field in U_i such that curl $\mathbf{V} = 0$, then $\mathbf{V} = \text{grad } F$ for some function F. The basic reason is that any closed curve C in U_i is the boundary of some oriented surface Σ in U_i in each of the cases $i = 1, 2$, or 3.

There is another problem for 3-dimensional space which involves closed surfaces rather than closed curves. It may be phrased as follows: Let U be a connected open set in $\mathbf{R}^3$ and let $\mathbf{V}$ be a continuously differentiable vector field in U such that div $\mathbf{V} = 0$. Is the integral of (the normal component of) $\mathbf{V}$ over any closed, orientable piece-wise smooth surface Σ in U equal to 0?

If not, what are the possible values of the integral of **V** over any such closed surface? If U is a convex open set, then any such integral of 0. One proves this by the use of Gauss's theorem (also called the divergence theorem):

$$\iint_{\Sigma} \mathbf{V} = \iiint_{D} (\text{div } \mathbf{V}) \, dx \, dy \, dz.$$

Here D is a domain in U with piece-wise smooth boundary Σ (the boundary may have several components). The main point is that a closed orientable surface Σ contained in a convex open set U is always the boundary of a domain D contained in U. However, if the open set U has holes in it, this may not be true, and the situation is more complicated. For example, suppose that U is the complement of the origin in 3-space, and **V** is the vector field in U with components $P = x/r^3$, $Q = y/r^3$, and $R = z/r^3$, where $r = (x^2 + y^2 + z^2)^{1/2}$ is the distance from the origin. It is readily verified that div $\mathbf{V} = 0$; on the other hand, the integral of **V** over any sphere with center at the origin is readily calculated to be $\pm 4\pi$; the sign depends on the orientation conventions. The set of all possible values of the surface integral $\iint_{\Sigma} \mathbf{V}$ for all closed, orientable surfaces Σ in U is precisely the set of all integral multiples of 4π.

On the other hand, if U is the complement of the z-axis in 3-space, then the situation is exactly the same as in the case where U is convex. The reason is that any closed, orientable surface in U bounds a domain D in U; the existence of the hole in U does not matter.

There is a whole series of analogous problems in Euclidean spaces of dimension four or more. Also, one could consider similar problems on curved submanifolds of Euclidean space. Although there would doubtless be interesting new complications, we have already presented enough examples to give the flavor of the subject.

At some point in the nineteenth century certain mathematicians tried to set up general procedures to handle problems such as these. This led them to introduce the following terminology and definitions. The closed curves, surfaces, and higher dimensional manifolds over which one integrates vector fields, etc. were called *cycles*. In particular, a closed curve is a 1-dimensional cycle, a closed surface is a 2-dimensional cycle, and so on. To complete the picture, a 0-dimensional cycle is a point. It is understood, of course, that cycles of dimension > 0 always have a definite orientation, i.e., a 2-cycle is an oriented closed surface. Moreover, it is convenient to attach to each cycle a certain integer which may be thought of as its "multiplicity." To integrate a vector field over a 1-dimensional cycle or closed curve with multiplicity $+3$ means to integrate it over a path going around the curve 3 times; the result will be 3 times the value of the integral going around it once. If the multiplicity is -3, then one integrates 3 times around the curve in the opposite direction. If the symbol c denotes a 1-dimensional cycle, then the symbol $3c$ denotes this cycle with the multiplicity $+3$, and $-3c$ denotes the same cycle with multiplicity -3. It is also convenient to allow formal sums and linear combinations of cycles (all of the same dimension), that is, expressions like $3c_1 + 5c_2 - 10c_3$, where c_1, c_2, and c_3 are cycles. With this definition of addition, the set of all n dimensional cycles in an open set U of Euclidean space becomes an abelian group; in fact it is a free abelian group. It is customary to denote this group by $Z_n(U)$. There is one further convention that is understood here: If c is the 1-dimensional cycle determined by a certain oriented closed curve, and c' denotes the cycle determined by the same curve with the opposite orientation, then $c = -c'$. This is consistent with the fact that the integral of a vector field over c' is the negative of the integral over c. Of course, the same convention also holds for higher dimensional cycles.

It is important to point out that 1-dimensional cycles are only assumed to be closed curves, they are not assumed to be *simple* closed curves. Thus they may have various self-intersections or singularities. Similarly, a 2-dimensional cycle in U is an oriented surface in U which is allowed to have various self-intersections or singularities. It is really a continuous (or differentiable) mapping of a compact, connected, oriented 2-manifold into U. On account of the possible existence of self-intersections or singularities, these cycles are often called *singular cycles*.

Once one knows how to define the integral of a vector field (or differential form) over a cycle, it is obvious how to define the integral over a formal linear combination of cycles. If $c_1, \ldots, c_k$ are cycles in U and

$$z = n_1 c_1 + \cdots + n_k c_k$$

where $n_1, n_2, \ldots, n_k$ are integers, then

$$\int_z \mathbf{V} = \sum_{i=1}^k n_i \int_{c_i} \mathbf{V}$$

for any vector field $\mathbf{V}$ in U.

The next step is to define an equivalence relation between cycles. This equivalence relation is motivated by the following considerations. Assume that U is an open set in 3-space.

(a) Let u and w be 1-dimensional cycles in U, i.e., u and w are elements of the group $Z_1(U)$. Then we wish to define $u \sim w$ so that this implies

$$\int_u \mathbf{V} = \int_w \mathbf{V}$$

for *any* vector field $\mathbf{V}$ in U such that curl $\mathbf{V} = 0$.

(b) Let u and w be elements of the group $Z_2(U)$. Then we wish to define $u \sim w$ so that this implies

$$\int_u \mathbf{V} = \int_w \mathbf{V}$$

for any vector field $\mathbf{V}$ in U such that div $\mathbf{V} = 0$.

Note that the condition

$$\int_u \mathbf{V} = \int_w \mathbf{V}$$

can be rewritten as follows, in view of our conventions:

$$\int_{u-w} \mathbf{V} = 0.$$

Thus $u \sim w$ if and only if $u - w \sim 0$.

In Case (a), Stokes's theorem suggests the proper definition, while in Case (b) the divergence theorem points the way.

We will discuss Case (a) first. Suppose we have an oriented surface in U whose boundary consists of the oriented closed curves $c_1, c_2, \ldots, c_k$. The orientations of the boundary curves are determined according to the conventions used in the statement of Stokes's theorem. Then the 1-dimensional cycle

$$z = c_1 + c_2 + \cdots + c_k$$

is defined to be *homologous to zero*, written

$$z \sim 0.$$

More generally, any linear combination of cycles homologous to zero is also defined to be homologous to zero. The set of all cycles homologous to zero is a subgroup of $Z_1(U)$ which is denoted by $B_1(U)$. We define z and z' to be homologous (written $z \sim z'$) if and only if $z - z' \sim 0$. Thus the set of equivalence classes of cycles, called *homology classes*, is nothing other than the quotient group

$$H_1(U) = Z_1(U)/B_1(U)$$

which is called the 1-dimensional *homology group* of U.

Analogous definitions apply to Case (b). Let D be a domain in U whose boundary consists of the connected oriented surfaces $s_1, s_2, \ldots, s_k$. The orientation of the boundary surfaces is determined by the conventions used for the divergence theorem. Then the 2-dimensional cycle

$$z = s_1 + s_2 + \cdots + s_k$$

is by definition homologous to zero, written $z \sim 0$. As before, any linear combination of cycles homologous to zero is also defined to be homologous to 0, and the set of cycles homologous to 0 constitutes a subgroup, $B_2(U)$, of $Z_2(U)$. The quotient group

$$H_2(U) = Z_2(U)/B_2(U)$$

is called the 2-dimensional *homology group* of U.

Let us consider some examples. If U is an open subset of the plane, then $H_1(U)$ is a free abelian group, and it has a basis (or minimal set of generators) in 1-1 correspondence with the holes in U. If U is an open subset of 3-space then both $H_1(U)$ and $H_2(U)$ are free abelian groups, and each hole in U contributes generators to $H_1(U)$ or $H_2(U)$, or perhaps to both. This helps explain the different kinds of holes in this case.

In principle, there is nothing to stop us from generalizing this procedure, and defining for any topological space X and nonnegative integer n the group $Z_n(X)$ of n-dimensional cycles in X, the subgroup $B_n(X)$ consisting of cycles which are homologous to zero, and the quotient group

$$H_n(X) = Z_n(X)/B_n(X),$$

called the n-dimensional homology group of X. However, there are difficulties in formulating the definitions rigorously in this generality; the reader may have noticed that some of the definitions in the preceding pages were lacking in precision. Actually, it took mathematicians some years to surmount these difficulties. The key idea was to think of an n-dimensional cycle as made up of small n-dimensional pieces which fit together in the right way, in much the same way that bricks fit together to make a wall. In this book, we will use n-dimensional cycles that consist of n-dimensional cubes which fit together in a nice way. To be more precise, the "singular" cycles will be built from "singular" cubes; a singular n-cube in a topological space X is simply a continuous map $T : I^n \to X$, where I^n denotes the unit n-cube in Euclidean n-space.

There is another complication which should be pointed out. We mentioned in connection with the examples above that if U is an open subset of the plane or 3-space, then the homology groups of U are free abelian groups. However, there exist open subsets U of Euclidean n-space for all $n > 3$ such that the group $H_1(U)$ contains elements of finite order (compare the discussion of the homology groups of nonorientable surfaces in §III.4). Suppose that $u \in H_1(U)$ is a homology class of order $k \neq 0$. Let z be a 1-dimensional cycle in the homology class u. Then z is not homologous to 0, but $k \cdot z$ is homologous to 0. This implies that if $\mathbf{V}$ is any vector field in U such that curl $\mathbf{V} = 0$, then

$$\int_z \mathbf{V} = 0.$$

To see this, let $\int_z \mathbf{V} = r$. Then $\int_{kz} \mathbf{V} = k \cdot r$; but $\int_{kz} \mathbf{V} = 0$ since $kz \sim 0$. Therefore $r = 0$. It is not clear that this phenomenon was understood in the nineteenth century; at least there seems to have been some confusion in Poincaré's early papers on topology about this point. Of course one source of difficulty is the fact that this phenomenon eludes our ordinary geometric intuition, since it does not occur in 3-dimensional space. Nevertheless it is a phenomenon of importance in algebraic topology.

Before ending this account, we should make clear that we do not claim that the nineteenth century development of homology theory actually proceeded along the lines we have just described. For one thing, the nineteenth century mathematicians involved in this development were more interested in complex analysis than real analysis. Moreover, many of their false starts and tentative attempts to establish the subject can only be surmised from reading the published papers which have survived to the present. The reader who wants to go back to the original sources is referred to the papers by

Riemann [7], E. Betti [1], and Poincaré [6]. Betti was a professor at the University of Pisa who became acquainted with Riemann in the last years of the latter's life. Presumably he was strongly influenced by Riemann's ideas on this subject.

§4. References to Further Articles on the Background and Motivation for Homology Theory

The student will probably find it helpful to read further articles on this subject. The following are recommended (most of them are easy reading): Seifert and Threlfall [8], Massey [5], Wallace [9], and Hocking and Young [4]. The bibliographies in Blackett [2] and Frechet and Fan [3] list many additional articles which are helpful and interesting.

Bibliography for Chapter I

[1] E. Betti, Sopra gli spazi di un numero qualunque di dimensioni, *Ann. Mat. Pura Appl.* **4** (1871), 140–158.
[2] D. W. Blackett, *Elementary Topology, A Combinatorial and Algebraic Approach*, Academic Press, New York, 1967, p. 219.
[3] M. Frechet and K. Fan, *Initiation to Combinatorial Topology*, Prindle, Weber, and Schmidt, Boston, 1967, 113–119.
[4] J. G. Hocking and G. S. Young, *Topology*, Addison-Wesley, Reading, 1961, 218–222.
[5] W. S. Massey, *Algebraic Topology: An Introduction*, Springer-Verlag, New York, 1977, Chapter 8.
[6] H. Poincaré, Analysis situs; 1ere complement a l' analysis situs; 5^{ieme} complement a l' analysis situs, *Collected Works*, Gauthier-Villars, Paris 1953, vol. VI.
[7] G. F. B. Riemann, Fragment aus der Analysis Situs, *Gesammelte Mathematische Werke* (2nd edition), Dover, New York, 1953, 479–483.
[8] H. Seifert and W. Threlfall, *Lehrbuch der Topologie*, Chelsea Publishing Co., New York, 1947, Chapter I.
[9] A. H. Wallace, *An Introduction to Algebraic Topology*, Pergamon Press, Elmsford, 1957, 92–95.

CHAPTER II

Definitions and Basic Properties of Homology Theory

§1. Introduction

This chapter gives formal definitions of the basic concepts of homology theory, and rigorous proofs of their basic properties. For the most part, examples and applications are postponed to Chapter III and subsequent chapters.

§2. Definition of Cubical Singular Homology Groups

First, we list some terminology and notation which will be used from here on:

$\mathbf{R}$ = real line.
I = closed unit interval, $[0,1]$.
$\mathbf{R}^n = \mathbf{R} \times \mathbf{R} \times \cdots \times \mathbf{R}$ (n factors, $n > 0$) Euclidean n-space.
$I^n = I \times I \times \cdots \times I$ (n factors, $n > 0$) unit n-cube.

By definition, I^0 is a space consisting of a single point.

Any topological space homeomorphic to I^n may be called an n-dimensional cube.

Definition 2.1. A *singular n-cube* in a topological space X is a continuous map $T: I^n \to X$ ($n \geq 0$).

Note the special cases $n = 0$ and $n = 1$.

$Q_n(X)$ denotes the free abelian group generated by the set of all singular n-cubes in X. Any element of $Q_n(X)$ has a unique expression as a finite linear combination with integral coefficients of n-cubes in X.

Definition 2.2. A singular n-cube $T:I^n \to X$ is *degenerate* if there exists an integer i, $1 \leq i \leq n$, such that $T(x_1,x_2,\ldots,x_n)$ does not depend on x_i.

Note that a singular 0-cube is never degenerate; a singular 1-cube $T:I \to X$ is degenerate if and only if T is a constant map.

Let $D_n(X)$ denote the subgroup of $Q_n(X)$ generated by the degenerate singular n-cubes, and let $C_n(X)$ denote the quotient group $Q_n(X)/D_n(X)$. The latter is called the group of *cubical singular n-chains in X*, or just *n-chains in X* for simplicity.

Remarks. If $X = \emptyset$, the empty set, then $Q_n(X) = D_n(X) = C_n(X) = \{0\}$ for all $n \geq 0$.

If X is a space consisting of a single point, then there is a unique singular n-cube in X for all $n \geq 0$; this unique n-cube is degenerate if $n \geq 1$. Hence $C_0(X)$ is an infinite cyclic group and $C_n(X) = \{0\}$ for $n > 0$ in this case.

For any space X, $D_0(X) = \{0\}$, hence $C_0(X) = Q_0(X)$.

For any space X, it is readily verified that for $n \geq 1$, $C_n(X)$ is a free abelian group on the set of all nondegenerate n-cubes in X (or, more precisely, their cosets mod $D_n(X)$).

The Faces of a Singular n-cube $(n > 0)$

Let $T:I^n \to X$ be a singular n-cube in X. For $i = 1, 2, \ldots, n$, we will define singular $(n-1)$-cubes

$$A_i T, B_i T : I^{n-1} \to X$$

by the formulas

$$A_i T(x_1, \ldots, x_{n-1}) = T(x_1, \ldots, x_{i-1}, 0, x_i, \ldots, x_{n-1}),$$
$$B_i T(x_1, \ldots, x_{n-1}) = T(x_1, \ldots, x_{i-1}, 1, x_i, \ldots, x_{n-1}).$$

$A_i T$ is called the *front i-face* and $B_i T$ is called the *back i-face* of T.

These face operators satisfy the following identities, where $T:I^n \to X$ is an n-cube, $n > 1$, and $1 \leq i < j \leq n$:

$$
\begin{aligned}
A_i A_j(T) &= A_{j-1} A_i(T), \\
B_i B_j(T) &= B_{j-1} B_i(T), \\
A_i B_j(T) &= B_{j-1} A_i(T), \\
B_i A_j(T) &= A_{j-1} B_i(T).
\end{aligned}
\tag{2.1}
$$

We now define the *boundary operator*; it is a homomorphism $\partial_n : Q_n(X) \to Q_{n-1}(X)$, $n \geq 1$. To define such a homomorphism, it is only necessary to define it on the basis elements, the singular cubes, by the basic property of free abelian groups. Usually we will write ∂ rather than ∂_n for brevity.

Definition 2.3. For any n-cube T, $n > 0$,

$$\partial_n(T) = \sum_{i=1}^{n} (-1)^i [A_i T - B_i T].$$

The reader should write out this formula explicitly for the cases $n = 1, 2,$ and 3, and by drawing pictures convince himself that it does in some sense represent the *oriented* boundary of an n-cube T. The following are the two most important properties of the boundary operator:

$$\partial_{n-1}(\partial_n(T)) = 0 \qquad (n > 1) \tag{2.2}$$

$$\partial_n(D_n(X)) \subset D_{n-1}(X) \qquad (n > 0). \tag{2.3}$$

The proof of (2.2) depends on Identities (2.1); the proof of (2.3) is easy.

As a consequence of (2.3), ∂_n *induces* a homomorphism $C_n(X) \to C_{n-1}(X)$, which we denote by the same symbol, ∂_n. Note that this new sequence of homomorphisms $\partial_1, \partial_2, \ldots, \partial_n, \ldots$, satisfies Equation (2.2): $\partial_{n-1}\partial_n = 0$.

We now define

$$Z_n(X) = \text{kernel } \partial_n = \{u \in C_n(X) \,|\, \partial(u) = 0\} \qquad (n > 0)$$
$$B_n(X) = \text{image } \partial_{n+1} = \partial_{n+1}(C_{n+1}(X)) \qquad (n \geq 0).$$

Note that as a consequence of the equation $\partial_{n-1}\partial_n = 0$, it follows that

$$B_n(X) \subset Z_n(X) \quad \text{for } n > 0.$$

Hence we can define

$$H_n(X) = Z_n(X)/B_n(X) \quad \text{for } n > 0.$$

It remains to define $H_0(X)$ and $H_n(X)$ for $n < 0$, which we will do in a minute. $H_n(X)$ is called the *n-dimensional singular homology group* of X, or the *n-dimensional homology group* of X for short. These groups $H_n(X)$ will be our main object of study. The groups $C_n(X)$, $Z_n(X)$, and $B_n(X)$ are only of secondary importance. More terminology: $Z_n(X)$ is called the *group of n-dimensional singular cycles* of X, or *group of n-cycles*. $B_n(X)$ is called the *group of n-dimensional boundaries* or *group of n-dimensional bounding cycles*.

To define $H_0(X)$, we will first define $Z_0(X)$, then set $H_0(X) = Z_0(X)/B_0(X)$ as before. It turns out that there are actually two slightly different candidates for $Z_0(X)$, which give rise to slightly different groups $H_0(X)$. In some situations one definition is more advantageous, while in other situations the other is better. Hence we will use both. The difference between the two is of such a simple nature that no trouble will result.

First Definition of $\tilde{H}_0(X)$

This definition is very simple. We define $Z_0(X) = C_0(X)$ and

$$H_0(Z) = Z_0(X)/B_0(X) = C_0(X)/B_0(X).$$

There is another way we could achieve the same result: we could define $C_n(X) = \{0\}$ for $n < 0$, define $\partial_n : C_n(X) \to C_{n-1}(X)$ in the only possible way for $n \le 0$ (i.e., $\partial_n = 0$ for $n \le 0$), and then define $Z_0(X) = $ kernel ∂_0. More generally, we could then define $Z_n(X) = $ kernel ∂_n for all integers n, positive or negative, $B_n(X) = \partial_{n+1}(C_{n+1}(X)) \subset Z_n(X)$, and $H_n(X) = Z_n(X)/B_n(X)$ for all n. Of course we then obtain $H_n(X) = \{0\}$ for $n < 0$.

Note that $H_0(X)$ is defined even in case X is empty.

Second Definition—The Reduced 0-dimensional Homology Group, $\tilde{H}_0(X)$

For this purpose, we define a homomorphism $\varepsilon : C_0(X) \to \mathbf{Z}$, where $\mathbf{Z}$ denotes the ring of integers. This homomorphism is often called the *augmentation*. Since $C_0(X) = Q_0(X)$ is a free group on the set of 0-cubes, it suffices to define $\varepsilon(T)$ for any 0-cube T in X. The definition is made in the simplest possible nontrivial way: $\varepsilon(T) = 1$. It then follows that if $u = \sum_i n_i T_i$ is any 0-chain, $\varepsilon(u) = \sum_i n_i$ is just the sum of the coefficients. One now proves the following important formula:

$$\varepsilon \circ \partial_1 = 0. \tag{2.4}$$

To prove this formula, it suffices to verify that for any singular 1-cube T in X, $\varepsilon(\partial_1(T)) = 0$, and this is a triviality.

We now define $\tilde{Z}_0(X) = $ kernel ε. Formula (2.4) assures us that $B_0(X) \subset \tilde{Z}_0(X)$, hence we can define

$$\tilde{H}_0(X) = \tilde{Z}_0(X)/B_0(X).$$

$\tilde{H}_0(X)$ is called the *reduced 0-dimensional homology group* of X. To avoid some unpleasantness later, we agree to only consider the reduced group $\tilde{H}_0(X)$ in case the space X is nonempty. It is often convenient to set $\tilde{H}_n(X) = H_n(X)$ for $n > 0$.

We will now discuss the relation between the groups $H_0(X)$ and $\tilde{H}_0(X)$. First of all, note that $\tilde{Z}_0(X)$ is a subgroup of $Z_0(X) = C_0(X)$, hence $\tilde{H}_0(X)$ is a subgroup of $H_0(X)$. Let $\xi : \tilde{H}_0(X) \to H_0(X)$ denote the inclusion homomorphism. Secondly, from Formula (2.4), it follows that $\varepsilon(B_0(X)) = 0$, hence the augmentation ε induces a homomorphism.

$$\varepsilon_* : H_0(X) \to \mathbf{Z}.$$

Proposition 2.1. *The following sequence of groups and homomorphisms*

$$0 \to \tilde{H}_0(X) \xrightarrow{\xi} H_0(X) \xrightarrow{\varepsilon_*} \mathbf{Z} \to 0$$

is exact. Thus we may identify $\tilde{H}_0(X)$ with the kernel of ε_*. (The space X is assumed nonempty.)

The proof is easy. It follows that $H_0(X)$ is the direct sum of $\tilde{H}_0(X)$ and an infinite cyclic subgroup; however, this direct sum decomposition is not natural or canonical; the infinite cyclic summand can often be chosen in many different ways.

Some Examples

EXAMPLE 2.1. $X =$ space consisting of a single point. Then we find that

$$H_0(X) \approx Z$$
$$H_n(X) = \{0\} \quad \text{for } n \neq 0$$
$$\tilde{H}_0(X) = \{0\}$$

$\varepsilon_* : H_0(X) \to Z$ is an isomorphism.

EXAMPLE 2.2. 0-dimensional homology group of an arcwise connected space, X. We then see that $\varepsilon : C_0(X) \to Z$ is an epimorphism, and $B_0(X) =$ kernel ε (proof left to reader). It follows that $\varepsilon_* : H_0(X) \to Z$ is an isomorphism, and $\tilde{H}_0(X) = \{0\}$. (Note: X is assumed nonempty.)

EXAMPLE 2.3. Let X be a space with many arc-components; denote the arc-components by $X_\gamma, \gamma \in \Gamma$. Note that each singular n-cube lies entirely in one of the arc-components. Hence $Q_n(X)$ breaks up naturally into a direct sum,

$$Q_n(X) = \sum_{\gamma \in \Gamma} Q_n(X_\gamma).$$

Similarly, with $D_n(X)$:

$$D_n(X) = \sum_{\gamma \in \Gamma} D_n(X_\gamma),$$

hence on passing to quotient groups we see that

$$C_n(X) = \sum_{\gamma \in \Gamma} C_n(X_\gamma) \quad \text{(direct sum)}.$$

Next, note that if a singular n-cube is entirely contained in the arc-component X_γ, then its faces are also entirely contained in X_γ. It follows that the boundary $\partial_n : C_n(X) \to C_{n-1}(X)$ maps $C_n(X_\gamma)$ into $C_{n-1}(X_\gamma)$. Therefore we have the following direct sum decompositions

$$Z_n(X) = \sum_{\gamma \in \Gamma} Z_n(X_\gamma),$$

$$B_n(X) = \sum_{\gamma \in \Gamma} B_n(X_\gamma),$$

and hence

$$H_n(X) = \sum_{\gamma \in \Gamma} H_n(X_\gamma).$$

In words, *the nth homology group of X is the direct sum of the nth homology groups of its arc-components.*

We can apply this last result and Example 2.2 to determine the structure of $H_0(X)$ for any space X. The result is that $H_0(X)$ is a direct sum of infinite cyclic groups, with one summand for each arc-component of X.

Note that such a simple direct sum theorem does not hold for $\tilde{H}_0(X)$. For example, if X has exactly two arcwise connected components, what is the structure of $\tilde{H}_0(X)$?

EXERCISE

2.1. Determine the structure of the homology group $H_n(X)$, $n \geq 0$, if X is (a) the set of rational numbers with their usual topology. (b) a countable, discrete space.

These examples show the relation between the structure of $H_0(X)$ and certain topological properties of X (the number of arcwise connected components). In an analogous way, the algebraic structure of the groups $H_n(X)$ for $n > 0$ express certain topological properties of the space X. Naturally, these will be properties of a more subtle nature. One of our principal aims will be to develop methods of determining the structure of the groups $H_n(X)$ for various spaces X.

§3. The Homomorphism Induced by a Continuous Map

Homology theory associates with every topological space X the sequence of groups $H_n(X)$, $n = 0, 1, 2, \ldots$. Equally important, it associates with every continuous map $f : X \to Y$ between spaces a sequence of homomorphisms $f_* : H_n(X) \to H_n(Y)$, $n = 0, 1, 2, \ldots$. Certain topological properties of the continuous map f are reflected in algebraic properties of the homomorphisms f_*. We will now give the definition of f_*, which is very simple.

First of all, we define homomorphisms $f_\# : Q_n(X) \to Q_n(Y)$ by the simple rule

$$f_\#(T) = f \circ T$$

for any singular n-cube $T : I^n \to X$, $n = 0, 1, 2, \ldots$. We now list the main properties of this homomorphism $f_\#$:

(3.1) If T is a degenerate singular n-cube, so is $f_\#(T)$. Hence $f_\#$ maps $D_n(X)$ into $D_n(Y)$, and induces a homomorphism of $C_n(X)$ into $C_n(Y)$. We will denote this induced homomorphism by the same symbol,

$$f_\# : C_n(X) \to C_n(Y), \qquad n = 0, 1, 2, \ldots$$

to avoid an undue proliferation of notation.

(3.2) The following diagram is commutative for $n = 1, 2, 3, \ldots$:

$$
\begin{array}{ccc}
Q_n(X) & \xrightarrow{\ f_\# \ } & Q_n(Y) \\
\downarrow{\scriptstyle \partial_n} & & \downarrow{\scriptstyle \partial_n} \\
Q_{n-1}(X) & \xrightarrow{\ f_\# \ } & Q_{n-1}(Y).
\end{array}
$$

This fact can also be expressed by the equation $\partial_n \circ f_\# = f_\# \circ \partial_n$, or by the statement that $f_\#$ commutes with the boundary operator. [To prove this, one observes that $f_\#(A_i T) = A_i(f_\# T)$ and $f_\#(B_i T) = B_i(f_\#(T))$.] It follows that the following diagram is commutative for $n = 1, 2, 3, \ldots$:

$$
\begin{array}{ccc}
C_n(X) & \xrightarrow{\ f_\# \ } & C_n(Y) \\
\downarrow{\scriptstyle \partial_n} & & \downarrow{\scriptstyle \partial_n} \\
C_{n-1}(X) & \xrightarrow{\ f_\# \ } & C_{n-1}(Y).
\end{array}
$$

Hence $f_\#$ maps $Z_n(X)$ into $Z_n(Y)$ and $B_n(X)$ into $B_n(Y)$ for all $n \geq 0$ and induces a homomorphism of quotient groups, denoted by

$$f_* : H_n(X) \to H_n(Y) \qquad n = 0, 1, 2, \ldots.$$

This is our desired definition.

(3.3) The following diagram is also readily seen to be commutative:

$$
\begin{array}{ccc}
C_0(X) & & \\
\downarrow{\scriptstyle f_\#} & \searrow^{\varepsilon} & Z. \\
C_0(Y) & \nearrow_{\varepsilon} &
\end{array}
$$

Hence $f_\#$ also maps $\tilde{Z}_0(X)$ into $\tilde{Z}_0(Y)$ and induces a homomorphism of $\tilde{H}_0(X)$ into $\tilde{H}_0(Y)$ which is denoted by the same symbol:

$$f_* : \tilde{H}_0(X) \to \tilde{H}_0(Y).$$

The student should verify that the following two diagrams are also commutative:

$$
\begin{array}{ccc}
\tilde{H}_0(X) & \xrightarrow{\ \xi \ } & H_0(X) \\
\downarrow{\scriptstyle f_*} & & \downarrow{\scriptstyle f_*} \\
\tilde{H}_0(Y) & \xrightarrow{\ \xi \ } & H_0(Y)
\end{array}
\qquad
\begin{array}{ccc}
H_0(X) & & \\
\downarrow{\scriptstyle f_*} & \searrow^{\varepsilon_*} & Z. \\
H_0(Y) & \nearrow_{\varepsilon_*} &
\end{array}
$$

Here the notation is that of Proposition 2.1.

(3.4) Let $f: X \to X$ denote the identity map. It is easy to verify successively that the following homomorphisms are identity maps:

$$f_\#: Q_n(X) \to Q_n(X),$$
$$f_\#: C_n(X) \to C_n(X),$$
$$f_*: H_n(X) \to H_n(X),$$

and

$$f_*: \tilde{H}_0(X) \to \tilde{H}_0(X).$$

Of course, the real interest lies in the fact that f_* is the identity.

(3.5) Let X, Y, and Z be topological spaces, and $g: X \to Y$, $f: Y \to Z$ continuous maps. We will denote by $fg: X \to Z$ the composition of the two maps. Under these conditions, we have the homomorphisms $f_* g_*$ and $(fg)_*$ from $H_n(X)$ to $H_n(Z)$ for all $n \geq 0$, and from $\tilde{H}_0(X)$ to $\tilde{H}_0(Z)$. We assert that these two homomorphisms are the same in all cases:

$$(fg)_* = f_* g_*.$$

To prove this assertion, one verifies first that $(fg)_\#$ and $f_\# g_\#$ are the same homomorphisms from $Q_n(X)$ to $Q_n(Z)$, then that $(fg)_\#$ and $f_\# g_\#$ are the same homomorphisms from $C_n(X)$ to $C_n(Z)$. From this the assertion follows.

Since Properties (3.4) and (3.5) are so obvious, the reader may wonder why we even bothered to mention them explicitly. These properties will be used innumerable times in the future, and it is in keeping with the customs of modern mathematics to make explicit any axiom or theorem that one uses.

CAUTION: If $f: X \to Y$ is a 1-1 map, it does *not* necessarily follow that $f_*: H_n(X) \to H_n(Y)$ is 1-1; similarly, the fact that f is onto does *not* imply that f_* is onto. There will be plenty of examples to illustrate this point later on.

EXERCISES

3.1. Let X and Y be spaces having a finite number of arcwise connected components, and $f: X \to Y$ a continuous map. Describe the induced homomorphism $f_*: H_0(X) \to H_0(Y)$. Generalize to the case where X or Y have an infinite number of arc-components.

3.2. Let X_y be an arc-component of X, and $f: X_y \to X$ the inclusion map. Prove that $f_*: H_n(X_y) \to H_n(X)$ is a monomorphism, and the image is the direct summand of $H_n(X)$ corresponding to X_y, as described in Example 2.3. Consequence: the direct sum decomposition of Example 2.3 can be described completely in terms of such homomorphisms which are induced by inclusion maps.

3.3. (Application to Retracts) A subset A of a topological space X is called a *retract* of X if there exists a *continuous* map $r: X \to A$ such that $r(a) = a$ for any $a \in A$. This is a rather strong condition on the subspace A. (a) Construct examples of pairs (X, A) such that A is a retract of X, and such that A is *not* a retract of X.

(b) Let A be a retract of X with retracting map $r: X \to A$, and let $i: A \to X$ denote the inclusion map. Prove that $r_*: H_n(X) \to H_n(A)$ is an epimorphism,

$i_*: H_n(A) \to H_n(X)$ is a monomorphism, and that $H_n(X)$ is the direct sum of the image of i_* and the kernel of r_*.

§4. The Homotopy Property of the Induced Homomorphisms

In this section we will prove a basic property of the homomorphism induced by a continuous map. This property is to a large extent responsible for the distinctive character of a homology theory, and is one of the factors making possible the computation of the homology groups $H_n(X)$ for many spaces X.

Definition 4.1. Two continuous maps $f, g: X \to Y$ are *homotopic* (notation: $f \simeq g$) if there exists a continuous map $F: I \times X \to Y$ such that $F(0,x) = f(x)$ and $F(1,x) = g(x)$ for any $x \in X$.

Intuitively speaking, $f \simeq g$ if and only if it is possible to "continuously deform" the map f into the map g. The reader should prove that $\simeq$ is an equivalence relation on the set of all continuous maps from X into Y. The equivalence classes are called *homotopy classes*. The classification of continuous maps into homotopy classes is often very convenient; for example, usually there will be uncountably many continuous maps from X into Y, but if X and Y are reasonable spaces, there will often only be finitely many or countably many homotopy classes.

Theorem 4.1. *Let f and g be continuous maps of X into Y. If f and g are homotopic, then the induced homomorphisms, f_* and g_*, of $H_n(X)$ into $H_n(Y)$ are the same. Also, $f_* = g_*: \tilde{H}_0(X) \to \tilde{H}_0(Y)$.*

PROOF: Let $F: I \times X \to Y$ be a continuous map such that $F(0,x) = f(x)$ and $F(1,x) = g(x)$. We will use the continuous map F to construct a sequence of homomorphisms

$$\varphi_n: C_n(X) \to C_{n+1}(Y), \qquad n = 0, 1, 2, \ldots$$

such that the following relation holds:

$$-f_\# + g_\# = \partial_{n+1} \circ \varphi_n + \varphi_{n-1} \circ \partial_n, \qquad n = 0, 1, 2, \ldots. \tag{4.1}$$

[For $n = 0$, we will interpret this equation as follows: $C_{-1}(X) = C_{-1}(Y) = \{0\}$, ∂_0 is the 0 homomorphism, and $\varphi_{-1}: C_{-1}(X) \to C_0(Y)$ is (of necessity) the 0 homomorphism.] We assert that the theorem follows immediately from Equation (4.1). To see this, let $u \in H_n(X)$; choose a representative cycle $u' \in Z_n(X)$ for the homology class u. Since $\partial_n(u') = 0$, it follows from Equation (4.1) that

$$-f_\#(u') + g_\#(u') = \partial_{n+1}(\varphi_n(u')).$$

Hence $-f_\#(u') + g_\#(u') \in B_n(Y)$, and therefore $f_*(u) = g_*(u)$. The proof in case $u \in \tilde{H}_0(X)$ is left to the student.

This is a typical procedure in algebraic topology; from the continuous map F we construct homomorphisms (algebraic maps) φ_n which reflect properties of F.

To construct the homomorphisms φ_n, we define a sequence of homomorphisms

$$\Phi_n : Q_n(X) \to Q_{n+1}(Y), \qquad n = 0, 1, 2, \ldots$$

as follows. For any singular n-cube $T : I^n \to X$, define a singular $(n+1)$-cube $\Phi_n(T) : I^{n+1} \to Y$ by the formula

$$(\Phi_n T)(x_1, \ldots, x_{n+1}) = F(x_1, T(x_2, \ldots, x_{n+1})). \tag{4.2}$$

We wish to compute $\partial_{n+1}\Phi_n(T)$. For this purpose, observe that

$$A_1 \Phi_n(T) = f_\#(T),$$
$$B_1 \Phi_n(T) = g_\#(T),$$
$$A_i \Phi_n(T) = \Phi_{n-1} A_{i-1}(T) \qquad (2 \le i \le n+1),$$
$$B_i \Phi_n(T) = \Phi_{n-1} B_{i-1}(T) \qquad (2 \le i \le n+1).$$

We now compute:

$$\partial_{n+1}\Phi_n(T) = \sum_{i=1}^{n+1} (-1)^i [A_i \Phi_n(T) - B_i \Phi_n(T)]$$

$$= -[f_\#(T) - g_\#(T)] + \sum_{i=2}^{n+1} (-1)^i \Phi_{n-1}(A_{i-1}(T) - B_{i-1}(T))$$

$$= -f_\#(T) + g_\#(T) + \sum_{j=1}^{n} (-1)^{j+1} \Phi_{n-1}(A_j(T) - B_j(T))$$

$$= -f_\#(T) + g_\#(T) - \Phi_{n-1}\partial_n(T).$$

Therefore we conclude that for any $u \in Q_n(X)$,

$$-f_\#(u) + g_\#(u) = \partial_{n+1}\Phi_n(u) + \Phi_{n-1}\partial_n(u). \tag{4.3}$$

Next, observe that if T is a *degenerate* singular n-cube, $n > 0$, then $\Phi_n(T)$ is a degenerate $(n+1)$-cube. Hence

$$\Phi_n(D_n(X)) \subset D_{n+1}(Y)$$

and therefore Φ_n induces a homomorphism

$$\varphi_n : C_n(X) \to C_{n+1}(Y).$$

From (4.3) it follows that φ_n satisfies Equation (4.1), as desired.   Q.E.D.

Some terminology. The function F above is called a *homotopy* between the continuous maps f and g. The homomorphisms φ_n, $n = 0, 1, 2, \ldots$, constitute a *chain homotopy* or *algebraic homotopy* between the *chain maps* $f_\#$ and $g_\#$.

We will now discuss some applications of this theorem. Later on when we are able to actually determine the structure of some homology groups and compute some induced homomorphisms, we will be able to use it to prove that certain maps are *not* homotopic. For example, it can be shown that there are infinitely many homotopy classes of maps of an n-sphere onto itself if $n > 0$.

Homotopy Type of Spaces

Definition 4.2. Two spaces X and Y are of the same *homotopy type* if there exist continuous maps $f: X \to Y$ and $g: Y \to X$ such that gf is homotopic to the identity map $X \to X$, and fg is homotopic to the identity map $Y \to Y$. The maps f and g occurring in this definition are called *homotopy equivalences*.

For example, if X and Y are homeomorphic, then they are of the same homotopy type (but not conversely).

Theorem 4.2. *If $f: X \to Y$ is a homotopy equivalence, then $f_*: H_n(X) \to H_n(Y)$, $n = 0, 1, 2, \ldots$, and $f_*: \tilde{H}_0(X) \to \tilde{H}_0(Y)$ are isomorphisms.*

The proof, which is simple, is left to the reader.

Definition 4.3. A space X is *contractible to a point* if there exists a continuous map $F: I \times X \to X$ such that $F(0,x) = x$ and $F(1,x) = x_0$ for any $x \in X$ (here x_0 is a fixed point of X).

For example, any convex subset of Euclidean n-space is contractible to a point (proof to be supplied by the reader). If a space X is contractible to a point, then it has the same homotopy type as a space consisting of a single point, and its homology groups are as follows:

$$H_0(X) \approx \mathbf{Z}, \qquad \tilde{H}_0(X) = 0,$$
$$H_n(X) = 0 \quad \text{for } n \neq 0.$$

Definition 4.4. A subset A of a space X is a *deformation retract* of X if there exists a retraction $r: X \to A$ (i.e., A is a retract of X) and a continuous map $F: I \times X \to X$ such that $F(0,x) = x$, $F(1,x) = r(x)$ for any $x \in X$.

For example, in Definition 4.3, the set $\{x_0\}$ is a deformation retract of X.

If A is a deformation retract of X, then the inclusion map $i: A \to X$ is a homotopy equivalence; the proof is left to the reader. Hence the induced homomorphism $i_*: H_n(A) \to H_n(X)$ is an isomorphism. This is a useful principle to remember when trying to determine the homology groups of a space.

§5. The Exact Homology Sequence of a Pair

In order to be able to use homology groups effectively, it is necessary to be able to determine their structure for various spaces; so far we can only do this for a few spaces, such as those which are contractible. In most cases, the definition of $H_n(X)$ is useless as a means of computing its structure. In order to make further progress, it seems to be necessary to have some general theorems which give relations between the homology groups of a space X and those of any subspace A contained in X. If $i: A \to X$ denotes the inclusion map, then there is defined the induced homomorphism $i_*: H_n(A) \to H_n(X)$ for $n = 0, 1, 2, \ldots$. As was mentioned earlier, i_* need not be either an epimorphism or monomorphism.

In this section we will generalize our earlier definition of homology groups, by defining relative homology groups for any pair (X,A) consisting of a topological space X and a subspace A; these groups are denoted by $H_n(X,A)$, where $n = 0, 1, 2, \ldots$. There is a nice relation between these relative homology groups and the homomorphisms $i_*: H_n(A) \to H_n(X)$, which is expressed by something called the *homology sequence of the pair* (X,A). Thus it will turn out that knowledge of the structure of the groups $H_n(X,A)$ will give rise to information about the homomorphisms $i_*: H_n(A) \to H_n(X)$ and vice-versa. In the next section we will take up various properties of the relative homology groups, such as the *excision property*; this will enable us to actually determine these relative homology groups in certain cases.

The relative homology groups are true generalizations of the homology groups defined earlier in the sense that if A is the empty set, then $H_n(X,A) = H_n(X)$. Nevertheless, the primary interest in algebraic topology centers on the nonrelative homology groups $H_n(X)$ for any space X. Our point of view is that the relative groups $H_n(X,A)$ are introduced mainly for the purpose of making possible the computation of the "absolute" homology groups $H_n(X)$, even though in certain circumstances the relative groups are of independent interest.

The Definition of Relative Homology Groups

Let A be a subspace of the topological space X, and let $i: A \to X$ denote the inclusion map. It is readily verified that the induced homomorphism $i_\#: C_n(A) \to C_n(X)$ is a monomorphism, hence we can consider $C_n(A)$ to be a subgroup of $C_n(X)$; it is the subgroup generated by all nondegenerate singular cubes in A. We will use the notation $C_n(X,A)$ to denote the quotient group $C_n(X)/C_n(A)$; it is called the group of n-dimensional chains of the pair (X,A). The boundary operator $\partial_n: C_n(X) \to C_{n-1}(X)$ has the property that $\partial_n(C_n(A)) \subset C_{n-1}(A)$, hence it induces a homomorphism of quotient groups

$$\partial_n': C_n(X,A) \to C_{n-1}(X,A)$$

which we will usually denote by ∂_n, or ∂, for simplicity. In analogy with the definitions in §2, we define the group of *n-dimensional cycles of* (X,A) for $n > 0$ by

$$Z_n(X,A) = \text{kernel } \partial_n = \{u \in C_n(X,A) \mid \partial(u) = 0\}$$

and for $n \geq 0$ the *group of n-dimensional bounding cycles* by

$$B_n(X,A) = \text{image } \partial_{n+1} = \partial_{n+1}(C_{n+1}(X,A)).$$

Since $\partial_n \partial_{n+1} = 0$, it follows that

$$B_n(X,A) \subset Z_n(X,A)$$

and hence we can define

$$H_n(X,A) = Z_n(X,A)/B_n(X,A).$$

In case $n = 0$, we define $Z_0(X,A) = C_0(X,A)$ and $H_0(X,A) = C_0(X,A)/B_0(X,A)$.

Intuitively speaking, the relative homology group $H_n(X,A)$ is defined in the same way as $H_n(X)$, except that one neglects anything in the subspace A. For example, let $u \in C_n(X)$; then the coset of u in the quotient group, $C_n(X,A)$, is a cycle mod A if and only if $\partial(u) \in C_n(A)$, i.e., $\partial(u)$ is a chain in the subspace of A.

EXERCISE

5.1. Prove that $C_n(X,A)$ is a free abelian group generated by the (cosets of) the non-degenerate singular n-cubes of X which are not contained in A.

It is convenient to display the chain groups $C_n(A)$, $C_n(X)$, and $C_n(X,A)$ together with their boundary operators in one large diagram as follows:

$$(5.1)$$

Here the vertical arrows denote the appropriate boundary operator, ∂, and $j_{\#}$ denotes the natural epimorphism of $C_n(X)$ onto its quotient group $C_n(X,A)$. It is clear that each square in this diagram is commutative. In order to avoid having to consider the case $n = 0$ as exceptional, we will define for any integer $n < 0$,

$$C_n(A) = C_n(X) = C_n(X,A) = \{0\}.$$

Thus this diagram extends infinitely far upwards and downwards.

As was pointed out in §3, the homomorphisms $i_{\#}$ induce homomorphisms i_* of $H_n(A)$ into $H_n(X)$ for all n. Similarly, the homomorphisms $j_{\#}$ induce homomorphisms

$$j_* : H_n(X) \to H_n(X,A), \qquad n = 0, 1, 2, \ldots.$$

We will now define a third sequence of homomorphisms

$$\partial_* : H_n(X,A) \to H_{n-1}(A)$$

for all integral values of n by a somewhat more elaborate procedure, as follows. Let $u \in H_n(X,A)$; we wish to define $\partial_*(u) \in H_{n-1}(A)$. Choose a representative n-dimensional cycle $u' \in C_n(X,A)$ for the homology class u. Because $j_{\#}$ is an epimorphism, we can choose a chain $u'' \in C_n(X)$ such that $j_{\#}(u'') = u'$. Consider the chain $\partial(u'') \in C_{n-1}(X)$; using the commutativity of Diagram (5.1) and the fact that u' is a cycle, we see that $j_{\#}\partial(u'') = 0$; hence $\partial(u'')$ actually belongs to the subgroup $C_{n-1}(A)$ of $C_{n-1}(X)$. Also $\partial(u'')$ is easily seen to be a cycle; we define $\partial_*(u)$ to be the homology class of the cycle $\partial(u'')$.

To justify this definition of ∂_*, one must verify that it does not depend on the choice of the representative cycle u' or of the chain u'' such that $j_{\#}(u'') = u'$. In addition, it must be proved that ∂_* is a homomorphism, i.e., $\partial_*(u + v) = \partial_*(u) + \partial_*(v)$. These verifications should be carried out by the reader.

The homomorphism ∂_* is called the *boundary operator of the pair* (X,A).

It is natural to consider the following infinite sequence of groups and homomorphisms for any pair (X,A):

$$\cdots \xrightarrow{j_*} H_{n+1}(X,A) \xrightarrow{\partial_*} H_n(A) \xrightarrow{i_*} H_n(X) \xrightarrow{j_*} H_n(X,A) \xrightarrow{\partial_*} \cdots .$$

This sequence will be called the *homology sequence of the pair* (X,A). Once again, in order to avoid having to consider the case $n = 0$ as exceptional, we will make the convention that for $n < 0$, $H_n(A) = H_n(X) = H_n(X,A) = \{0\}$. Thus the homology sequence of a pair extends to infinity in both the right and left directions.

The following is the main theorem of this section:

Theorem 5.1. *The homology sequence of any pair* (X,A) *is exact.*

In order to prove this theorem, it obviously suffices to prove the following six inclusion relations:

$$\text{image } i_* \subset \text{kernel } j_*, \qquad \text{image } i_* \supset \text{kernel } j_*,$$
$$\text{image } j_* \subset \text{kernel } \partial_*, \qquad \text{image } j_* \supset \text{kernel } \partial_*,$$
$$\text{image } \partial_* \subset \text{kernel } i_*, \qquad \text{image } \partial_* \supset \text{kernel } i_*.$$

We strongly urge the reader to carry out these six proofs, none of which is difficult. It is only by working through such details that one can acquire familiarity with the techniques of this subject.

EXERCISES

5.1. For any pair (X,A), prove the following assertions:
(a) $i_*: H_n(A) \to H_n(X)$ is an isomorphism for all n if and only if $H_n(X,A) = 0$ for all n.
(b) $j_*: H_n(X) \to H_n(X,A)$ is an isomorphism for all n if and only if $H_n(A) = 0$ for all n.
(c) $H_n(X,A) = 0$ for $n \le q$ if and only if $i_*: H_n(A) \to H_n(X)$ is an isomorphism for $n < q$ and an epimorphism for $n = q$.

5.2. Let X_γ, $\gamma \in \Gamma$, denote the arcwise connected components of X. Prove that $H_n(X,A)$ is isomorphic to the direct sum of the groups $H_n(X_\gamma, X_\gamma \cap A)$ for all $\gamma \in \Gamma$. Also, determine the structure of $H_0(X_\gamma, X_\gamma \cap A)$. (*Hint*: There are two cases to consider.)

5.3. For any pair (X,A), prove there are natural isomorphisms, as follows: Let $Z_n(X \bmod A) = \{x \in C_n(X) \mid \partial(x) \in C_n(A)\}$. Then

$$Z_n(X,A) \approx Z_n(X \bmod A)/C_n(A),$$

$$B_n(X,A) \approx [B_n(X) + C_n(A)]/C_n(A)$$
$$\approx B_n(X)/[B_n(X) \cap C_n(A)],$$

$$H_n(X,A) \approx Z_n(X \bmod A)/[B_n(X) + C_n(A)].$$

[Note: The notation $B_n(X) + C_n(A)$ denotes the least subgroup of $C_n(X)$ which contains both $B_n(X)$ and $C_n(A)$; it need not be isomorphic to their direct sum.]

5.4. Give a discussion of the exact sequence of a pair (X,A) in case the subspace A is empty.

5.5. Let (X,A) be a pair with A nonempty, and let us agree to consider the reduced homology groups $\tilde{H}_0(A)$ and $\tilde{H}_0(X)$ as subgroups of $H_0(A)$ and $H_0(X)$ respectively (cf. Proposition 2.1). Show that the boundary operator $\partial_*: H_1(X,A) \to H_0(A)$ sends $H_1(X,A)$ into the subgroup $\tilde{H}_0(A)$, and that the following sequence is exact:

$$\cdots \overset{j_*}{\to} H_1(X,A) \overset{\partial_*}{\to} \tilde{H}_0(A) \overset{i_*}{\to} \tilde{H}_0(X) \overset{j_*}{\to} H_0(X,A) \to 0.$$

(This result may be paraphrased as follows: If $A \ne \varnothing$, we may replace $H_0(A)$ and $H_0(X)$ by $\tilde{H}_0(A)$ and $\tilde{H}_0(X)$ in the homology sequence of (X,A), and the resulting sequence will still be exact.)

5.6. Let X be a totally disconnected topological space, and let A be an arbitrary subset of X. Determine the various groups and homomorphisms in the homology sequence of (X,A).

§6. The Main Properties of Relative Homology Groups

In order to determine the structure of the relative homology groups of a pair, we need to know the general properties of these newly defined homology groups. First we will consider some properties that are strictly analogous to those discussed in §§3 and 4 for "absolute" homology groups.

Let (X,A) and (Y,B) be pairs consisting of a topological space and a subspace. We will say that a continuous function f mapping X into Y is *a map of the pair* (X,A) *into the pair* (Y,B) if $f(A) \subset B$; we will use the notation $f:(X,A) \to (Y,B)$ to indicate that f is such a map.

Our first observation is that *any map of pairs* $f:(X,A) \to (Y,B)$ *induces a homomorphism* $f_*:H_n(X,A) \to H_n(Y,B)$ *of the corresponding relative homology groups*. This induced homomorphism is defined as follows.

The continuous map f induces a homomorphism $f_\#:C_n(X) \to C_n(Y)$ for all n, as described in §3. Since $f(A) \subset B$, it follows that $f_\#$ sends the subgroup $C_n(A)$ into the subgroup $C_n(B)$, and hence there is induced a homomorphism of quotient groups $C_n(X,A) \to C_n(Y,B)$ which we will also denote by $f_\#$. These induced homomorphisms commute with the boundary operators, in the sense that the following diagram is commutative for each n:

$$\begin{array}{ccc} C_n(X,A) & \xrightarrow{\ f_\#\ } & C_n(Y,B) \\ \downarrow{\scriptstyle\partial} & & \downarrow{\scriptstyle\partial} \\ C_{n-1}(X,A) & \xrightarrow{\ f_\#\ } & C_{n-1}(Y,B). \end{array}$$

It now follows exactly as in §3 that $f_\#$ induces a homomorphism $f_*: H_n(X,A) \to H_n(Y,B)$ of the corresponding homology groups for all n.

The reader should formulate and verify the analogs for maps of pairs of the properties described in (3.4) and (3.5) for maps of spaces.

Note that the homomorphism $j_*:H(X) \to H_n(X,A)$ which is part of the homology sequence of the pair (X,A) (as explained in the preceding section) is actually a homomorphism of the kind we have just described. For, we can consider that the identity map of X into itself defines a map $j:(X,\varnothing) \to (X,A)$ of pairs, and then it is easily checked that the homomorphism $j_*:H_n(X) \to H_n(X,A)$ defined in the preceding section is the homomorphism induced by j.

Next, we will consider the homotopy relation for maps of pairs. The appropriate generalization of Definition 4.1 is the following: Two maps

$f, g:(X,A) \to (Y,B)$ are *homotopic* (as maps of pairs) if there exists a continuous map $F:(I \times X, I \times A) \to (Y,B)$ such that $F(0,x) = f(x)$ and $F(1,x) = g(x)$ for any $x \in X$. The point is that we are requiring that $F(I \times A) \subset B$ in addition to the conditions of Definition 4.1. This additional condition enables one to prove the following result:

Theorem 6.1. *Let $f, g:(X,A) \to (Y,B)$ be maps of pairs. If f and g are homotopic (as maps of pairs), then the induced homomorphisms f_* and g_* of $H_n(X,A)$ into $H_n(Y,B)$ are the same.*

The proof proceeds along the same lines as that of Theorem 4.1. Because of the stronger hypothesis on the homotopy F, it follows that the homomorphisms φ_n constructed in the proof of 4.1 satisfy the following condition:

$$\varphi_n(C_n(A)) \subset C_{n+1}(B).$$

Hence φ_n induces a homomorphism of quotient groups

$$\varphi_n: C_n(X,A) \to C_{n+1}(Y,B).$$

The details are left to the reader.

EXERCISE

6.1. Formulate the appropriate definition of two pairs, (X,A) and (Y,B), being of the same homotopy type, and prove an analog of Theorem 4.2 for such pairs. Similarly, generalize the concepts of retract and deformation retract from spaces to pairs of spaces, and prove the analogs of the properties stated in §§3 and 4 for these concepts.

Next, we will consider the effect of a map $f:(X,A) \to (Y,B)$ on the exact homology sequences of the pairs (X,A) and (Y,B). We can conveniently arrange the two exact sequences and the homomorphisms induced by f in a ladderlike diagram, as follows:

$$
\begin{array}{ccccccccc}
\cdots \longrightarrow & H_n(A) & \xrightarrow{i_*} & H_n(X) & \xrightarrow{j_*} & H_n(X,A) & \xrightarrow{\partial_*} & H_{n-1}(A) & \longrightarrow \cdots \\
& \downarrow & & \downarrow & & \downarrow & & \downarrow & \\
\cdots \longrightarrow & H_n(B) & \xrightarrow{i'_*} & H_n(Y) & \xrightarrow{j'_*} & H_n(Y,B) & \xrightarrow{\partial'_*} & H_{n-1}(B) & \longrightarrow \cdots .
\end{array}
\tag{6.1}
$$

We assert that *each square of this diagram is commutative.* For the left-hand square and the middle square, this assertion is a consequence of Property (3.5) and its analog for pairs. For the right-hand square, which involves ∂_* and ∂'_*, the asertion of commutativity is the statement of a new property of the homology of pairs. To prove it, one must go back to the basic definitions of the concepts involved. Since the proof is absolutely straightforward, the details are best left to the reader.

The commutativity of Diagram (6.1) helps to give us new insight into the significance of the relative homology groups. From a strictly algebraic point of view, there are usually many different ways that we could define groups $H_n(X,A)$ for each integer n in such a way that we would obtain an exact sequence involving the homomorphism $i_*: H_n(A) \to H_n(X)$ at every third step. The fact that Diagram (6.1) is commutative for *any* map f of pairs means that we have chosen a *natural* way to define the groups $H_n(X,A)$ and the exact homology sequence of a pair.

EXERCISE

6.2. Let A be an infinite cyclic group and let B be a cyclic group of order n, $n > 1$. How many solutions are there to the following algebraic problem (up to isomorphism): Determine an abelian group G and homomorphisms $\varphi: A \to G$ and $\psi: G \to B$ such that the following sequence is exact:

$$0 \to A \xrightarrow{\varphi} G \xrightarrow{\psi} B \to 0.$$

We now come to what is perhaps the most important and at the same time the most subtle property of the relative homology groups, called the *excision property*. There is no analogue of this property for absolute homology groups. It will give us some indication as to what the relative homology groups depend on. Ideally, we would like to be able to say that $H_n(X,A)$ depends only on $X - A$, the complement of A in X. While this statement is true under certain rather restrictive hypotheses, in general it is false. Another rough way of describing the situation is to say that *under certain hypotheses*, $H_n(X,A)$ is isomorphic to $H_n(X/A)$ for $n > 0$, and $H_0(X,A) \approx \tilde{H}_0(X/A)$, where X/A denotes the quotient space obtained from X by shrinking the subset A to a point. In any case, the true statement is somewhat weaker.

Theorem 6.2. *Let (X,A) be a pair, and let W be a subset of A such that $\overline{W}$ is contained in the interior of A. Then the inclusion map $(X - W, A - W) \to (X,A)$ induces an isomorphism of relative homology groups:*

$$H_n(X - W, A - W) \approx H_n(X,A), \qquad n = 0, 1, 2, \ldots.$$

The statement of this theorem can be paraphrased as follows: Under the given hypotheses, we can *excise* the set W without affecting the relative homology groups.

The proof of this theorem depends on the fact that in the definition of homology groups we can restrict our consideration to singular cubes which are arbitrarily small, and this will not change anything. For example, if X is a metric space, and ε is a small positive number, we can insist that only singular cubes of diameter less than ε be used in the definition of $H_n(X,A)$ if we wish. If X is not a metric space, we can prescribe an "order of smallness" by choosing an open covering of X, and then using only singular cubes which are small enough to be contained in a single set of the given open covering. For technical reasons, it is convenient to allow a slightly more general type of covering of X in our definition.

Definition 6.1. Let $\mathcal{U} = \{U_\lambda | \lambda \in \Lambda\}$ be a family of subsets of the topological space X such that *the interiors of the set* U_λ *cover* X (we may think of such a family as a generalization of the notion of an open covering of X). A singular n-cube $T: I^n \to X$ is said to be *small of order* $\mathcal{U}$ if there exists an index $\lambda \in \Lambda$ such that $T(I^n) \subset U_\lambda$.

For example, if X is a metric space and ε is small positive number, we could choose $\mathcal{U}$ to be the covering of X by all spheres of radius ε.

We can now go through our preceding definitions and systematically modify them by allowing only singular cubes which are small of order $\mathcal{U}$. This procedure works, because if $T: I^n \to X$ is a singular n-cube which is small of order $\mathcal{U}$, then $\partial_n(T)$ is a linear combination of singular $(n-1)$-cubes, all of which are also small of order $\mathcal{U}$.

Notation. $Q_n(X,\mathcal{U})$ denotes the subgroup of $Q_n(X)$ generated by the singular n-cubes which are small of order $\mathcal{U}$, $D_n(X,\mathcal{U}) = Q_n(X,\mathcal{U}) \cap D_n(X)$, and $C_n(X,\mathcal{U}) = Q_n(X,\mathcal{U})/D_n(X,\mathcal{U})$. Similarly, for any subspace A of X, $Q_n(A,\mathcal{U}) = Q_n(A) \cap Q_n(X,\mathcal{U})$, $D_n(A,\mathcal{U}) = D_n(A) \cap Q_n(A,\mathcal{U})$, and $C_n(A,\mathcal{U}) = Q_n(A,\mathcal{U})/D_n(A,\mathcal{U})$. Finally, for the relative chain groups we let $C_n(X,A,\mathcal{U}) = C_n(X,\mathcal{U})/C_n(A,\mathcal{U})$.

Note that ∂_n maps $Q_n(X,\mathcal{U})$ into $Q_{n-1}(X,\mathcal{U})$, and hence induces homomorphisms

$$C_n(X,\mathcal{U}) \to C_{n-1}(X,\mathcal{U}),$$
$$C_n(A,\mathcal{U}) \to C_{n-1}(A,\mathcal{U}),$$

and

$$C_n(X,A,\mathcal{U}) \to C_{n-1}(X,A,\mathcal{U}),$$

all of which we will continue to denote by the same symbol, ∂_n. Thus we can define exactly as before

$$Z_n(X,A,\mathcal{U}) = \{u \in C_n(X,A,\mathcal{U}) | \partial_n(u) = 0\},$$
$$B_n(X,A,\mathcal{U}) = \partial_{n+1}(C_{n+1}(X,A,\mathcal{U})).$$

Then since $B_n(X,A,\mathcal{U}) \subset Z_n(X,A,\mathcal{U})$, we can define the homology group

$$H_n(X,A,\mathcal{U}) = Z_n(X,A,\mathcal{U})/B_n(X,A,\mathcal{U}).$$

Notice what happens for $n = 0$: $Q_0(X,\mathcal{U}) = Q_0(X)$, and hence it follows that

$$C_0(X,A,\mathcal{U}) = C_0(X,A),$$
$$Z_0(X,A,\mathcal{U}) = C_0(X,A),$$
$$H_0(X,A,\mathcal{U}) = C_0(X,A)/B_0(X,A,\mathcal{U}).$$

Next, note that the inclusion $Q_n(X,\mathcal{U}) \subset Q_n(X)$ induces homomorphisms

$$\sigma_n : C_n(X,A,\mathcal{U}) \to C_n(X,A)$$

(actually, σ_n is a monomorphism, although this fact seems to be of no great importance). Obviously, the homomorphism σ commutes with the boundary operator ∂, i.e., the following diagram is commutative:

$$
\begin{array}{ccc}
C_n(X,A,\mathcal{U}) & \xrightarrow{\ \partial\ } & C_{n-1}(X,A,\mathcal{U}) \\
\downarrow{\scriptstyle \sigma} & & \downarrow{\scriptstyle \sigma} \\
C_n(X,A) & \xrightarrow{\ \partial\ } & C_{n-1}(X,A).
\end{array}
$$

Hence σ maps $Z_n(X,A,\mathcal{U})$ into $Z_n(X,A)$ and $B_n(X,A,\mathcal{U})$ into $B_n(X,A)$, and thus induces a homomorphism

$$\sigma_* : H_n(X,A,\mathcal{U}) \to H_n(X,A)$$

for all n.

Theorem 6.3. *Assume that $\mathcal{U}$ satisfies the above hypotheses. Then the induced homomorphisms $\sigma_* : H_n(X,A,\mathcal{U}) \to H_n(X,A)$ are isomorphisms for all n.*

This theorem is the precise formulation of the assertion made earlier that we can restrict our consideration to singular cubes which are small of order $\mathcal{U}$ in defining $H_n(X,A)$. The proof, which is rather long, is given in the next section.

We will now give the proof of Theorem 6.2, the excision property, using Theorem 6.3.

Let (X,A) and W satisfy the conditions of Theorem 6.2. The hypotheses imply that

$$\text{Interior } (A) \cup \text{Interior } (X - W) = X,$$

hence $\mathcal{U} = \{A, X - W)$ is a generalized open covering of the kind that occurs in Theorem 6.3. Note that for each n,

$$C_n(X,\mathcal{U}) = C_n(A) + C_n(X - W)$$

by the definition of $C_n(X,\mathcal{U})$ (N.B., this is *not* a direct sum).

To prove the excision property, consider the following commutative diagram for each integer n:

$$
\begin{array}{ccc}
C_n(X - W, A - W) & \xrightarrow{\ 1\ } & C_n(X,A) \\
& \searrow{\scriptstyle 2} & \uparrow{\scriptstyle \sigma_n} \\
& & C_n(X,A,\mathcal{U}).
\end{array}
\qquad (6.2)
$$

Each of the homomorphisms indicated in this diagram is induced by an inclusion relation. On passing to homology groups, we obtain the following commutative diagram:

$$
\begin{array}{ccc}
H_n(X - W, A - W) & \xrightarrow{\ 3\ } & H_n(X,A) \\
& \searrow{}^{4} & \Big\uparrow{}_{\sigma_*} \\
& & H_n(X,A,\mathscr{U}).
\end{array}
\tag{6.3}
$$

We wish to prove that the homomorphism indicated by arrow 3 is an isomorphism. Since σ_* is an isomorphism (by Theorem 6.3), it suffices to prove that arrow 4 is an isomorphism. Now the homomorphism designated by arrow 4 is induced by homomorphism designated by arrow 2; therefore let us consider this homomorphism in more detail. By definition,

$$
\begin{aligned}
C_n(X - W, A - W) &= C_n(X - W)/C_n(A - W) \\
&= C_n(X - W)/[C_n(X - W) \cap C_n(A)]
\end{aligned}
$$

since $C_n(A - W) = C_n(X - W) \cap C_n(A)$. Similarly,

$$
\begin{aligned}
C_n(X,A,\mathscr{U}) &= C_n(X,\mathscr{U})/C_n(A,\mathscr{U}) \\
&= [C_n(X - W) + C_n(A)]/C_n(A).
\end{aligned}
$$

Thus the homomorphism denoted by arrow 2 consists of homomorphisms

$$
\frac{C_n(X - W)}{C_n(X - W) \cap C_n(A)} \rightarrow \frac{C_n(X - W) + C_n(A)}{C_n(A)}
\tag{6.4}
$$

for $n = 0, 1, 2, \ldots$, which are induced by the obvious inclusion relations. But according to the first isomorphism theorem a homomorphism such as that in (6.4) is an isomorphism. Hence arrow 2 in (6.2) designates an isomorphism, and it follows that the induced homomorphism, arrow 4 in (6.3), is also an isomorphism. This completes the proof of Theorem 6.2.

We will give examples of the use of the excision property and other properties of relative homology groups in the next chapter.

§7. The Subdivision of Singular Cubes and the Proof of Theorem 6.3

In this section, we introduce the technique of subdivision of singular cubes and use it to prove Theorem 6.3. Although this technique is based on a rather simple and natural geometric idea, the actual proof is rather long and involved. For that reason it may be advisable to skip this section on a first reading and return to it later.

Actually, we will first prove Theorem 6.3 for the easier case of absolute homology groups (the case where $A = \emptyset$ in the statement of the theorem). The general case will then follow by an easy argument using a purely algebraic proposition called the *five-lemma*.

The first step in the proof of Theorem 6.3 is to introduce the so-called *subdivision operator*, and prove its properties. This will involve some lengthy formulas, tedious verifications, etc. The reader must not let those obscure the essentially simple geometric ideas behind the proof.

First, we will consider the process of subdividing a (singular) cube. Probably the simplest way to subdivide the cube I^n is to divide it into 2^n cubes each of side $\frac{1}{2}$, by means of the hyperplanes $x_i = \frac{1}{2}$, $i = 1, 2, \ldots, n$. This leads to the following definitions. Let $\mathscr{E}_n$ denote the set of all vertices of the cube I^n; an n-tuple of real numbers $e = (e_1, e_2, \ldots, e_n)$ belong to $\mathscr{E}_n$ if and only if $e_i = 0$ or 1 for all i. For any singular n-cube $T: I^n \to X$ and any $e \in \mathscr{E}_n$, define

$$F_e(T): I^n \to X$$

by

$$(F_e T)(x) = T(\tfrac{1}{2}(x + e)) \tag{7.1}$$

for all $x = (x_1, \ldots, x_n) \in I^n$. Then define $\mathrm{Sd}_n: Q_n(X) \to Q_n(X)$ by

$$\mathrm{Sd}_n(T) = \sum_{e \in E_n} F_e(T). \tag{7.2}$$

All this is for $n \geq 1$; if T is a singular 0-cube, we define

$$\mathrm{Sd}_0(T) = T.$$

We will now list some properties of the homomorphism Sd_n.

(a) If T is a degenerate cube, then so is $F_e(T)$. Hence Sd_n maps $D_n(X)$ into $D_n(X)$ and induces a homomorphism

$$\mathrm{sd}_n: C_n(X) \to C_n(X).$$

(b) The homomorphisms Sd_n commute with the boundary operator, i.e.,

$$\partial_n \circ \mathrm{Sd}_n = \mathrm{Sd}_{n-1} \circ \partial_n.$$

In order to prove this, one verifies the following three identities regarding the operators F_e.

(b.1) Assume e and $e' \in \mathscr{E}_n$ are such that $e_i = e_i'$ for $i \neq j$, $e_j = 1$, $e_j' = 0$. Then

$$A_j F_e = B_j F_{e'}.$$

(b.2) Assume $e \in \mathscr{E}_n$, $e_j = 0$, and $e' \in \mathscr{E}_{n-1}$ is defined by

$$e' = (e_1, \ldots, e_{j-1}, e_{j+1}, \ldots, e_n).$$

Then

$$A_j F_e = F_{e'} A_j.$$

(b.3) Assume $e \in \mathscr{E}_n$, $e_j = 1$, and $e' \in \mathscr{E}_{n-1}$ is defined by

$$e' = (e_1, \ldots, e_{j-1}, e_{j+1}, \ldots, e_n).$$

Then

$$B_j F_e = F_{e'} B_j.$$

These three identities are exactly what one needs to verify that

$$\partial_n \, \mathrm{Sd}_n(T) = \mathrm{Sd}_{n-1} \partial_n(T).$$

Naturally, it follows that the induced homomorphism $\mathrm{sd}_n : C_n(X) \to C_n(X)$ also commutes with the boundary operator.

(c) If $u \in C_0(X) = Q_0(X)$, then $\varepsilon(\mathrm{Sd}_0(u)) = \varepsilon(u)$, This is a triviality, since $\mathrm{Sd}_0 = \mathrm{sd}_0$ is the identity map. We can summarize this property by stating that the operator Sd_0 is *augmentation preserving*.

(d) For any n-chain $u \in C_n(X)$, there exists an integer $q \geq 0$ such that

$$\mathrm{sd}_n^q(u) \in C_n(X, \mathscr{U}),$$

where sd_n^q denotes the homomorphism obtained by q-fold iteration of sd_n. In order to prove this assertion, it suffices to prove that for each singular n-cube $T : I^n \to X$, there exists an integer $q(T)$ such that $\mathrm{Sd}_n^{q(T)}(T)$ is a sum of cubes which are small of order $\mathscr{U}$, i.e., such that $\mathrm{Sd}^{q(T)}(T) \in Q_n(X, \mathscr{U})$. Then if u is a linear combination of the singular n-cubes $T_1, T_2, \ldots, T_k$, it suffices to choose q to be the largest of the integers $q(T_1), q(T_2), \ldots, q(T_k)$.

To prove that such an integer $q(T)$ exists, consider the open covering of the compact metric space I^n by the inverse images under T of the *interiors* of the sets of the covering $\mathscr{U}$; let ε denote the Lebesgue number[1] of this covering. Then if we choose $q(T)$, so that

$$2^{-q(T)} < \varepsilon / \sqrt{n},$$

the required condition will be satisfied (the $\sqrt{n}$ occurs in the denominator because that is the ratio of the length of the diagonal to the length of the side for an n-dimensional cube).

Next, we are going to define homomorphisms

$$\varphi_n : C_n(X) \to C_{n+1}(X), \qquad n = 0, 1, \ldots$$

such that for any $u \in C_n(X)$,

$$\mathrm{sd}_n(u) - u = \partial_{n+1} \varphi_n(u) + \varphi_{n-1} \partial_n(u). \tag{7.3}$$

In the terminology of §4, the φ_n's are a chain homotopy between the subdivision operator, sd, and the identity map. In order to define φ_n, we first

---

[1] We say ε is a *Lebesgue number* of a covering of a metric space S if the following condition holds: any subset of S of diameter $< \varepsilon$ is contained in some set of the covering. It is a theorem that any open covering of a compact metric space has a Lebesgue number. The reader may either prove this as an exercise or look up a proof in a general topology book.

define two auxiliary functions $\eta_0, \eta_1 : I^2 \to I^1$ by the formulas

$$\eta_0(x_1, x_2) = \frac{x_1}{2 - x_2}$$

$$\eta_1(x_1, x_2) = \begin{cases} \dfrac{x_1 + 1}{2 - x_2} & \text{if } x_1 + x_2 \le 1, \\ 1 & \text{if } x_1 + x_2 \ge 1. \end{cases}$$

To gain a better understanding of η_0 and η_1, note that η_0 maps the square I^2 onto the interval $[0, \tfrac{1}{2}]$ and that the curves

$$\eta_0(x_1, x_2) = \text{constant}$$

are straight lines through the point $(0, 2)$. Also, η_1 maps the square I^2 onto the interval $[\tfrac{1}{2}, 1]$, and the curves

$$\eta_1(x_1, x_2) = \text{constant}$$

are straight lines through the point $(-1, 2)$, provided $x_1 + x_2 \le 1$.

Now for any $e \in \mathscr{E}_n$ and any singular n-cube $T : I^n \to X$, $n > 0$, define a singular $(n + 1)$-cube $G_e(T) : I^{n+1} \to X$ by the formula $(G_e T)(x_1, \ldots, x_{n+1}) = T(\eta_{e_1}(x_1, x_{n+1}), \eta_{e_2}(x_2, x_{n+1}), \ldots, \eta_{e_n}(x_n, x_{n+1}))$. Define

$$\Phi_n : Q_n(X) \to Q_{n+1}(X), \qquad n > 0$$

by

$$\Phi_n(T) = (-1)^{n+1} \sum_{e \in \mathscr{E}_n} G_e(T).$$

We will complete the definition by defining $\Phi_0 : Q_0(X) \to Q_1(X)$ to be the zero map. The motivation for the definition of Φ_n is indicated in Figure 1 for the case $n = 1$. We will now prove some properties of the homomorphisms Φ_n.

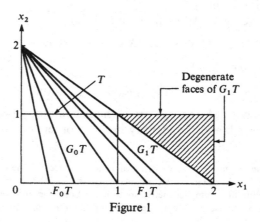

Figure 1

(e) If T is a degenerate cube, then so is $G_e(T)$. Hence Φ_n maps $D_n(X)$ into $D_{n+1}(X)$ and induces the desired homomorphism

$$\varphi_n: C_n(X) \to C_{n+1}(X), \qquad n = 0, 1, \ldots.$$

(f) For any singular n-cube $T: I^n \to X$, we have

$$\partial_{n+1}\Phi_n(T) = \mathrm{Sd}_n(T) - T - \Phi_{n-1}\partial_n(T) + \text{degenerate cubes.} \qquad (7.4)$$

Equation (7.3) follows from this. Of course Formula (7.4) is a triviality if $n = 0$. Therefore we will concentrate on the case $n > 0$. To compute $\partial_{n+1}\Phi_n(T)$, one needs the following identities:

(f.1) $A_{n+1}G_e(T) = F_e(T)$

(f.2) $B_{n+1}G_e(T) = T$ if $e = (0,0, \ldots ,0)$ and $B_{n+1}G_e(T)$ is a degenerate cube otherwise.

(f.3) Assume $e, e' \in \mathscr{E}_n, j \le n, e_j = 1, e'_j = 0$, and $e_i = e'_i$ for all $i \ne j$. Then

$$A_j G_e(T) = B_j G_{e'}(T)$$

for any n-cube T.

(f.4) Assume $e \in \mathscr{E}_n, n \ge 2$, and $e' \in \mathscr{E}_{n-1}$ is defined by

$$e' = (e_1, \ldots ,e_{j-1},e_{j+1}, \ldots ,e_n), \qquad j \le n,$$

If $e_j = 0$, then

$$A_j G_e(T) = G_{e'} A_j(T),$$

while if $e_j = 1$, then

$$B_j G_e(T) = G_{e'} B_j(T).$$

In case $n = 1$, $A_1 G_0 T$ and $B_1 G_1 T$ are degenerate.

By using Identities (f.1)–(f.4), it is a straightforward matter to verify Formula (7.4) and hence (7.3).

(g) If $u \in C_n(X,\mathscr{U})$, then $\varphi_n(u) \in C_{n+1}(X,\mathscr{U})$ also. To prove this, observe that if a cube T is small of order $\mathscr{U}$, then so is $G_e(T)$. Hence $\Phi_n(T) \in Q_{n+1}(X,\mathscr{U})$, and φ_n has the required property.

We have now defined the operators sd_n and φ_n, and proved their principal properties. For the sake of simplicity, we will write sd rather than sd_n and φ rather than φ_n from now on.

We also need the following formulas. For any integer $q > 0$, define

$$\psi_q: C_n(X) \to C_{n+1}(X), \qquad n = 0, 1, 2, \ldots$$

by

$$\psi_q(u) = \sum_{i=0}^{q-1} \mathrm{sd}^i(\varphi(u)).$$

The following equation now readily follows from Equation (7.3):

$$\mathrm{sd}^q(u) - u = \partial\psi_q(u) + \psi_q\partial(u). \qquad (7.5)$$

Note that Statement (g) above leads to the following:

(g') If $u \in C_n(X,\mathscr{U})$, then $\psi_q(u) \in C_{n+1}(X,\mathscr{U})$ for any integer $q > 0$.

With these preliminaries out of the way, we can now prove directly that

$$\sigma_* : H_n(X, \mathcal{U}) \to H_n(X)$$

is an isomorphism.

First, we prove that σ_* is an epimorphism. Let $x \in H_n(X)$; we will prove there exists an element $y \in H_n(X, \mathcal{U})$ such that $\sigma_*(y) = x$. Let $u \in C_n(X)$ be a representative cycle for x. By Statement (d) above, there exists an integer q such that

$$\mathrm{sd}^q(u) \in C_n(X, \mathcal{U}).$$

Since u is a cycle and sd commutes with the boundary operator, it follows that $\mathrm{sd}^q(u)$ is also a cycle. If we apply Equation (7.5), we see that u and $\mathrm{sd}^q(u)$ belong to the same homology class. Let y be the homology class of $\mathrm{sd}^q(u)$ in $H_n(X, \mathcal{U})$. Then $\sigma_*(y) = x$, as desired.

Next, we will prove that σ_* is a monomorphism. Assume $x \in H_n(X, \mathcal{U})$ and $\sigma_*(x) = 0$. We will show that $x = 0$. Let $v \in C_n(X, \mathcal{U})$ be a representative cycle for x. Since $\sigma_*(x) = 0$, there exists an element $u \in C_{n+1}(X)$ such that

$$\partial(u) = v.$$

Apply Statement (d) above to obtain an integer q such that

$$\mathrm{sd}^q(u) \in C_{n+1}(X, \mathcal{U}).$$

Now apply Equation (7.5),

$$\mathrm{sd}^q(u) - u = \partial \psi_q(u) + \psi_q(v).$$

Apply the boundary operator to both sides to obtain

$$\partial(\mathrm{sd}^q u) - v = \partial \psi_q(v)$$

or

$$v = \partial(\mathrm{sd}^q u - \psi_q(v)).$$

Since $v \in C_n(X, \mathcal{U})$, $\psi_q(v) \in C_{n+1}(X, \mathcal{U})$ by (g') above. Thus

$$\mathrm{sd}^q u - \psi_q(v) \in C_{n+1}(X, \mathcal{U})$$

and hence v is the boundary of a chain which is small of order $\mathcal{U}$. Therefore $x = 0$.

This completes the proof of Theorem 6.3 in the case $A = \varnothing$.

Next, we will prove Theorem 6.3 in the general case, where A is an arbitrary subset of X. Observe that for each integer n we have the following commutative diagram:

$$
\begin{array}{ccccccccc}
0 & \longrightarrow & C_n(A, \mathcal{U}) & \longrightarrow & C_n(X, \mathcal{U}) & \longrightarrow & C_n(X, A, \mathcal{U}) & \longrightarrow & 0 \\
 & & \downarrow{\sigma''} & & \downarrow{\sigma'} & & \downarrow{\sigma} & & \\
0 & \longrightarrow & C_n(A) & \xrightarrow{i_*} & C_n(X) & \xrightarrow{j_*} & C_n(X, A) & \longrightarrow & 0.
\end{array}
$$

Both of the rows in this diagram are exact sequences of chain groups. On passing to the corresponding homology groups, we obtain the following ladder-like diagram involving two long exact sequences:

$$\begin{array}{ccccccccc}
\longrightarrow & H_{n+1}(X,A,\mathscr{U}) & \longrightarrow & H_n(A,\mathscr{U}) & \longrightarrow & H_n(X,\mathscr{U}) & \longrightarrow & H_n(X,A,\mathscr{U}) & \longrightarrow \cdots \\
 & \downarrow{\sigma_*} & & \downarrow{\sigma''_*} & & \downarrow{\sigma'_*} & & \downarrow{\sigma_*} & \\
\longrightarrow & H_{n+1}(X,A) & \xrightarrow{\partial_*} & H_n(A) & \xrightarrow{i_*} & H_n(X) & \xrightarrow{j_*} & H_n(X,A) & \longrightarrow \cdots .
\end{array}$$

Each square in this diagram is commutative; the proof of this fact is exactly the same as the proof of the commutativity of Diagram (6.1). By what we have already proved, the homomorphisms σ'_* and σ''_* are isomorphisms. It now follows from the so-called *five-lemma* that the homomorphism σ_* is also an isomorphism, as was to be proved.

It remains to state and prove the *five-lemma*:

Lemma 7.1. *Consider the following diagram of abelian groups and homomorphisms.*

$$\begin{array}{ccccccccc}
A_1 & \xrightarrow{i_1} & A_2 & \xrightarrow{i_2} & A_3 & \xrightarrow{i_3} & A_4 & \xrightarrow{i_4} & A_5 \\
\downarrow{f_1} & & \downarrow{f_2} & & \downarrow{f_3} & & \downarrow{f_4} & & \downarrow{f_5} \\
B_1 & \xrightarrow{j_1} & B_2 & \xrightarrow{j_2} & B_3 & \xrightarrow{j_3} & B_4 & \xrightarrow{j_4} & B_5
\end{array}$$

Assume that each row is exact, that each square is commutative, that f_1 is an epimorphism, f_2 and f_4 are isomorphisms, and f_5 is a monomorphism. Then f_3 is also an isomorphism.

PROOF: It suffices to prove the following two assertions:
(a) For any $x \in A_3$, if $f_3(x) = 0$ then $x = 0$.
(b) Given any $x \in B_3$, there exists an element $y \in A_3$ such that $f_3(y) = x$.
The proof of each of these two assertions is carried out by a technique called "diagram chasing." For the reader who has seen this technique used before, the proof of this lemma will be very easy. For those who are unfamiliar with the technique, the proof of this lemma is an ideal exercise, and such readers are urged to work out the details of the proof. The proof of a proposition such as the five-lemma by diagram chasing requires practically no cleverness or ingenuity. At each stage of the proof there is only one possible "move"; one does not have to make any choices.

Determination of the Homology Groups of Certain Spaces: Applications and Further Properties of Homology Theory

§1. Introduction

In this chapter, we will actually determine the homology groups of various spaces: the n-dimensional sphere, finite graphs, and compact 2-dimensional manifolds. We also use homology theory to prove some classical theorems of topology, most of which are due to L. E. J. Brouwer. In addition, we prove some more basic properties of homology groups.

§2. Homology Groups of Cells and Spheres—Applications

We will now use the exact homology sequence and the excision property to determine the homology groups of a noncontractible space, namely, the n-sphere

$$S^n = \{x \in \mathbf{R}^{n+1} \, | \, \|x\| = 1\}.$$

This example is not only interesting in its own right; it is also basic to much that follows.

Theorem 2.1. *For any integer* $n \geq 0$,

$$\tilde{H}_i(S^n) = \begin{cases} \mathbf{Z} & \text{if } i = n, \\ \{0\} & \text{if } i \neq n. \end{cases}$$

Hence

$$H_0(S^n) = \begin{cases} \mathbf{Z} \oplus \mathbf{Z} & \text{if } n = 0, \\ \mathbf{Z} & \text{if } n > 0. \end{cases}$$

It is clear that the second statement is equivalent to the first statement for $i = 0$, in view of the relation between reduced and nonreduced homology groups.

PROOF OF THEOREM 2.1. The proof is by induction on n. The theorem is true for $n = 0$, because S^0 is a space consisting of exactly two points. In order to make the inductive step, we will identify S^n with the "equator" of S^{n+1}, i.e.,

$$S^n = \{x = (x_1, \ldots, x_{n+2}) \in S^{n+1} | x_{n+2} = 0\}.$$

We also need to consider the following two subsets of S^{n+1}:

$$E_+^{n+1} = \{(x_1, \ldots, x_{n+2}) \in S^{n+1} | x_{n+2} \geq 0\},$$
$$E_-^{n+1} = \{(x_1, \ldots, x_{n+2}) \in S^{n+1} | x_{n+2} \leq 0\}.$$

These may be referred to as the upper and lower hemispheres of S^{n+1}. These hemispheres are obviously homeomorphic to the set

$$E^{n+1} = \{(x_1, \ldots, x_{n+2}) \in R^{n+2} | |x| \leq 1 \text{ and } x_{n+2} = 0\},$$

hence they are contractible. The reader should draw a picture illustrating these sets for the case $n = 1$. Now consider the following diagram of homology groups:

$$\tilde{H}_i(S^n) \xleftarrow{\partial_*} H_{i+1}(E_-^{n+1}, S^n) \xrightarrow{k_*} H_{i+1}(S^{n+1}, E_+^{n+1}) \xleftarrow{j_*} \tilde{H}_{i+1}(S^{n+1}).$$

In this diagram, $j: S^{n+1} \to (S^{n+1}, E_+^{n+1})$ and $k: (E_-^{n+1}, S^n) \to (S^{n+1}, E_+^{n+1})$ denote inclusion maps. Consideration of the homology sequence of the pair (E_-^{n+1}, S^n) shows that ∂_* is an isomorphism, because E_-^{n+1} is contractible; similarly, it follows from the exactness of the homology sequence of the pair (S^{n+1}, E_+^{n+1}) and the contractibility of E_+^{n+1} that j_* is an isomorphism. To complete the proof, it suffices to prove that k_* is an isomorphism. Now the pair (E_-^{n+1}, S^n) is obtained from the pair (S^{n+1}, E_+^{n+1}) by excising the set $E_+^{n+1} - S^n$. However, we can not invoke the excision property (Theorem II.6.2) because the closure of $E_+^{n+1} - S^n$ is not contained in the interior of E_+^{n+1}. There is a way around this difficulty, however. Let

$$W = \{(x_1, \ldots, x_{n+2}) \in S^{n+1} | x_{n+2} \geq \tfrac{1}{2}\}.$$

Now consider the following diagram:

$$
\begin{array}{ccc}
H_{i+1}(E_-^{n+1}, S^n) & \xrightarrow{k_*} & H_{i+1}(S^{n+1}, E_+^{n+1}) \\
 & \searrow{h_*} \qquad e_* \nearrow & \\
 & H_{i+1}(S^{n+1} - W, E_+^{n+1} - W) &
\end{array}
$$

Here the symbols e and h denote inclusion maps. This diagram is obviously commutative. Now we *can* invoke the excision property to conclude that e_* is an isomorphism. Moreover, h_* is also an isomorphism, because the map h

is a homotopy equivalence of pairs; there is an obvious deformation retraction of the pair $(S^{n+1} - W, E_+^{n+1} - W)$ onto the pair (E_-^{n+1}, S^n). It follows from the commutativity of the diagram that k_* is also an isomorphism, as desired.   Q.E.D.

This proof illustrates the strategy that frequently has to be employed in applying the excision property. The situation is reminiscent of that often encountered in trying to apply the Seifert–Van Kampen theorem to determine the structure of the fundamental group of a space.

In §5 we will indicate an alternative proof of this theorem using the Mayer–Vietoris sequence.

We will now state some applications and corollaries of this result.

Proposition 2.2. *The sphere S^n is not contractible to a point.*

For the statement of the next two propositions, we will use the notation E^n to denote the set $\{x \in \mathbf{R}^n \,|\, |x| \leq 1\}$, called the unit *disc* or *ball* in $\mathbf{R}^n$ (the proofs are left to the reader).

Proposition 2.3. *S^n is not a retract of E^{n+1}.*

Proposition 2.4. *The relative homology groups of the pair (E^n, S^{n-1}) are as follows (for $n \geq 1$)*

$$H_i(E^n, S^{n-1}) = \begin{cases} 0 & i \neq n, \\ \mathbf{Z} & i = n. \end{cases}$$

Proposition 2.5 (*Brouwer fixed point theorem*). *Any continuous map $f : E^n \to E^n$ has at least one fixed point, i.e., a point x such that $f(x) = x$.*

PROOF: Assume to the contrary that $f(x) \neq x$ for all $x \in E^n$. Then the two distinct points x and $f(x)$ determine a unique straight line which intersects S^{n-1} in two points. Let $v(x)$ denote that point of the intersection which is such that x is between $v(x)$ and $f(x)$, or x is equal to $v(x)$. Then v is a map of E^n onto S^{n-1}. It is a nice technical exercise for the student to prove that v is continuous. It is obvious from the definition that v is a retraction. But this contradicts Proposition 8.3.   Q.E.D.

For a discussion of the significance of the Brouwer fixed point theorem, see *Algebraic Topology: An Introduction*, Chapter 2, §6.

We will use the knowledge we have gained about the homology groups of S^n to study continuous maps of S^n into itself. Let $f : S^n \to S^n$ be such a continuous map; consider the induced homomorphism

$$f_* : \tilde{H}_n(S^n) \to \tilde{H}_n(S^n).$$

Since $\tilde{H}_n(S^n)$ is an infinite cyclic group, there exists a unique integer d such that $f_*(u) = du$ for any $u \in \tilde{H}_n(S^n)$. This integer d is called the *degree* of f. It has the following basic properties:

(a) It is a homotopy invariant, i.e., if f_0 and f_1 are homotopic maps of S^n into itself, then f_0 and f_1 have the same degree. This fact is a direct consequence of the homotopy property of the induced homomorphism. It is proved in books on homotopy theory that the converse statement is also true, i.e., if f_0 and f_1 have the same degree, then they are homotopic.

(b) The degree of the composition of two maps is the product of the degrees. To be precise, if f and g are continuous maps $S^n \to S^n$, then degree $(gf) = ($degree $g)($degree $f)$.

Given any map $f : S^n \to S^n$, we will define a new map $\Sigma f : S^{n+1} \to S^{n+1}$, called the *suspension* of f by the following formula:

$$(\Sigma f)(x_1, x_2, \ldots, x_{n+2}) = \begin{cases} (0, \ldots, 0, x_{n+2}) & \text{if } |x_{n+2}| = 1, \\ \left(tf\left(\frac{x_1}{t}, \ldots, \frac{x_{n+1}}{t}\right), x_{n+2} \right) & \text{if } |x_{n+2}| < 1. \end{cases}$$

where $t = (1 - x_{n+2}^2)^{1/2}$. The geometric idea behind this formula may be described as follows: Σf maps the north pole of S^{n+1} to the north pole, the south pole of S^{n+1} to the south pole, and the equator into the equator according to the given map f. The meridian of S^{n+1} through the point x on the equator is mapped homeomorphically onto the meridian through the point $f(x)$.

(c) The degree of the suspension, Σf, is the same as that of the original map f. The proof of this property is left to the reader; it depends on the diagram used to prove Theorem 2.1 and the following two inclusions:

$$(\Sigma f)(E_+^{n+1}) \subset E_+^{n+1} \quad \text{and} \quad (\Sigma f)(E_-^{n+1}) \subset E_-^{n+1}.$$

In order to make use of this notion of degree, it is necessary to know the degree of certain explicit maps. The following are some propositions along this line. The proofs are left to the reader as exercises for the most part.

(d) The degree of the identity map is $+1$.

(e) The degree of a constant map is 0.

(f) Any map $f : S^0 \to S^0$ has degree ± 1 or 0.

(g) Let $v : S^n \to S^n$ denote the map which is reflection in a hyperplane through the origin of $\mathbf{R}^{n+1}$; then v has degree -1. To prove this, note first of all that we may choose our coordinate system so that the hyperplane in question has the equation $x_{n+1} = 0$. Then it is an easy task to prove this formula by induction on n, starting with the case $n = 0$.

(h) Let $f : S^n \to S^n$ denote the antipodal map, defined by $f(x) = -x$. Then the degree of f is $(-1)^{n+1}$. (*Hint:* Represent f as a composition of reflections.)

(i) Let $f : S^n \to S^n$ be a map which is fixed point free, i.e., $f(x) \neq x$ for all x. Then f is homotopic to the antipodal map, and hence has degree $(-1)^{n+1}$.

We will now use these facts to discuss the existence of continuous tangent vector fields on S^n. By a *tangent vector field* on S^n we mean a function v which assigns to each point $x \in S^n$ a vector $v(x)$ which is tangent to S^n at the point x. The tangency condition means that the vector $v(x)$ must be perpendicular to the unit vector x for all $x \in S^n$. The vector field v is said to be continuous (or differentiable) if the components of v are continuous (or differentiable) real-valued functions. When we speak of a *nonzero* vector field v, we mean that $v(x) \neq 0$ for *all* $x \in S^n$. The main theorem about such vector fields is the following:

Theorem 2.6. *There exists a continuous nonzero tangent vector field on S^n if and only if n is odd.*

It is easy to give an example of a continuous nonzero tangent vector field on S^n for n odd: One defines

$$v(x_1, \ldots, x_{n+1}) = (-x_2, x_1, -x_4, x_3, \ldots, -x_{n+1}, x_n).$$

To prove that such a vector field does not exist on S^n for n even, one proves the following statement: If there exists a continuous nonzero tangent vector field v on S^n, then the identity map of S^n onto itself is homotopic to a fixed-point-free map $f : S^n \to S^n$. In fact, one may define f by the formula

$$f(x) = \frac{x + v(x)}{|x + v(x)|}$$

and the homotopy by

$$f_t(x) = \frac{x + tv(x)}{|x + tv(x)|}, \qquad 0 \leq t \leq 1.$$

Theorem 2.6 now follows from this statement and Property (i) above.

Later on we will prove that there exist maps $S^n \to S^n$ of every possible degree provided $n \geq 1$.

The discussion of the degree of a map that we have just given applies only to maps of S^n into itself. These considerations may be extended to a slightly more general situation as follows. Let X and Y be topological spaces which are homeomorphic to S^n ($n \geq 1$) or more generally, have the same homotopy type as S^n. Then $H_n(X)$ and $H_n(Y)$ are infinite cyclic groups, hence there are two different choices possible for a generator of each of these groups. If definite choices of a generator have been made in each case, we will say that the spaces X and Y have been *oriented*. Assume that the chosen generators are denoted by $x \in H_n(X)$ and $y \in H_n(Y)$ respectively. Let $f : X \to Y$ be a continuous map; then there exists a unique integer d such that $f_*(x) = dy$. This integer d is called the *degree* of f. Note that changing the orientation of either X or Y changes the sign of the degree. It is a homotopy invariant of f, and has properties analogous to those discussed above.

EXAMPLE 2.1. Let $X = S^1$ and $Y = \mathbf{R}^2 - \{0\}$. We leave it to the reader to prove that S^1 is a deformation retract of $\mathbf{R}^2 - \{0\}$. A continuous map $S^1 \to \mathbf{R}^2 - \{0\}$ may be interpreted as a closed, continuous curve in the plane $\mathbf{R}^2$ which does not pass through the origin. The degree of such a map is essentially the same thing as the winding number of the closed path around the origin, as described in books on analysis.

EXERCISES

2.1. Prove that S^{n-1} is a deformation retract of $\mathbf{R}^n - \{0\}$.

2.2. Prove that the complement of a point in S^n is homeomorphic to $\mathbf{R}^n$ (stereographic projection).

2.3. Prove by two different methods that $\mathbf{R}^m$ and $\mathbf{R}^n$ are not homeomorphic if $m \neq n$: (a) Prove that their Alexandroff 1-point compactifications are not homeomorphic, and (b) Prove that the complement of a point in $\mathbf{R}^m$ is not homeomorphic to the complement of a point in $\mathbf{R}^n$.

2.4. Prove that any homeomorphism h of E^n onto itself maps S^{n-1} onto S^{n-1}. (*Hint*: Consider the complement of a point.)

2.5. Let $f : S^n \to S^n$ be a continuous map whose degree is nonzero. Prove that f maps S^n onto S^n.

2.6. If X is a Hausdorff space and $x \in X$, then $H_n(X, X - \{x\})$ is called the n-dimensional *local homology group* of X at x. Justify this name by showing that it only depends on arbitrarily small neighborhoods of x in X.

2.7. Determine the local homology groups at various points of the closed n-dimensional ball, E^n. Use this computation to give another solution of Exercise 2.4.

2.8. Use local homology groups to prove that an n-dimensional and an m-dimensional manifold are not homeomorphic if $m \neq n$.

2.9. Prove that a Möbius strip is *not* homeomorphic to the annulus $\{x \in \mathbf{R}^2 \,|\, 1 \le |x| \le 2\}$, although they have the same homotopy type and both are compact. (*Suggestion*: As a first step, determine local homology groups at various points of both spaces.)

§3. Homology of Finite Graphs

In this section we will use the properties of relative homology groups to develop a systematic procedure for computing the homology groups of a rather simple type of topological space called a graph. The results obtained are not very profound; however, they are illustrative of the techniques we will use later to determine the homology groups of more general spaces.

Definition 3.1. A *finite, regular graph* (or just a *graph* for short) is a pair consisting of a Hausdorff space X and a finite subspace X^0 (points of X^0 are called *vertices*) such that the following conditions hold:

(a) $X - X^0$ is the disjoint union of a finite number of open subsets e_1, $e_2, \ldots, e_k$, called *edges*. Each e_i is homeomorphic to an open interval of the real line.

(b) The point set boundary, $\bar{e}_i - e_i$, of the edge e_i consists of two distinct vertices, and the pair $(\bar{e}_i, e_i)$ is homeomorphic to the pair $([0,1],(0,1))$.

One could also consider infinite graphs, and nonregular graphs, i.e., those for which $\bar{e}_i - e_i$ may consist of one *or* two vertices; (cf. *Algebraic Topology: An Introduction*, Chapter VI). However, we will not do this for the present.

Note that a graph is compact, since it is the union of a finite number of compact subsets (the closed edges $\bar{e}_i$ and the vertices). It may be either connected or disconnected, and it may have isolated vertices. If a vertex v belongs to the closure of an edge e_i, it is customary to say that e_i and v are *incident*.

It is easy to give many examples of graphs. It can be shown that every graph, as defined here, can be embedded homeomorphically in Euclidean 3-space, and many can be embedded in the plane. A famous theorem of Kuratowski (1920) gives necessary and sufficient conditions for a graph to be embedded in the plane.

If a space X can be given a structure of a graph by specifying a set of vertices X^0 then we can specify additional graph structures on X by *subdividing*, i.e., inserting additional vertices (provided the set of edges is nonempty).

We will now show how to determine the structure of the homology groups of a graph X. First, we will determine the relative homology groups of the pair (X,X^0) and then use the exact homology sequence of (X,X^0) to achieve our goal. Let $e_1, e_2, \ldots, e_k$ denote the edges of the given graph (X,X^0). We will consistently use the notation $\dot{e}_i = \bar{e}_i - e_i$ to denote the boundary of the edge e_i. It follows from Proposition 2.4 and the definition of a graph that

$$H_q(\bar{e}_i, \dot{e}_i) = \begin{cases} \mathbf{Z} & \text{for } q = 1, \\ 0 & \text{for } q \neq 1. \end{cases} \tag{3.1}$$

Theorem 3.1. *Let (X,X^0) be a finite, regular graph with edges $e_1, e_2, \ldots, e_k$. Then the inclusion map $(\bar{e}_i, \dot{e}_i) \to (X,X^0)$ induces a monomorphism $H_q(\bar{e}_i, \dot{e}_i) \to H_q(X,X^0)$ for $i = 1, 2, \ldots, k$ and $H_q(X,X^0)$ is the direct sum of the image subgroups. It follows that $H_1(X,X^0)$ is a free abelian group of rank k, and $H_q(X,X^0) = 0$ for $q \neq 1$.*

(Note: The *rank* of a free abelian group is the number of elements in a basis; it is proved in books on linear algebra that it is an invariant of the group).

PROOF: The third sentence of the theorem is a consequence of the two preceding sentences, in view of Equation (3.1) above. Therefore we will concentrate our attention on the first two sentences of the theorem.

According to the definition of a graph, the set $\bar{e}_i$ is homeomorphic to the unit interval $I = [0,1]$; choose a definite homeomorphism of $\bar{e}_i$ with I for $i = 1, 2, \ldots, k$ and let a_i denote the point which corresponds to $\frac{1}{2} \in I$; it is the midpoint of the edge e_i. Similarly, let d_i denote the subset of e_i which corresponds to the closed subinterval $[\frac{1}{4}, \frac{3}{4}]$, and $D = d_1 \cup d_2 \cup \cdots \cup d_k$, $A = \{a_1, a_2, \ldots, a_k\}$. Our proof of the theorem is based on the consideration of the following diagram:

$$
\begin{array}{ccccc}
H_q(D, D - A) & \xrightarrow{\ 1\ } & H_q(X, X - A) & \xleftarrow{\ 2\ } & H_q(X, X^\circ) \\
\Big\uparrow{\scriptstyle 5} & & \Big\uparrow & & \Big\uparrow{\scriptstyle 6} \\
H_q(d_i, d_i - \{a_i\}) & \xrightarrow{\ 3\ } & H_q(\bar{e}_i, \bar{e}_i - \{a_i\}) & \xleftarrow{\ 4\ } & H_q(\bar{e}_i, \dot{e}_i).
\end{array}
$$

All homomorphisms in this diagram are induced by inclusion maps of the corresponding pairs. It follows that each square of this diagram is commutative. We assert that *all four horizontal arrows in this diagram denote isomorphisms*. For arrow 4, this follows from the fact that $\dot{e}_i$ is a deformation retract of $\bar{e}_i - \{a_i\}$, together with the five-lemma (Lemma II.7.1). Exactly the same kind of argument shows that Arrow 2 is an isomorphism. It follows from the excision property that Arrows 1 and 3 are isomorphisms.

The theorem now follows from the fact that the space D is disconnected and its components are $d_1, d_2, \ldots, d_k$, and $d_i - \{a_i\} = d_i \cap (D - A)$ (cf. Exercise II.5.2). Q.E.D.

We will now consider the exact homology sequence of the pair (X, X°). The structure of the relative homology groups $H_q(X, X^\circ)$ is described by the theorem just proved. Since X° is a finite space with the discrete topology, $H_q(X^\circ) = 0$ for $q \neq 0$, and $H_0(X^\circ)$ is a free abelian group whose rank is equal to the number of vertices. From this it follows easily that $H_q(X) = 0$ for $q > 1$, and the only nontrivial portion of the homology sequence of the pair (X, X°) is the following:

$$0 \to H_1(X) \xrightarrow{j_*} H_1(X, X^\circ) \xrightarrow{\partial_*} H_0(X^\circ) \xrightarrow{i_*} H_0(X) \to 0. \qquad (3.2)$$

We already know that $H_0(X)$ is a free abelian group whose rank is equal to the number of arc-components of the topological space X. For a finite, regular graph, it is readily proved that the components and arc-components are the same.

Thus we know the structure of all the groups in the homology sequence of the pair (X, X°), with the exception of $H_1(X)$. To determine the structure of this one remaining group, we need the following two results from linear algebra:

(A) *Any subgroup of a free abelian group is also free abelian.*

(B) *Let $f:A \to F$ be an epimorphism of an abelian group A onto the free abelian group F. Then the kernel of f is a direct summand of A; the other summand is isomorphic to F.*

The proofs of these propositions may be found in textbooks on linear algebra. The proof of (B) is especially simple.

Definition 3.2. The *Euler characteristic* of a graph is the number of vertices minus the number of edges.

We can now state the main theorem about the homology groups of a graph:

Theorem 3.2. *Let (X,X^0) be a finite, regular graph. Then $H_q(X) = 0$ for $q > 1$, $H_1(X)$ is a free abelian group, and*

$$\text{rank}(H_0(X)) - \text{rank}(H_1(X)) = \text{Euler characteristic.}$$

We leave it to the reader to prove this theorem, using the homology sequence of the pair (X,X^0) and the two results from linear algebra stated above.

This theorem gives a simple method for determining the structure for $H_1(X)$. For we can determine the rank of $H_0(X)$ by counting the number of components, and we can determine the Euler characteristic by counting the number of vertices and edges. For certain purposes it is necessary to go more deeply into the structure of $H_1(X)$, and actually give some sort of concrete representation of the elements of this group. This we will now proceed to do.

The exact sequence (3.2) shows that $H_1(X)$ and $H_0(X)$ are the kernel and cokernel respectively of the homomorphism $\partial_*:H_1(X,X^0) \to H_0(X^0)$. Our procedure will be to choose convenient bases for the free abelian groups $H_1(X,X^0)$ and $H_0(X^0)$, and then express ∂_* in terms of these bases. The edges of the graph X will be denoted by $e_1, \ldots, e_k$ and the vertices by $v_1, \ldots, v_m$.

It is easy to choose a natural basis for the group $H_0(X^0)$. Since X^0 is a discrete space, $H_0(X^0)$ is naturally isomorphic to the direct sum of the groups $H_0(v_i)$ for $i = 1, 2, \ldots, m$. The augmentation homomorphism $\varepsilon:H_0(v_i) \to \mathbf{Z}$ is an isomorphism; therefore it is natural to choose as a generator of $H_0(v_i)$ the element a_i such that $\varepsilon(a_i) = 1$. Then $\{a_1, \ldots, a_m\}$ is a basis for $H_0(X^0)$. To avoid proliferation of notation, it is convenient to use the same symbol v_i for the basis element $a_i \in H_0(v_i)$. This abuse of notation will hardly ever lead to confusion, and it is sanctioned by many decades of use. Thus we will denote our basis of $H_0(X^0)$ by $\{v_1, \ldots, v_m\}$.

Choosing a basis for $H_1(X,X^0)$ is only slightly more complicated.

According to Theorem 3.1, $H_1(X,X^0)$ decomposes into the direct sum of infinite cyclic subgroups, which correspond to the edges $e_1, \ldots, e_k$. Thus to choose a basis for $H_1(X,X^0)$ it suffices to choose a generator for the infinite cyclic group $H_1(\bar{e}_i,\dot{e}_i)$ for $i = 1, 2, \ldots, m$. It turns out that such a choice is purely arbitrary; there is no natural or preferred choice of a generator. In order to understand the meaning of such a choice, consider the following commutative diagram (cf. Exercise II.5.5):

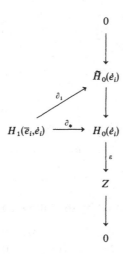

The homomorphism ∂_1 is an isomorphism; thus choosing a generator for $H_1(\bar{e}_i,\dot{e}_i)$ is equivalent to choosing a generator for $\tilde{H}_0(\dot{e}_i)$. The set $\dot{e}_i$ consists of two vertices; let us denote them by v_α and v_β. Using the convention introduced in the preceding paragraph, we may use the same symbols, v_α and v_β, to denote a basis for $H_0(\dot{e}_i)$. With this convention, the two possible choices of a generator for the infinite cyclic subgroup $\tilde{H}_0(\dot{e}_i)$ are $v_\alpha - v_\beta$ and $v_\beta - v_\alpha$. Thus we see that a choice of basis for $H_1(\bar{e}_i,\dot{e}_i)$ corresponds to an ordering of the vertices of the edge e_i. For this reason, we will say that we *orient* the edge e_i when we make such a choice. To make things precise, we lay down the following rule: *Orient the edge e_i by choosing an ordering of its two vertices. If $v_\beta > v_\alpha$, then this ordering of vertices corresponds to the generator $\partial_1^{-1}(v_\beta - v_\alpha)$ of the group $H_1(\bar{e}_i,\dot{e}_i)$.*

We can now give the following recipe for the homomorphism ∂_*: $H_1(X,X^0) \to H_0(X^0)$:

(a) A basis for $H_0(X^0)$ consists of the set of vertices.
(b) Orient the edges by choosing an order for the vertices of each edge. On a diagram or drawing of the given graph, it is convenient to indicate the orientation by an arrow on each edge pointing from the first vertex to the second.
(c) A basis for $H_1(X,X^0)$ consists of the set of oriented edges.
(d) If e_i is any edge, with vertices v_α and v_β and orientation determined by the relation $v_\beta > v_\alpha$, then

$$\partial_*(e_i) = v_\beta - v_\alpha.$$

EXAMPLE 3.1. Figure 2 shows a graph with six vertices and nine edges which can not be imbedded in the plane. (This graph comes up in the well-known problem of the three houses and the three utilities). We have oriented all the edges by placing arrows on them which point upwards. According to the preceding rules, the homomorphism ∂_* is given by the following formulas:

$$
\begin{aligned}
\partial_*(e_1) &= v_1 && -v_4 \\
\partial_*(e_2) &= \quad v_2 && -v_5 \\
\partial_*(e_3) &= \quad\quad v_3 && -v_6 \\
\partial_*(e_4) &= \quad v_2 && -v_4 \\
\partial_*(e_5) &= \quad\quad v_3 && -v_5 \\
\partial_*(e_6) &= v_1 && -v_5 \\
\partial_*(e_7) &= \quad v_2 && -v_6 \\
\partial_*(e_8) &= \quad\quad v_3 \quad -v_4 \\
\partial_*(e_9) &= v_1 && -v_6.
\end{aligned}
$$

In other words, ∂_* is represented by the following matrix:

$$
\begin{bmatrix}
1 & 0 & 0 & -1 & 0 & 0 \\
0 & 1 & 0 & 0 & -1 & 0 \\
0 & 0 & 1 & 0 & 0 & -1 \\
0 & 1 & 0 & -1 & 0 & 0 \\
0 & 0 & 1 & 0 & -1 & 0 \\
1 & 0 & 0 & 0 & -1 & 0 \\
0 & 1 & 0 & 0 & -0 & -1 \\
0 & 0 & 1 & -1 & 0 & 0 \\
1 & 0 & 0 & 0 & 0 & -1
\end{bmatrix}
$$

There remains the problem of determining the kernel and cokernel of ∂_*. In books on linear algebra there is an algorithm described for introducing new

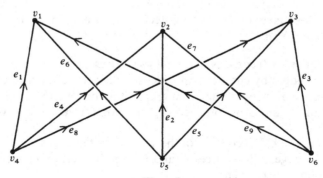

Figure 2

bases in the domain and range of such a homomorphism so that the corresponding matrix is a diagonal matrix. Then generators of the kernel and cokernel can be read off with ease. Unfortunately, this algorithm is rather lengthy and tedious. As a practical alternative, one can proceed as follows.

The Euler characteristic of this graph is $6 - 9 = -3$. Since it is connected, $H_0(X)$ has rank 1. Hence $H_1(X)$ has rank 4, by Theorem 3.2. Therefore we should be able to find four linearly independent elements in the kernel of ∂_*, and then hope to prove that they form a basis for the kernel of ∂_*. Consider the following four elements of $H_1(X, X^0)$:

$$z_1 = e_1 - e_6 + e_2 - e_4,$$
$$z_2 = e_2 - e_4 + e_8 - e_5,$$
$$z_3 = e_3 - e_5 + e_6 - e_9,$$

and

$$z_4 = e_7 - e_2 + e_5 - e_3.$$

These four elements (which we may as well call *cycles*) were determined by inspection of the above diagram. They correspond in an obvious way to certain oriented closed paths in the diagram. It is readily verified that all four of these cycles actually belong to the kernel of ∂_*, and that they are linearly independent. Finally, it is a nice exercise in linear algebra to check that the set $\{e_1, e_2, e_3, e_4, e_5, z_1, z_2, z_3, z_4\}$ is also a basis for $H_1(X, X^0)$. These facts suffice to prove that $\{z_1, z_2, z_3, z_4\}$ is actually a basis for the kernel of ∂_*, or what is equivalent, for the homology group $H_1(X)$. We leave it to the reader to carry through the details of the proof. The reader is strongly urged to make diagrams of several graphs and determine a set of linearly independent cycles which constitute a basis for the 1-dimensional homology group of each graph. It is only by such exercises that one can gain an adequate understanding and intuitive feeling for homology theory. The idea that a 1-dimensional homology class is represented by a linear combination of cycles is very important.

Next we will discuss the problem of determining the homomorphism induced on the 1-dimensional homology groups by a continuous map from one graph to another. This problem is probably just as important as the problem of determining the structure of the 1-dimensional homology groups. Let (X, X^0) and (Y, Y^0) be finite regular graphs and $f: X \to Y$ a continuous map. In order to have an effective procedure for determining the induced homomorphism $f_*: H_1(X) \to H_1(Y)$, it is necessary to impose some conditions of f. The following will be convenient for our purposes:

(A) $f(X^0) \subset Y^0$, i.e., f maps vertices into vertices.

(B) Given any edge e_i of X, either f maps $\bar{e}_i$ homeomorphically onto some closed edge $\bar{e}'_j$ of Y, or f maps $\bar{e}_i$ onto a vertex of Y.

Of course most continuous maps f do not satisfy these conditions. However, it can be shown that one can deform any map f homotopically into one which *does* satisfy them, provided one subdivides (X,X^0) first. In view of the invariance of f_* under homotopies, this is allowable for our purposes.

Since $f(X^0) \subset Y^0$, we may consider f as a map of pairs: $(X,X^0) \to (Y,Y^0)$. Hence we obtain the following commutative diagram involving the exact homology sequences of the pairs (X,X^0) and (Y,Y^0):

$$
\begin{array}{ccccccccc}
0 & \longrightarrow & H_1(X) & \xrightarrow{j_*} & H_1(X,X^0) & \xrightarrow{\partial_*} & H_0(X^0) & \xrightarrow{i_*} & H_0(X) & \longrightarrow & 0 \\
 & & \downarrow{f_*} & & \downarrow{f_1} & & \downarrow{f_0} & & \downarrow{f_*} & & \\
0 & \longrightarrow & H_1(Y) & \xrightarrow{j_*} & H_1(Y,Y^0) & \xrightarrow{\partial_*} & H_0(Y^0) & \xrightarrow{i_*} & H_0(Y) & \longrightarrow & 0.
\end{array}
$$

From this diagram, it is clear that the homomorphism $f_*: H_1(X) \to H_1(Y)$ is completely determined by the homomorphism labelled f_1. To determine the homomorphism f_1, it suffices to describe its effect on the basis we have chosen for $H_1(X,X^0)$, i.e., on the oriented edges. Suppose first that f maps $\bar{e}_i$ homeomorphically onto the closed edge $\bar{e}'_j$ of Y, as stated in condition B above. We assume that the edges e_i and e'_j have both been oriented by choosing an order of their vertices. Then two cases arise, according as the map f is orientation preserving, or orientation reversing (the meaning of these terms is obvious). We leave it to the reader to prove that

$$
f_1(e_i) = \begin{cases} +e'_j & \text{if } f \text{ preserves orientation,} \\ -e'_j & \text{if } f \text{ reverses orientation.} \end{cases}
$$

Here f_1 denotes the homomorphism $H_1(X,X^0) \to H_1(Y,Y^0)$ induced by f, while $e_i \in H_1(X,X^0)$ and $e'_j \in H_1(Y,Y^0)$ denote the basis elements represented by the corresponding oriented edges. Suppose next that f maps the edge e_i of X onto the vertex v'_j of Y. Then

$$
f_1(e_i) = 0.
$$

To prove this equation, consider the following commutative diagram:

$$
\begin{array}{ccc}
H_1(X,X^0) & \xrightarrow{f_1} & H_1(Y,Y^0) \\
\uparrow & & \uparrow \\
H_1(\bar{e}_i,\dot{e}_i) & \xrightarrow{f_*} & H_1(v'_j,v'_j).
\end{array}
$$

The vertical arrows denote homomorphisms induced by inclusion maps. Since $H_1(v'_j,v'_j) = 0$, the assertion follows.

EXAMPLE 3.2. By subdividing into short arcs, the circle S^1 may be considered as a graph in various different ways. Let us consider S^1 as the unit circle in the complex plane, $\mathbf{C}$:

$$
S^1 = \{z \in \mathbf{C} \,|\, |z| = 1\}.
$$

Let $f : S^1 \to S^1$ be the continuous map defined by $f(z) = z^3$. We wish to determine the induced homomorphism $f_* : H_1(S^1) \to H_1(S^1)$. In order to solve this problem, we need to subdivide S^1 into a regular graph in two different ways. The first subdivision is into 6 equal arcs by means of the vertices

$$v_j = \exp\left(j \frac{\pi \sqrt{-1}}{3}\right), \qquad j = 0, 1, \ldots, 5.$$

The corresponding (oriented) edges $e_0, e_1, \ldots, e_5$ are shown in Figure 3. The second subdivision is into two semicircles by the vertices $u_0 = +1$ and $u_1 = -1$; the corresponding (oriented) edges, denoted by e'_0 and e'_1 are also shown in the diagram. Let $X^0 = \{v_0, v_1, \ldots, v_5\}$ and $Y^0 = \{u_0, u_1\}$. Then we can consider f as a map of pairs, $(S^1, X^0) \to (S^1, Y^0)$, and the conditions A and B above are fulfilled, with $X = Y = S^1$. The induced homomorphism $f_1 : H_1(S^1, X^0) \to H_1(S^1, Y^0)$ is described by the following equation,

$$f_1(e_j) = \begin{cases} -e'_0 & \text{if } j = 0, 2, \text{ or } 4, \\ -e'_1 & \text{if } j = 1, 3, \text{ or } 5, \end{cases} \tag{3.3}$$

in view of our choice of orientations. The kernels of the homomorphisms

$$\partial_* : H_1(S^1, X^0) \to H_0(X^0),$$
$$\partial_* : H_1(S^1, Y^0) \to H_0(Y^0)$$

are both of rank 1, and they are generated by the cycles

$$x = \sum_{j=0}^{5} e_j \quad \text{and} \quad y = e'_0 + e'_1 \tag{3.4}$$

respectively. We can consider each of these cycles as a representative of a generator of the infinite cyclic group $H_1(S^1)$; in view of the way the orientations of the edges were chosen, it seems likely that the generators so represented are the negatives of each other. It follows readily from Equations

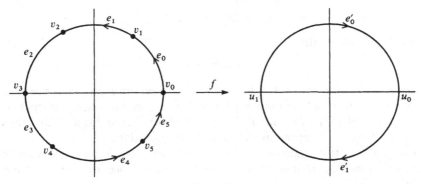

Figure 3

(3.3) and (3.4) that

$$f_1(x) = -3y;$$

thus the map f has degree ± 3. Actually, its degree is $+3$.

EXAMPLE 3.3. The preceding example raises the following question: suppose we subdivide a given space X into a finite regular graph in two different ways. Using each of these subdivision, we can determine cycles which represent elements of the homology group $H_1(X)$. How can we compare representative cycles from the two different subdivisions? The following example shows how this problem can be solved. Consider two different subdivisions of the unit circle S^1 in the complex plane; for example, consider the two subdivisions considered in the previous example with vertex sets

$$X^0 = \{v_0, \dots, v_5\} \quad \text{and} \quad Y^0 = \{u_0, u_1\}$$

respectively. We will define a continuous map $g: S^1 \to S^1$ such that g is homotopic to the identity map, and so that g is a map of pairs $(S^1, X^0) \to (S^1, Y^0)$ such that conditions A and B above hold. The easiest way to define g is to define it separately on each closed cell $\bar{e}_j$, taking care that the various mappings so defined agree on the end points of the cells. We list the definitions as follows:

(a) g shall map $\bar{e}_0$ homeomorphically onto $\bar{e}_0'$ with $g(v_0) = u_0$ and $g(v_1) = u_1$.
(b) $g(\bar{e}_1) = g(\bar{e}_2) = u_1$.
(c) g maps $\bar{e}_3$ homeomorphically onto e_1' with $g(v_3) = u_1$ and $g(v_4) = u_0$.
(d) $g(\bar{e}_4) = g(\bar{e}_5) = u_0$.

We leave it to the reader to verify that g is actually homotopic to the identity map of S^1 onto itself. The induced homomorphism $g_1: H_1(S^1, X^0) \to H_1(S^1, Y^0)$ is described by the following equations, using the same orientations of edges as in the preceding example:

$$g_1(e_0) = -e_0'$$
$$g_1(e_3) = -e_1'$$
$$g_1(e_j) = 0 \quad \text{for } j = 1, 2, 4, \text{ or } 5.$$

From this it follows that

$$g_1(x) = -y,$$

where x and y are the cycles defined in the previous example. Since g is homotopic to the identity map, we know that the induced homomorphism $g_*: H_1(S^1) \to H_1(S^1)$ is the identity homomorphism. From this it follows that the cycles x and $-y$ represent the same homology class.

The point of this example is not so much to prove rigorously what is intuitively obvious, as it is to illustrate a general procedure for handling questions of this kind.

EXERCISES

3.1. Determine the degree of the mapping $f:S^1 \to S^1$ defined by $f(z) = z^k$ for any integer k.

3.2. Prove that for any integer k and any positive integer n there exists a continuous map $f:S^n \to S^n$ of degree k.

3.3. Identify S^2 with the Alexandroff 1-point compactification of the complex plane $\mathbf{C}$, obtained by adjoining a point to $\mathbf{C}$, called the *point at infinity*. Let $f(z)$ be a polynomial of positive degree with complex coefficients; we may consider f to be a continuous nonconstant map $f:\mathbf{C} \to \mathbf{C}$. Prove that we may extend f to a *continuous* map $\bar{f}:S^2 \to S^2$ by mapping the point at infinity onto itself.

3.4 Let $f(z) = z^k$, $k > 0$. Determine the degree of the extension $\bar{f}:S^2 \to S^2$ of f defined according to the procedure of the preceding exercise.

3.5. Let $f(x)$ be a polynomial of degree $k > 0$ with complex coefficients. Determine the degree of the extension $\bar{f}:S^2 \to S^2$ of f defined as above.

3.6. Let $f(z)$ be a polynomial of degree $k > 0$ with complex coefficients. Prove that the equation $f(z) = 0$ has at least one root in the field of complex numbers, $\mathbf{C}$ (this is the so-called *fundamental theorem of algebra*).

3.7. Let $X = \{(x,y,z) \in \mathbf{R}^3 \,|\, xyz = 0\}$ i.e., X is the union of the three coordinate planes. Prove that any homeomorphism of X onto itself must have the origin, $(0,0,0)$, as a fixed point. (*Suggestion:* Determine the local homology groups at various points.)

§4. Homology of Compact Surfaces

A compact surface is homeomorphic to one of the following: the 2-sphere, S^2; the torus, $S^1 \times S^1$; the real projective plane; a connected sum of tori; or, a connected sum of projective planes. For a description of these various surfaces, see *Algebraic Topology: An Introduction*, Chapter I. The main fact that we will use is that every connected surface can be obtained from some polygonal disc by identifying the edges in pairs according to a certain scheme.

EXAMPLE 4.1 (The torus). We can think of a torus as obtained from a rectangle by identification of the opposite edges, as shown in Figure 4. Under the

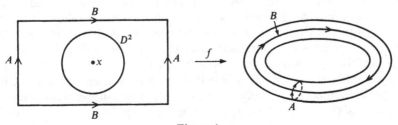

Figure 4

identification, each pair of edges becomes a circle, and the two circles, labelled A and B in the diagram, intersect in a single point. We will use the following notation:

E^2 = the rectangle
X = the torus
$f: E^2 \to X$, the identification map,
$\dot{E}^2$ = boundary of the rectangle
$X^1 = f(\dot{E}^2) = A \cup B$.

The homology groups of X^1 can be determined by the methods of the preceding section. If we knew the relative homology groups $H_q(X,X^1)$, then we could hope to determine the homology groups of X by studying the exact homology sequence of the pair (X,X^1).

Proposition 4.1. *The identification map* $f:(E^2,\dot{E}^2) \to (X,X^1)$ *induces an isomorphism* $f_*: H_q(E^2,\dot{E}^2) \to H_q(X,X^1)$ *of relative homology groups for all* q. *Hence* $H_q(X,X^1) = 0$ *for* $q \neq 2$, *and* $H_2(X,X^1)$ *is infinite cyclic.*

PROOF: The last sentence is a consequence of the first sentence and Proposition 2.4. The pattern of proof of the first sentence of the proposition, using the excision property, deformation retracts, etc., is one that we have used before a couple of times.

Let x denote the center point of the rectangle E^2, and let D^2 denote a closed disc with center at the point x whose radius is small enough so that it is contained entirely in the interior of the rectangle E^2. Consider the following diagram of relative homology groups:

$$
\begin{array}{ccccc}
H_q(E^2,\dot{E}^2) & \xrightarrow{\;1\;} & H_q(E^2, E^2 - \{x\}) & \xleftarrow{\;3\;} & H_q(D^2, D^2 - \{x\}) \\
\downarrow{\scriptstyle f_*} & & \downarrow & & \downarrow{\scriptstyle 5} \\
H_q(X,x^1) & \xrightarrow{\;2\;} & H_q(X, X - \{fx\}) & \xleftarrow{\;4\;} & H_q(fD^2, fD^2 - \{fx\}).
\end{array}
$$

In this diagram the horizontal arrows all denote homomorphisms induced by inclusion maps, and the vertical arrows denote homomorphisms induced by f. Each square in the diagram is commutative.

We assert that the four homomorphisms denoted by horizontal arrows are all isomorphisms. For arrows 3 and 4 this assertion follows from the excision property. For arrow 1, it follows from the fact that $\dot{E}^2$ is a deformation retract of $E^2 - \{x\}$; one must also use the five-lemma. By a similar argument, the assertion can be proved for arrow 2.

To complete the proof, observe that arrow 5 is an isomorphism, because f maps D^2 homeomorphically onto $f(D^2)$. It now follows from the commutativity of the diagram that f_* is also an isomorphism.   Q.E.D.

The subset X^1 of X can be subdivided so as to be a finite, regular graph; it is obviously connected, and its Euler characteristic is -1. Therefore $H_0(X^1) = \mathbf{Z}$, $H_1(X^1) = \mathbf{Z} \oplus \mathbf{Z}$, and $H_q(X^1) = 0$ for $q > 1$. If we put this information about the homology groups of (X,X^1) and X^1 into the exact homology sequence of the pair (X,X^1), we see that $H_q(X) = 0$ for $q > 2$, and the only nontrivial part of this homology sequence is the following:

$$0 \to H_2(X) \xrightarrow{j_*} H_2(X,X^1) \xrightarrow{\partial_*} H_1(X^1) \xrightarrow{i_*} H_1(X) \to 0. \qquad (4.1)$$

From this sequence, we see that $H_2(X)$ and $H_1(X)$ are the kernel and cokernel respectively of the homomorphism ∂_*. Thus it is necessary to determine ∂_*. For this purpose consider the following commutative diagram:

$$
\begin{array}{ccc}
H_2(X,X^1) & \xrightarrow{\ \partial_*\ } & H_1(X^1) \\
\uparrow{\scriptstyle f_*} & & \uparrow{\scriptstyle f_{1*}} \\
H_2(E^2,\dot{E}^2) & \xrightarrow{\ \partial'_*\ } & H_1(\dot{E}^2).
\end{array}
$$

By the proposition just proved, f_* is an isomorphism. It follows from consideration of the homology sequence of the pair $(E^2,\dot{E}^2)$ that ∂'_* is an isomorphism. The homomorphism f_{1*} is induced by the identification maps $f_1 : \dot{E}^2 \to X^1$; this is a map of finite, regular graphs of the type discussed in Section 3. Using the techniques of that section, it is a routine matter to calculate that f_{1*} is the zero homomorphism; we leave the details to the reader. From this it follows that ∂_* is also the zero homomorphism.

Going back to the exact sequence (4.1) we see that both j_* and i_* are isomorphisms. Thus we have completely determined the structure of the homology groups of the torus, as follows:

$$H_0(X) = \mathbf{Z} \qquad (X \text{ is connected}),$$
$$H_1(X) = \mathbf{Z} \oplus \mathbf{Z},$$
$$H_2(X) = \mathbf{Z},$$

and

$$H_q(X) = 0 \quad \text{for } q > 2.$$

The fact that the inclusion map $i : X^1 \to X$ induces an isomorphism $i_* : H_1(X^1) \to H_1(X)$ is also significant. This means that elements of $H_1(X)$ can be represented by cycles on the graph X^1. Note also that this statement is still true if the inclusion map $i : X^1 \to X$ is deformed homotopically into some other map.

EXAMPLE 4.2 (The connected sum of n tori, $n > 1$ (an orientable surface of genus n)). This example is completely analogous to the torus. Such a surface can be obtained from a polygonal disc having $4n$ edges by identifying the edges of pairs according to the scheme shown in Figure 5. Under the identification, each pair of edges becomes a circle on the surface X, and these $2n$ circles, which may be denoted $A_1, A_2, \ldots, A_n, B_1, B_2, \ldots, B_n$ all intersect in a single point. The union of these circles may be denoted by the symbol X^1, by analogy with the case of the torus. Let $(E^2, \dot{E}^2)$ denote the pair consisting of the polygonal disc and its boundary circle. One can prove that the identification map $f : (E^2, \dot{E}^2) \to (X, X^1)$ induces isomorphisms $f_* : H_q(E^2, \dot{E}^2) \to H_q(X, X^1)$ for all q; the proof of Proposition 4.1 applies without any essential change. Then one completes the determination of the homology groups of X by studying the homology sequence of the pair (X, X^1). The final results are the following:

$$H_0(X) \text{ and } H_2(X) \text{ are infinite cyclic,}$$

$$H_1(X) \text{ is free abelian of rank } 2n,$$

and

$$H_q(X) = 0 \quad \text{for } q > 2.$$

Exactly as in the case of the torus, the inclusion map $i : X^1 \to X$ induces an isomorphism $i_* : H_1(X^1) \to H_1(X)$.

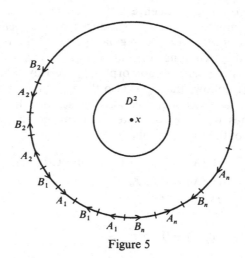

Figure 5

EXAMPLE 4.3 (The projective plane). The projective plane may be obtained from a circular disc by identifying diametrically opposite points on the boundary. It is harder to visualize than the surfaces we have considered so far because it can not be imbedded homeomorphically in Euclidean 3-space. It is a nonorientable surface, and this results in a somewhat different structure for its homology groups, as we shall see.

As in the previous cases, denote the disc by E^2, the projective plane by X, and let $f : (E^2, \dot{E}^2) \to (X, X^1)$ be the identification map. Here $\dot{E}^2$ denotes the boundary circle of E^2, and $X^1 = f(\dot{E}^2)$ is also a circle. The induced map

$f_1 : \dot{E}^2 \to X^1$ is a 2-to-1 map, i.e., it has degree ± 2. Exactly as before, we can prove that $f_* : H_q(E^2, \dot{E}^2) \to H_q(X, X^1)$ is an isomorphism for all q. The only nontrivial part of the homology sequence of the pair (X, X^1) is the following:

$$0 \xrightarrow{} H_2(X) \xrightarrow{\ j_* \ } H_2(X, X^1) \xrightarrow{\ \partial_* \ } H_1(X^1) \xrightarrow{\ i_* \ } H_1(X) \xrightarrow{} 0$$

$$\Big\uparrow{\scriptstyle f_*} \qquad\qquad \Big\uparrow{\scriptstyle f_{1*}}$$

$$H_2(E^2, \dot{E}^2) \xrightarrow{\ \partial'_* \ } H_1(\dot{E}^2)$$

Since f_* and ∂'_* are isomorphisms, and f_{1*} has degree ± 2, we conclude that ∂_* also has degree ± 2. It now follows from exactness of the homology sequence of (X, X^1) that

$$H_2(X) = 0,$$

and

$$H_1(X) \text{ is cyclic of order 2.}$$

Of course $H_0(X) = \mathbf{Z}$ and $H_q(X) = 0$ for $q > 2$ exactly as before.

This is our first example of a space whose homology groups have an element of finite order; in fact it is probably the simplest example of such a space. It can be proved that if X is any reasonable subset of Euclidean 3-space, its homology groups have no elements of finite order.

EXAMPLE 4.4 (The Klein bottle, K). We have two different ways of obtaining a Klein bottle by identifying edges of a square: That indicated on the left in Figure 6, in which opposite edges are to be identified, or that indicated on the right in Figure 6, in which adjacent edges are to be identified. It is interesting to use both representations to compute the homology groups of K, and then compare the results. The details are left to the reader. In either case, it is readily seen that $H_2(K) = 0$. What is the structure of $H_1(K)$? How can one prove algebraically that both methods lead to the same result?

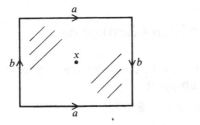

Figure 6

EXAMPLE 4.5 (An arbitrary nonorientable compact surface). An arbitrary nonorientable surface X is the connected sum of n projective planes, $n \geq 1$. If n is odd, it can be considered as the connected sum of a projective plane and an orientable surface, while if n is even, it can be considered as the connected sum of a Klein bottle and an orientable surface. The integer n is sometimes called the *genus* of the nonorientable surface. Whether n is odd or even, we obtain two distinct ways of representing X as the quotient space of disc; for details, see *Algebraic Topology: An Introduction*, Chapter IV, Example 5.4. The reader should use at least one of these ways (and preferably both) to determine the homology groups of X. The final result is that $H_2(X) = 0$, and $H_1(X)$ is the direct sum of a free abelian group of rank $n - 1$ and a cyclic group of order 2.

Note that for the orientable surfaces, $H_2(X)$ is infinite cyclic and $H_1(X)$ is a free abelian group, while for nonorientable surfaces $H_2(X) = 0$ and $H_1(X)$ has a subgroup which is cyclic of order 2. Later on we will see that analogous results hold for compact, connected n-dimensional manifolds for any positive integer n.

EXERCISES

4.1. Compute the homology groups of a space obtained by identifying the three edges of a triangle to a single edge as shown in Figure 7. (Note: This space is not a manifold.)

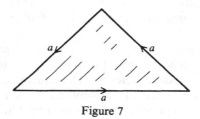

Figure 7

4.2. Given any integer $n > 1$, show how to construct a space X such that $H_1(X)$ is cyclic of order n.

§5. The Mayer–Vietoris Exact Sequence

In this section we will be concerned with the following question: Suppose the space X is the union of two subspaces,

$$X = A \cup B.$$

What relations hold between the homology groups of the three subspaces A, B, $A \cap B$ and the homology groups of the whole space? If we make certain rather mild assumptions on the subspaces involved, we can give a rather nice answer to this question in the form of an exact sequence, called the Mayer–Vietoris sequence. This exact sequence plays the same role in homology theory that the Seifert–Van Kampen theorem plays in the study of the fundamental group (see *Algebraic Topology: An Introduction,* Chapter IV).

In order to describe this exact sequence, let

$$i_*: H_n(A \cap B) \to H_n(A),$$
$$j_*: H_n(A \cap B) \to H_n(B),$$
$$k_*: H_n(A) \to H_n(X),$$

and

$$l_*: H_n(B) \to H_n(X)$$

denote homomorphisms induced by inclusion maps. Using these homomorphisms, we define homomorphisms

$$\varphi: H_n(A \cap B) \to H_n(A) \oplus H_n(B),$$
$$\psi: H_n(A) \oplus H_n(B) \to H_n(X)$$

by the formulas

$$\varphi(x) = (i_*(x), j_*(x)), \qquad x \in H_n(A \cap B),$$
$$\psi(u,v) = k_*(u) - l_*(v), \qquad u \in H_n(A), v \in H_n(B).$$

Theorem 5.1. *Let A and B be subsets of the topological space X such that $X = (\text{interior } A) \cup (\text{interior } B)$. Then it is possible to define natural homomorphisms*

$$\Delta: H_n(X) \to H_{n-1}(A \cap B)$$

for all values of n such that the following sequence is exact:

$$\cdots \xrightarrow{\Delta} H_n(A \cap B) \xrightarrow{\varphi} H_n(A) \oplus H_n(B) \xrightarrow{\psi} H_n(X) \xrightarrow{\Delta} H_{n-1}(A \cap B) \xrightarrow{\varphi} \cdots.$$

If $A \cap B \neq \varnothing$, the sequence remains exact if we substitute reduced homology groups for ordinary homology groups in dimension 0.

This sequence is called the Mayer–Vietoris sequence. The statement that the homomorphism Δ is natural has the following precise technical meaning: Assume that the subspaces A' and B' of X' are such that

$$X' = (\text{interior } A') \cup (\text{interior } B')$$

and that $f: X \to X'$ is a continuous map such that $f(A) \subset A'$ and $f(B) \subset B'$. Then the following diagram is commutative for all n:

$$
\begin{array}{ccc}
H_n(X) & \xrightarrow{\Delta} & H_{n-1}(A \cap B) \\
\downarrow{\scriptstyle f_*} & & \downarrow{\scriptstyle f_*} \\
H_n(X') & \xrightarrow{\Delta} & H_{n-1}(A' \cap B').
\end{array}
$$

PROOF OF THEOREM 5.1. Let $\mathcal{U} = \{A,B\}$; in view of the hypotheses assumed on A and B we can apply Theorem II.6.3 to conclude that the inclusion homomorphisms $\sigma : C_n(X,\mathcal{U}) \to C_n(X)$ induces isomorphisms $\sigma_* : H_n(X,\mathcal{U}) \to H_n(X)$ for all n. Note that

$$C_n(X,\mathcal{U}) = C_n(A) + C_n(B),$$

where the group on the right is the least subgroup of $C_n(X)$ containing $C_n(A)$ and $C_n(B)$ (it is *not* a direct sum). Therefore the homomorphisms $k_\# : C_n(A) \to C_n(X)$ and $l_\# : C_n(B) \to C_n(X)$ have the property that their images are contained in the subgroup $C_n(X,\mathcal{U})$ in each case. Hence we have commutative diagrams as follows:

$$
\begin{array}{ccc}
 & \overset{k_\#'}{\nearrow} C_n(X,\mathcal{U}) & \qquad\qquad \overset{l_\#'}{\nearrow} C_n(X,\mathcal{U}) \\
C_n(A) & \Big\downarrow \sigma & \quad C_n(B) \qquad\qquad \Big\downarrow \sigma \\
 & \underset{k_\#}{\searrow} C^n(X) & \qquad\qquad \underset{l_\#}{\searrow} C_n(X).
\end{array}
$$

Our strategy will be to replace the group $H_n(X)$ by $H_n(X,\mathcal{U})$ in proving Theorem 5.1; when we do this, we must systematically replace k by k' and l by l'. We will assume this has been done, and from now on will drop the primes from the notation for these homomorphisms.

By analogy with the definition of the homomorphisms φ and ψ above, we define homomorphisms

$$\Phi : C_n(A \cap B) \to C_n(A) \oplus C_n(B),$$
$$\Psi : C_n(A) \oplus C_n(B) \to C_n(X,\mathcal{U})$$

by the following formulas:

$$\Phi(x) = (i_\# x, j_\# x)$$
$$\Psi(u,v) = k_\#(u) - l_\#(v).$$

Now consider the following diagram of chain groups and homomorphisms:

$$
\begin{array}{ccccccccc}
 & & \downarrow & & \downarrow & & \downarrow & & \\
0 & \longrightarrow & C_{n+1}(A \cap B) & \overset{\Phi}{\longrightarrow} & C_{n+1}(A) \oplus C_{n+1}(B) & \overset{\Psi}{\longrightarrow} & C_{n+1}(X,\mathcal{U}) & \longrightarrow & 0 \\
 & & \downarrow & & \downarrow & & \downarrow & & \\
0 & \longrightarrow & C_n(A \cap B) & \overset{\Phi}{\longrightarrow} & C_n(A) \oplus C_n(B) & \overset{\Psi}{\longrightarrow} & C_n(X,\mathcal{U}) & \longrightarrow & 0 \\
 & & \downarrow & & \downarrow & & \downarrow & & \\
0 & \longrightarrow & C_{n-1}(A \cap B) & \overset{\Phi}{\longrightarrow} & C_{n-1}(A) \oplus C_{n-1}(B) & \overset{\Psi}{\longrightarrow} & C_{n-1}(X,\mathcal{U}) & \longrightarrow & 0 \\
 & & \downarrow & & \downarrow & & \downarrow & & \\
 & & \vdots & & \vdots & & \vdots & &
\end{array}
$$

The vertical arrows denote the appropriate boundary operator in each case. In the case of the vertical arrows in the middle column, this means the direct sum of the boundary operators for A and B. The two most important facts about this diagram are the following:

(1) Each square of this diagram is commutative. This is practically obvious, in view of the way the homomorphisms Φ and Ψ were defined.

(2) Each horizontal line in this diagram is exact. The verification of this fact is left to the reader; it should not present any real difficulty.

The reader should now compare this diagram with Diagram (II.5.1) which was used to set up the exact homology sequence of a pair. The essential properties of the two diagrams are the same. By the same process that was used in II.5 to define the boundary operator of a pair, one can now define the homomorphisms

$$\Delta : H_n(X, \mathcal{U}) \to H_{n-1}(A \cap B)$$

for all values of n. Moreover, the methods used to prove the exactness of the homology sequence of a pair apply without change to give the exactness of the following sequence of groups and homomorphisms:

$$\cdots \xrightarrow{\Delta} H_n(A \cap B) \xrightarrow{\varphi} H_n(A) \oplus H_n(B) \xrightarrow{\psi} H_n(X, \mathcal{U}) \xrightarrow{\Delta} \cdots .$$

All that remains is to substitute $H_n(X)$ for $H_n(X, \mathcal{U})$, and we have proved the exactness of the Mayer–Vietoris sequence.

We leave it to the reader to verify that the homomorphism Δ is natural and to investigate what happens when one uses reduced homology groups in dimension zero (provided $A \cap B \neq \varnothing$). Q.E.D.

EXAMPLE 5.1. We will show how the Mayer–Vietoris sequence can be used to make the inductive step in the proof of Theorem 2.1. As the inductive hypothesis, assume that $\tilde{H}_n(S^n) = \mathbf{Z}$, and $\tilde{H}_i(S^n) = 0$ for $i \neq n$. We wish to determine $\tilde{H}_i(S^{n+1})$. Let A be the complement of the point $(0, \ldots, 0, -1)$ in S^{n+1} and let B be the complement of the point $(0, \ldots, 0, +1)$ in S^{n+1}. Then A and B are open subsets of S^{n+1}, and $A \cup B = S^{n+1}$. Therefore we can apply the Mayer–Vietoris sequence; in this case $A \cap B \neq \varnothing$, and it is convenient to used reduced homology groups in dimension 0. Consider the following portion of the sequence:

$$\tilde{H}_{i+1}(A) \oplus \tilde{H}_{i+1}(B) \to \tilde{H}_{i+1}(S^{n+1}) \xrightarrow{\Delta} \tilde{H}_i(A \cap B) \to \tilde{H}_i(A) \oplus H_i(B).$$

One proves by stereographic projection that A and B are both homeomorphic to $\mathbf{R}^{n+1}$, hence they are contractible. Therefore $\tilde{H}_i(A) = \tilde{H}_i(B) = 0$ for all i. It follows by exactness that Δ is an isomorphism. Now $A \cap B$ is homeo-morphic to $\mathbf{R}^{n+1}$ minus a point, and therefore it contains S^n as a deformation retract. Hence by the inductive hypothesis $\tilde{H}_n(A \cap B) = \mathbf{Z}$, and $\tilde{H}_i(A \cap B) = 0$ for $i \neq n$. Since Δ is an isomorphism, it follows that $\tilde{H}_{n+1}(S^{n+1}) = \mathbf{Z}$, and $\tilde{H}_i(S^{n+1}) = 0$ for $i \neq n + 1$, as was to be proved.

One final comment about the Mayer–Vietoris sequence: we could weaken the hypotheses of Theorem 5.1 to read $X = A \cup B$, provided we knew that the inclusion homomorphism $C_n(A) + C_n(B) \to C_n(X)$ induced isomorphisms on homology groups; this was the purpose of the assumption that the *interiors* of A and B cover X. We will come back to this point later.

EXERCISES

5.1. Assume that $X = U \cup V$, where U and V are open subsets of X, and $U \cap V$ is nonempty and contractible. Express the homology groups of X in terms of those of U and V.

5.2. Assume that $X = A \cup B$, where A and B are closed subsets of X, and $A \cap B = \{x_0\}$. Assume further that x_0 has an open neighborhood N in X such that $N \cap A$ and $N \cap B$ are both contractible, and that during the contraction the point x_0 remains fixed. Express the homology groups of X in terms of those of A and B.

5.3. Assume that the space X and the subspaces A and B satisfy the hypotheses of Theorem 5.1. (a) Prove that the inclusion maps $(A, A \cap B) \to (X, B)$ and $(B, A \cap B) \to (X, A)$ induce isomorphisms on homology. (b) Show that the homomorphism $\varDelta : H_n(X) \to H_{n-1}(A \cap B)$ is the composition of the following homomorphisms:

$$H_n(X) \xrightarrow{j_*} H_n(X, B) \approx H_n(A, A \cap B) \xrightarrow{\partial_*} H_{n-1}(A \cap B).$$

5.4. Use the result of Part (b) of the preceding exercise to *define* the homomorphism $\varDelta : H_n(X) \to H_{n-1}(A \cap B)$. Then prove directly (by diagram chasing, without going back to chain groups) that the Mayer–Vietoris sequence is exact (cf. Eilenberg and Steenrod, [2], Chapter I).

§6. The Jordan–Brouwer Separation Theorem and Invariance of Domain

The classical Jordan curve theorem may be stated as follows: Let C be a simple closed curve in the plane $\mathbf{R}^2$, i.e., C is a subset of $\mathbf{R}^2$ which is homeomorphic to S^1. Then $\mathbf{R}^2 - C$ has exactly two components, and C is the boundary of each component (in the sense of point set topology). It is our object in this section to prove a generalization of this theorem to $\mathbf{R}^n$, and derive various consequences. Most of the results of this section were first proved by the Dutch mathematician L. E. J. Brouwer.

The Mayer–Vietoris sequence will play an essential role in the proof. We will also need another general property of singular homology theory, which may be stated as follows:

Proposition 6.1. *Let (X, A) be a pair consisting of a topological space X and subspace A. (a) Given any homology class $u \in H_n(X, A)$, there exists a compact* *pair $(C,D) \subset (X,A)$ and a homology class $u' \in H_n(C,D)$ such that $i_*(u') = u$, where $i:(C,D) \to (X,A)$ is the inclusion map. (b) Let (C,D) be any compact pair such that $(C,D) \subset (X,A)$, and $v \in H_m(C,D)$ a homology class such that $i_*(v) = 0$. Then there exists a compact pair (C',D') such that $(C,D) \subset (C',D') \subset (X,A)$ and $j_*(v) = 0$, where $j:(C,D) \to (C',D')$ is the inclusion map.*

In the statement of this proposition, "a compact pair (C,D)" means a pair such that C is compact and D is a compact subset of C. An inclusion relation between pairs, such as $(C,D) \subset (X,A)$, means that $C \subset X$ and $D \subset A$. For the reader who is familiar with the concept of direct limit, this proposition may be restated as follows: $H_n(X,A)$ is the direct limit of the groups $H_n(C,D)$, where (C,D) ranges over all compact pairs contained in (X,A).

The proof of this proposition depends on the following fact: If $a \in Q_n(X)$, then there exists a compact set $C \subset X$ such that $a \in Q_n(C)$. In fact, if a is a linear combination of the singular n-cubes $T_1, T_2, \ldots, T_k$, then we may choose $C = T_1(I^n) \cup T_2(I^n) \cup \cdots \cup T_k(I^n)$. The proposition follows readily from this fact by choosing representative cycles for the homology classes involved, etc. The details can be easily worked out by the reader. We also leave it to the reader to verify that this proposition remains true if we replace ordinary homology groups by reduced homology groups everywhere in the statement.

In order to prove the Jordan–Brouwer separation theorem, we need the following lemma, which is of some interest in its own right:

Lemma 6.2. *Let Y be a subset of S^n which is homeomorphic to I^k, where $0 \le k \le n$. Then $\tilde{H}_i(S^n - Y) = 0$ for all i.*

PROOF: The proof is by induction on k. For $k = 0$, I^k is a single point (by definition), and $S^n - I^k$ is homeomorphic to $\mathbf{R}^n$, which is contractible.

In order to make the inductive step it is convenient to assume we have chosen a definite homeomorphism of Y with I^k; then we may as well identify Y with I^k by means of this homeomorphism. Let

$$Y_0 = \{(x_1, \ldots, x_k) \in Y \,|\, x_1 \le \tfrac{1}{2}\},$$
$$Y_1 = \{(x_1, \ldots, x_k) \in Y \,|\, x_1 \ge \tfrac{1}{2}\}.$$

Then

$$Y_0 \cup Y_1 = Y,$$
$$S^n - (Y_0 \cap Y_1) = (S^n - Y_0) \cup (S^n - Y_1),$$

and we may apply the Mayer–Vietoris sequence to this representation of $S^n - (Y_0 \cap Y_1)$ as the union of two open subsets. Note that $Y_0 \cap Y_1$ is homeomorphic to I^{k-1}, hence by the inductive hypothesis

$$\tilde{H}_i(S^n - (Y_0 \cap Y_1)) = 0$$

for all i. Therefore we conclude from the exactness of the Mayer–Vietoris sequence that

$$\varphi : \tilde{H}_i(S^n - Y) \to \tilde{H}_i(S^n - Y_0) \oplus \tilde{H}_i(S^n - Y_1)$$

is an isomorphism for all i.

Now recall the definition of the homomorphism φ from the preceding section:

$$\varphi(x) = (i_{0*}(x), i_{1*}(x)),$$

where $i_0 : S^n - Y \to S^n - Y_0$ and $i_1 : S^n - Y \to S^n - Y_1$ are inclusion maps. In order to complete the proof of the inductive step, we will assume that for some integer i, $\tilde{H}_i(S^n - Y) \neq 0$, and show that this assumption leads to a contradiction. As a first consequence of the assumption that $\tilde{H}_i(S^n - Y) \neq 0$, we see that we can find an element $a_0 \in \tilde{H}_i(S^n - Y)$ such that $i_{0*}(a_0) \neq 0$, or $i_{1*}(a_0) \neq 0$.

Let us first take up the case where $a_1 = i_{0*}(a_0) \neq 0$. Let

$$Y_{00} = \{(x_1, \ldots, x_k) \in Y_0 \,|\, 0 \le x_1 \le \tfrac{1}{4}\},$$
$$Y_{01} = \{(x_1, \ldots, x_k) \in Y_0 \,|\, \tfrac{1}{4} \le x_1 \le \tfrac{1}{2}\}.$$

Then

$$Y_0 = Y_{00} \cup Y_{01}.$$

Let

$$i_{00} : S^n - Y_0 \to S^n - Y_{00},$$
$$i_{01} : S^n - Y_0 \to S^n - Y_{01}.$$

Then by a repetition of the above argument using the Mayer–Vietoris sequence and the inductive hypothesis, we may prove that $i_{00*}(a_1) \neq 0$ or $i_{01*}(a_1) \neq 0$. In the other case where $i_{1*}(a_0) \neq 0$, we may represent Y_1 as the union of two subsets,

$$Y_1 = Y_{10} \cup Y_{11}$$

such that $i_{10*}(a_1) \neq 0$, or $i_{11*}(a_1) \neq 0$ where now $a_1 = i_{1*}(a_0) \neq 0$.

The reader will immediately see that we may continue this process *ad infinitum*. The net result is that we can construct an infinite decreasing sequence of subsets of Y each homeomorphic to I^k and denoted by

$$Y \supset Y' \supset Y'' \supset \cdots \supset Y^{(m)} \supset \cdots$$

such that the following two properties hold:

(a) Let Y^∞ denote the intersection of all the sets of this sequence; then Y^∞ is homeomorphic to I^{k-1}. Hence $\tilde{H}_j(S^n - Y^\infty) = 0$ for all j by our inductive hypothesis.

(b) Let us denote the complementary sets and their inclusion maps as follows:

$$S^n - Y \xrightarrow{i} S^n - Y' \xrightarrow{i'} S^n - Y'' \xrightarrow{i''} \cdots .$$

Using the element $a_0 \in \tilde{H}_i(S^n - Y)$, we may construct an infinite sequence

$$(a_0, a_1, a_2, \ldots)$$

of elements such that $a_m \in \tilde{H}_i(S^n - Y^{(m)})$ and $a_m \neq 0$ as follows:

$$a_1 = i_*(a_0)$$
$$a_2 = i'_*(a_1),$$
$$a_3 = i''_*(a_2), \quad \text{etc.}$$

We will complete the proof by showing that the existence of such an infinite sequence of nonzero elements contradicts Proposition 6.1. Apply Proposition 6.1(a) to obtain a compact set $C \subset S^n - Y$ and a homology class $a'_0 \in H_i(C)$ such that $a'_0 \to a_0$ under the inclusion map $C \to (S^n - Y)$. Since $\tilde{H}_i(S^n - Y^\infty) = 0$, we may apply Proposition 6.1(b) to the inclusion $C \subset S^n - Y^\infty$ to conclude that there exists a compact set C' such that

$$C \subset C' \subset S^n - Y^\infty$$

and $a'_0 \to 0$ under the homomorphism induced by the inclusion map $C \to C'$. Since C' is compact, there exists an integer m such that $C' \subset S^n - Y^{(m)}$. Now consider the following diagram of reduced homology groups:

$$
\begin{array}{ccc}
\tilde{H}_i(C) & \longrightarrow & \tilde{H}_i(C') \\
\downarrow & & \downarrow \\
\tilde{H}_i(S^n - Y) & \longrightarrow & \tilde{H}(S^n - Y^{(m)}).
\end{array}
$$

All homomorphisms in this diagram are induced by inclusion maps, hence the diagram is commutative. If we consider the element $a'_0 \in \tilde{H}_i(C)$ and chase it both ways around this diagram, we see that it must go to zero one way, while the other way it goes to $a_m \neq 0$. This is the desired contradiction, and hence the proof of the inductive step is complete. Q.E.D.

Perhaps the reader wonders who concocted such a complicated proof as this. The answer is that it is the work of many mathematicians; it has evolved over a relatively long portion of the history of algebraic topology. In order to appreciate why the proof of this lemma might have to be so complicated, the reader should consider some examples of subsets Y of S^3 which are homeomorphic to I^1 and such that $S^3 - I^1$ has a nontrivial fundamental group (cf. Artin and Fox, [1]).

Theorem 6.3. *Let A be a subset of S^n which is homeomorphic to S^k, $0 \leq k \leq n - 1$. Then $\tilde{H}_{n-k-1}(S^n - A) = \mathbf{Z}$, and $\tilde{H}_i(S^n - A) = 0$ for $i \neq n - k - 1$.*

PROOF: Once again the proof is by induction on k, using the Mayer–Vietoris sequence. If $k = 0$, then A consists of two points and $S^n - A$ is homeomorphic to $\mathbf{R}^n$ with one point removed. Hence $S^n - A$ has the homotopy type of S^{n-1}, and the theorem is true for this case.

Now we will make the inductive step. Since A is homeomorphic to S^k, it follows that $A = A_1 \cup A_2$, where A_1 and A_2 are subsets of A which are homeomorphic to I^k, and $A_1 \cap A_2$ is homeomorphic S^{k-1} (cf. the proof of Theorem 2.1). Therefore,

$$S^n - (A_1 \cap A_2) = (S^n - A_1) \cup (S^n - A_2),$$

and we may apply the Mayer–Vietoris sequence to this representation of $S^n - (A_1 \cap A_2)$ as the union of two open subsets. By the lemma just proved,

$$\tilde{H}_i(S^n - A_1) = \tilde{H}_i(S^n - A_2) = 0$$

for all i. It follows from the exactness of the Mayer–Vietoris sequence that

$$\Delta : \tilde{H}_{i+1}(S^n - (A_1 \cap A_2)) \to \tilde{H}_i(S^n - A))$$

is an isomorphism for all i. Since $A_1 \cap A_2$ is homeomorphic to S^{k-1}, this isomorphism suffices to prove the inductive step.   Q.E.D.

EXAMPLE 6.1. Suppose that A is a subset of S^3 which is homeomorphic to S^1, i.e., A is a simple closed curve in S^3. It follows from the theorem just proved that $\tilde{H}_1(S^3 - A)$ is infinite cyclic, and $\tilde{H}_i(S^3 - A) = 0$ for $i \neq 1$. It is well known that a simple closed curve in $\mathbf{R}^3$ or S^3 can be "knotted" in various different ways, or left unknotted. Thus the homology groups of $S^3 - A$ in this case are independent of how A is knotted. On the other hand, it may be shown that the fundamental group of $S^3 - A$ *does* depend on how A is knotted; cf. *Algebraic Topology: An Introduction*, Chapter IV, §6, and the references given there. The fact that the homology groups of $S^3 - A$ are independent of how A is knotted can be an advantage or a disadvantage, depending on what one is trying to do.

Corollary 6.4 (Jordan–Brouwer theorem). *Let A be a subset of S^n which is homeomorphic to S^{n-1}. Then $S^n - A$ has exactly two components.*

PROOF: Apply the case $k = n - 1$ of the preceding theorem to conclude that $H_0(S^n - A)$ has rank 2; hence $S^n - A$ has exactly two arc components. But it is readily seen that $S^n - A$ is locally arcwise connected, hence the components and arc-components are the same.

Proposition 6.5. *Let A be a subset of S^n which is homeomorphic to S^{n-1}. Then A is the boundary of each component of $S^n - A$.*

In order to better appreciate this proposition, consider the case where A is a subset of S^2 which is homeomorphic to $S^1 \times I$ (instead of S^1). Then $S^2 - A$ has two components, but the boundary of either component is a *proper* subset of A.

PROOF OF PROPOSITION 6.5. Since $S^n - A$ is locally connected, each component of $S^n - A$ is an open subset of $S^n - A$, and hence an open subset of S^n. Therefore the boundary of each component must be a subset of A. To complete the proof of the proposition, we must show that any point $a \in A$ is a boundary point of each component of $S^n - A$. Denote the components of $S^n - A$ by C_0 and C_1. Let N be any open neighborhood of a in S^n; we must show that $N \cap C_i \neq \varnothing$ for $i = 0$ and 1.

Note that $N \cap A$ is an open neighborhood of a in A. Since A is homeomorphic to S^{n-1}, we can find a decomposition

$$A = A_1 \cup A_2,$$

as in the proof of Theorem 6.3, such that A_1 and A_2 are homeomorphic to I^{n-1}, $A_1 \cap A_2$ is homeomorphic to S^{n-2}, and $A_2 \subset N \cap A$. It follows from Lemma 6.2 that $S^n - A_1$ is arcwise connected. Let $p_0 \in C_0$, and $p_1 \in C_1$; choose an arc in $S^n - A_1$ joining p_0 to p_1, i.e., a continuous map $f: I \to S^n - A_1$ such that $f(0) = p_0$, and $f(1) = p_1$. It follows from what we have already proved that $f(I) \cap A \neq \varnothing$, and hence $f(I) \cap A_2 \neq \varnothing$. Consider the subset $f^{-1}(A_2) \subset I$; this is a compact subset of I, and hence it must have a least point t_0 and a greatest point t_1. Obviously t_0 and t_1 are boundary points of $f^{-1}(A_2)$, and $f^{-1}(N)$ is an open subset of I which contains both t_1 and t_2. From this it follows by an easy argument that $f^{-1}(N) \cap f^{-1}(C_1)$ and $f^{-1}(N) \cap f^{-1}(C_2)$ are both nonempty. Hence $N \cap C_1 \neq \varnothing$ and $N \cap C_2 \neq \varnothing$, as desired.   Q.E.D.

Note that essential role that Lemma 6.2 plays in this proof.

In order to better appreciate the significance of Corollary 6.4 and Proposition 6.5, the reader should study the Alexander horned sphere or other wild imbeddings of S^2 in S^3, cf. Hocking and Young, [3] p. 176. For the case of imbeddings of S^1 in S^2, there is the so-called Schönflies theorem, which is stronger than the Jordan curve theorem (see E. Moise, [4]).

Next, we will prove another of L. E. J. Brouwer's theorems, usually referred to as "the theorem on invariance of domain."

Theorem 6.6. *Let U and V be homeomorphic subsets of S^n. If U is open, then so is V (and conversely).*

PROOF: Let $h: U \to V$ be a homeomorphism. For any point $x \in U$ we can find a closed neighborhood N of x in U such that N is homeomorphic to I^n and its boundary, $\dot{N}$, is homeomorphic to S^{n-1}. Let $y = h(x)$; then $N' = h(N)$ is a closed neighborhood of y in V with boundary $\dot{N}' = h(\dot{N})$. It follows from Lemma 6.2 that $S^n - N'$ is connected, and from Theorem 6.4 that $S^n - \dot{N}'$ has exactly two components. Note that $S^n - \dot{N}'$ is the disjoint union of $N' - \dot{N}'$ and $S^n - N'$; since both of these sets are connected, they are the components of $S^n - \dot{N}'$. Therefore both of them are open subsets of $S^n - \dot{N}'$ and hence of S^n. In particular, $N' - \dot{N}'$ is an open neighborhood of y which is entirely contained in V. Therefore y is an interior point of V. Since this argument obviously applies to any point $y \in V$, the proof is complete.   Q.E.D.

Brouwer's theorem on invariance of domain is a powerful theorem, and it deserves to be better known. It should be looked on as a very special topological property of S^n; or more generally, of n-dimensional manifolds. (See the exercises below.)

Corollary 6.7. *Let A and B be arbitrary subsets of S^n, and let $h: A \to B$ be a homeomorphism. Then h maps interior points onto interior points, and boundary points onto boundary points.*

This corollary shows that the property of being an interior or boundary point of a subset $A \subset S^n$ is an intrinsic property, independent of the imbedding.

EXERCISES

6.1. Let Y be a subset of $\mathbf{R}^n$ which is homeomorphic to I^k, $0 \le k \le n$. Determine the homology groups of $\mathbf{R}^n - Y$. (*Hint:* Consider $\mathbf{R}^n$ as the complement of a point in S^n.)

6.2. Let A be a subset of $\mathbf{R}^n$ which is homeomorphic to S^k, $0 \le k \le n - 1$. Determine the homology groups of $\mathbf{R}^n - A$. How many components does $\mathbf{R}^n - A$ have?

6.3. Does the analogue of Proposition 6.5 hold true for subsets of $\mathbf{R}^n$ which are homeomorphic to S^{n-1}?

6.4. Let A be a closed subset of $\mathbf{R}^n$ which is homeomorphic to $\mathbf{R}^{n-1}$. Prove that $\mathbf{R}^n - A$ has exactly two components.

6.5. Prove that Theorem 6.6 and Corollary 6.7 hold for subsets of $\mathbf{R}^n$. Then prove the following more general form of Brouwer's theorem. Assume M and N are n-dimensional manifolds; let U and V be subsets of M and N respectively such that U and V are homeomorphic. If U is an open subset of M, then V is an open subset of N. (Note: An n-dimensional manifold is a Hausdorff space such that each point has an open neighborhood which is homeomorphic to $\mathbf{R}^n$.)

6.6. Use Brouwer's theorem on invariance of domain to prove that $\mathbf{R}^m$ and $\mathbf{R}^n$ are not homeomorphic if $m \ne n$ (it is not necessary to use homology theory in this proof).

6.7. Prove that if $m > n$, then there is no subset of S^n which is homeomorphic to I^m.

6.8. Let U be an open subset of $\mathbf{R}^n$, and let $f: U \to \mathbf{R}^n$ be a map which is continuous and one-to-one. Prove that f is a homeomorphism of U onto $f(U)$.

6.9. Prove that no *proper* subset of S^n can be homeomorphic to S^n.

6.10. Prove that a continuous map $f: S^n \to \mathbf{R}^n$ cannot be one-to-one.

6.11. Let U be an open subset of $\mathbf{R}^m$. Prove that if $m > n$, there is no continuous, one-to-one map of U into $\mathbf{R}^n$. Generalize this statement by replacing $\mathbf{R}^m$ and $\mathbf{R}^n$ by manifolds of dimension m and n respectively.

6.12. Let A and B be subsets of S^n which are homeomorphic to S^p and S^q respectively, where $0 < p \le q \le n$. Determine the homology groups of $S^n - (A \cup B)$ in the following two cases:

(a) A and B are disjoint subsets of S^n.
(b) $A \cap B$ consists of exactly one point.

In case $p = q = n - 1$, determine the number of components of $S^n - (A \cup B)$.

6.13. Let A and B be homeomorphic subsets of $\mathbf{R}^n$. If A is closed, does it follow that B is closed?

6.14. Let X be a connected regular, finite graph and let A and B be subsets of S^3 which are homeomorphic to X. Prove that the (reduced) homology groups of $S^3 - A$ and $S^3 - B$ are isomorphic (*Hint*: Use induction on the number of edges of X. Any finite, connected graph has some edge whose removal does not disconnect it.)

§7. The Relation between the Fundamental Group and the First Homology Group*

The main theorem of this section asserts that for an arcwise connected space, the fundamental group completely determines the first homology group. The precise statement will be given after some preliminary definitions. It is assumed that the reader is familiar with the basic properties of the fundamental group; cf. *Algebraic Topology: An Introduction*, Chapter II.

First of all, for any topological space X and any base point $x_0 \in X$ we define a homomorphism

$$h_X : \pi(X, x_0) \to H_1(X)$$

as follows. Let $\alpha \in \pi(X, x_0)$; choose a closed path $f : I \to X$ belonging to the equivalence class α. We can think of f as a singular 1-cube, and hence as determining an element of the chain group $C_1(X)$. Since $f(0) = f(1) = x_0$, $\partial_1(f) = 0$; in other words, f is a cycle. We define $h_X(\alpha)$ to be the homology class of the cycle f. To see that $h_X(\alpha)$ is well defined, one must verify that if $g : I \to X$ is another closed path in the equivalence class α, then the cycles f and g belong to the same homology class. We leave this verification to the reader. Next, one should check that h_X is a homomorphism, i.e., $h_X(\alpha \cdot \beta) = h_X(\alpha) + h_X(\beta)$. This may be done as follows. Choose representatives $f : I \to X$ and $g : I \to X$ for α and β respectively. Then $f \cdot g : I \to X$ is a representative for $\alpha \cdot \beta$, where

$$(f \cdot g)t = \begin{cases} f(2t) & 0 \le t \le \tfrac{1}{2}, \\ g(2t - 1) & \tfrac{1}{2} \le t \le 1. \end{cases}$$

Now define a singular 2-cube $T : I^2 \to X$ by the formula

$$T(x_1, x_2) = \begin{cases} f(x_1 + 2x_2) & x_1 + 2x_2 \le 1, \\ g\left(\dfrac{x_1 + 2x_2 - 1}{x_1 + 1}\right) & x_1 + 2x_2 \ge 1. \end{cases}$$

---

* This section may be omitted by readers who are not familiar with the properties of the fundamental group.

The function T was chosen so that it is constant along the straight lines shown in Figure 8. It is readily checked that

$$\partial_2(T) = f + g - f \cdot g - c,$$

where c is a degenerate singular 1-cube. But this equation clearly implies that $h_X(\alpha \cdot \beta) = h_X(\alpha) + h_X(\beta)$, as required. In order to better understand the definition of the function T, it is suggested that the reader try to work out the formula for $T(x_1, x_2)$ himself so that it will have the required properties.

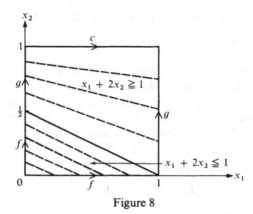

Figure 8

The homomorphism we have just defined satisfies the following obvious naturality condition. Let $\varphi: X \to Y$ be a continuous map such that $\varphi(x_0) = y_0$. Then the following diagram is commutative:

$$
\begin{array}{ccc}
\pi(X,x_0) & \xrightarrow{\ \varphi_* \ } & \pi(Y,y_0) \\
\Big\downarrow{\scriptstyle h_X} & & \Big\downarrow{\scriptstyle h_Y} \\
H_1(X) & \xrightarrow{\ \varphi_* \ } & H_1(Y).
\end{array}
$$

In addition, the following two rather obvious remarks apply to the homomorphism h.

(a) If the space X is not arcwise connected, $H_1(X)$ is the direct sum of the groups $H_1(X_\lambda)$, where $\{X_\lambda | \lambda \in A\}$ denotes the set of arc components of X. It is obvious that the image of the homomorphism h_X is entirely contained in the 1-dimensional homology group of the arccomponent of X which contains the basepoint x_0. Therefore the homomorphism h_X is mainly of interest in the case of arcwise connected spaces.

(b) Since $H_1(X)$ is abelian, the commutator subgroup of $\pi(X,x_0)$ is contained in the kernel of h_X. Let us use the notation $\pi'(X,x_0)$ to denote the "abelianized" fundamental group, i.e., the quotient group of $\pi(X,x_0)$ modulo its commutator subgroup. Then h_X induces a homomorphism $\pi'(X,x_0) \to H_1(X)$, which we will denote by the same symbol, h_X, or h for short.

With these properties out of the way, we can state the main result of this section:

Theorem 7.1. *Let X be an arcwise connected space. Then h is an isomorphism of the abelianized fundamental group $\pi'(X, x_0)$ onto $H_1(X)$.*

PROOF: In order to carry out the proof, it is convenient to show that one can compute the singular homology groups of an arcwise connected space X using only those singular cubes which have all their vertices mapped into the basepoint x_0. There is a certain analogy here with Theorem II.6.3.

Let $Q_n(X/x_0)$ denote the subgroup of $Q_n(X)$ generated by all singular n-cubes $T: I^n \to X$ such that $T(v) = x_0$ for any vertex v of the cube I^n. Define $D_n(X/x_0) = D_n(X) \cap Q_n(X/x_0)$ and $C_n(X/x_0) = Q_n(X/x_0)/D_n(X/x_0)$. Note that the boundary operator $\partial_n: Q_n(X) \to Q_{n-1}(X)$ obviously maps the subgroup $Q_n(X/x_0)$ into $Q_{n-1}(X/x_0)$, and hence it induces a boundary operator $\partial_n: C_n(X/x_0) \to C_{n-1}(X/x_0)$. As usual, we define the group of n-cycles, $Z_n(X/x_0)$, to be the kernel of ∂_n,

$$Z_n(X/x_0) = \{u \in C_n(X/x_0) \mid \partial_n(u) = 0\}$$

and the group of bounding cycles $B_n(X/x_0)$ to be $\partial_{n+1}(C_{n+1}(X/x_0))$. Then $B_n(X/x_0) \subset Z_n(X/x_0)$ and we define

$$H_n(X/x_0) = Z_n(X/x_0)/B_n(X/x_0).$$

The inclusion $Q_n(X/x_0) \subset Q_n(X)$ induces homomorphisms

$$\tau_n: C_n(X/x_0) \to C_n(X)$$

and

$$\tau_*: H_n(X/x_0) \to H_n(X).$$

Lemma 7.2. *If the space X is arcwise connected, then the homomorphism τ_* is an isomorphism for all n.*

PROOF OF LEMMA. The strategy of the proof is to show that the system of subgroups $C_n(X/x_0)$, $n = 0, 1, 2, \ldots$ is a "deformation retract" of the full chain groups $C_n(X)$, $n = 0, 1, 2, \ldots$ in some algebraic sense. To be precise, we will define a sequence of homomorphisms $\rho_n: C_n(X) \to C_n(X/x_0)$ such that the ρ_n's commute with the boundary operators and hence induce homomorphisms $\rho_*: H_n(X) \to H_n(X/x_0)$. It will turn out that $\rho_n \tau_n$ is the identity map of $C_n(X/x_0)$ for each n, hence $\rho_* \tau_*$ is the identity map of $H_n(X/x_0)$. Finally we will define a sequence of homomorphisms $\Phi_n: C_n(X) \to C_{n+1}(X)$ which will be a chain homotopy between the chain map $\tau_n \rho_n$ and the identity map of $C_n(X)$. Hence $\tau_* \rho_*$ is the identity map of $H_n(X)$, and the proof will be complete. Actually, we will only carry out this program for small values of n, because we only need to know that $\tau_*: H_1(X/x_0) \to H_1(X)$ is an isomorphism. The rest of the proof will be left as an exercise. Also, it turns out to be easiest to define the homomorphisms Φ_n first, and then define the homomorphisms ρ_n afterwards.

In order to define Φ_n, we will define homomorphisms $\varphi_n : Q_n(X) \to Q_{n+1}(X)$ such that $\varphi_n(D_n(X)) \subset D_{n+1}(X)$. We will do this in succession for $n = 0, 1, 2$. In each case, if T is a singular n-cube, $\varphi_n T$ will be a singular $n+1$ cube. We proceed as follows:

Case $n = 0$. We can identify the singular 0-cubes with the points of X. For each $x \in X$ such that $x \neq x_0$, choose a path $T : I \to X$ such that $T(0) = x_0$ and $T(1) = x$, and then define $\varphi(x) = T$. Complete the definition by defining $\varphi(x_0)$ to be the degenerate singular 1-cube at x_0. Note that

$$\partial_1 \varphi_0(x) = x - x_0$$

for any singular 0-cube x.

Case $n = 1$. Let $T : I \to X$ be a singular 1-cube; we have to define a singular 2-cube $\varphi_1 T : I^2 \to X$. We have already defined the chain homotopy φ_0 on the two faces $A_1 T$ and $B_1 T$, and we want the new definition to be consistent with what we have already defined. Therefore we impose the following three conditions on $\varphi_1 T$:

$$B_1 \varphi_1 T = T,$$
$$A_2 \varphi_2 T = \varphi_0 A_1 T,$$
$$B_2 \varphi_1 T = \varphi_0 B_1 T.$$

Note that these conditions imply that $A_1 \varphi_1 T \in Q_1(X/x_0)$. Given a singular 1-cube T, there always exist singular 2-cubes $\varphi_1 T$ satisfying these three conditions, because the subset of I^2 consisting of the union of any three edges is a retract of I^2. Therefore we may *define* φ_1 by choosing for each singular 1-cube T a singular 2-cube $\varphi_1 T$ satisfying these three conditions. We wish also to impose the following two additional conditions, which are consistent with the three we have already imposed, and with each other:

(a) If $T \in Q_1(X/x_0)$, i.e., if T maps both the vertices of I into x_0, define $\varphi_1 T$ by

$$(\varphi_1 T)(x_1, x_2) = T(x_2).$$

Then $\varphi_1 T$ is degenerate.

(b) If T is a degenerate 1-cube, i.e., $T(x) = $ constant, define

$$(\varphi_1 T)(x_1, x_2) = (\varphi_0 A_1 T)(x_1) = (\varphi_0 B_1 T)(x_1).$$

Then $\varphi_1 T$ is also degenerate.

Case $n = 2$. Given a singular 2-cube $T : I^2 \to X$ we wish to define $\varphi_2 T : I^3 \to X$ so that the definition is consistent with the definition of φ_1 on the four faces of T. Therefore we impose the following conditions on $\varphi_2 T$:

$$B_1 \varphi_2 T = T,$$
$$A_i \varphi_2 T = \varphi_1 A_{i-1} T \qquad i = 2, 3,$$
$$B_i \varphi_2 T = \varphi_1 B_{i-1} T \qquad i = 2, 3.$$

Given T, there will always exist singular 3-cubes $\varphi_2 T$ satisfying these five conditions, because the union of any five faces of I^3 is a retract of I^3. Define φ_2 by choosing for each 2-cube T a 3-cube $\varphi_2 T$ satisfying these five conditions: Note that $A_1 \varphi_2 T \in Q_2(X/x_0)$. We also impose the following two additional conditions, which are consistent with the previous five conditions and with each other:

(a) If $T \in Q_2(X/x_0)$, define $\varphi_2 T$ by

$$(\varphi_2 T)(x_1, x_2, x_3) = T(x_2, x_3).$$

Then $\varphi_2 T$ is degenerate in this case.

(b) If T is a degenerate 2-cube define $\varphi_2 T$ as follows. Since T is degenerate,

$$T(x_1, x_2) = A_1 T(x_2) = B_1 T(x_2)$$

or

$$T(x_1, x_2) = A_2 T(x_1) = B_2 T(x_1).$$

In the first case, define

$$\begin{aligned} \varphi_2 T(x_1, x_2, x_3) &= (\varphi_1 A_1 T)(x_1, x_3) \\ &= (\varphi_1 B_1 T)(x_1, x_3) \end{aligned}$$

while in the second case let

$$\begin{aligned} \varphi_2 T(x_1, x_2, x_3) &= (\varphi_1 A_2 T)(x_1, x_2) \\ &= (\varphi_1 B_2 T)(x_1, x_2). \end{aligned}$$

In either case, $\varphi_2 T$ is also degenerate.

The reader who so desires can define φ_n inductively, following the same pattern for the cases $n = 1$ and $n = 2$.

For $n = 1$ or 2 it is a routine matter to verify the following formula for any singular n-cube T:

$$\partial_{n+1} \varphi_n(T) = T - A_1 \varphi_n(T) - \varphi_{n-1} \partial_n(T);$$

while for $n = 0$ we have the simpler formula

$$\partial_1 \varphi_0(x) = x - x_0.$$

Therefore we define $\rho_n : Q_n(X) \to Q_n(X/x_0)$ as follows: For $n = 0$,

$$\rho_0(x) = x_0$$

for any singular 0-cube x. For $n = 1$ or 2,

$$\rho_n(T) = A_1 \varphi_n(T).$$

With this notation, the preceding formulas can be written as follows:

$$\partial_1 \varphi_0(u) = u - \rho_0(u), \qquad u \in Q_0(X) \tag{7.1}$$

$$\partial_{n+1} \varphi_n(u) + \varphi_{n-1} \partial_n(u) = u - \rho_n(u), \qquad u \in Q_n(X), n = 1 \text{ or } 2. \tag{7.2}$$

Note that ρ_n restricted to the subgroup $Q_n(X/x_0)$ is the identity map, and both φ_n and ρ_n map degenerate chains into degenerate chains. Therefore they define homomorphisms

$$\rho_n: C_n(X) \to C_n(X/x_0),$$
$$\Phi_n: C_n(X) \to C_{n+1}(X),$$

and analogues of Equations (7.1) and (7.2) hold. It remains to prove that ρ commutes with the boundary operator, i.e.,

$$\partial_n \rho_n(u) = \rho_{n-1}\partial_n(u).$$

This equation is an easy consequence of Equations (7.1) and (7.2): Apply ∂_n to both sides of Equation (7.2), and also substitute $\partial_{n+1}(u)$ for u in this equation, and compare the results.

This completes the proof of Lemma 7.2. This proof is conceptually quite simple, but the many details which need to be checked make it rather long.   Q.E.D.

We can now proceed with the proof of Theorem 7.1. First of all, note that $Z_1(X/x_0) = C_1(X/x_0)$; hence there is a natural epimorphism $k: C_1(X/x_0) \to H_1(X/x_0)$ and the kernel of k is $B_1(X/x_0)$.

Next, we will define a homomorphism $l: Q_1(X/x_0) \to \pi'(X/x_0)$ in a rather obvious way. Since $Q_1(X/x_0)$ is a free abelian group and $\pi'(X,x_0)$ is abelian, it suffices to define l on a basis for $Q_1(X/x_0)$, namely, on the singular 1-cubes. But each such basis element $T: I \to X$ with vertices at x_0 is a closed path and hence determines a unique element of $\pi(X,x_0)$. Note that l maps $D_1(X/x_0)$ trivially, and therefore induces a homomorphism $l': C_1(X/x_0) \to \pi'(X,x_0)$, which is obviously an epimorphism. Also, the following diagram is clearly commutative:

$$
\begin{array}{ccc}
C_1(X/x_0) & \xrightarrow{\ l'\ } & \pi'(X,x_0) \\
\downarrow{\scriptstyle k} & & \downarrow{\scriptstyle h} \\
H_1(X/x_0) & \xrightarrow{\ \tau_*\ } & H_1(X).
\end{array}
\qquad (7.3)
$$

Since τ_* is an isomorphism it follows from this diagram that

$$\text{kernel } l' \subset \text{kernel } k = B_1(X/x_0).$$

We will next show that

$$B_1(X/x_0) \subset \text{kernel } l'; \qquad (7.4)$$

from this it will follow that

$$\text{kernel } l' = \text{kernel } k,$$

and since both k and l' are epimorphisms, and Diagram (7.3) is commutative, h must be an isomorphism as desired.

To prove Inclusion (7.4), consider the following sequence of homomorphisms.

$$Q_2(X/x_0) \xrightarrow{\partial_2} Q_1(X/x_0) \xrightarrow{l} \pi'(X, x_0).$$

By using a basic property of the fundamental group (cf. Lemma 8.1 in Chapter II of *Algebraic Topology: An Introduction*) it is easy to prove that the composition $l\partial_2 = 0$. From this fact Inclusion (7.4) follows. Q.E.D.

This theorem should help to develop one's intuition about the first homology group $H_1(X)$.

EXERCISES

7.1. Assume that G is an arcwise connected topological space, $e \in G$, and there exists a continuous map $\mu : G \times G \to G$ such that $\mu(e, x) = \mu(x, e) = x$ for any $x \in G$. [Example: G is a topological group and e is the identity.] Prove that $\pi(X, e)$ is isomorphic to $H_1(X)$ (cf. Exercise 7.5 of Chapter II of *Algebraic Topology: An Introduction*).

7.2. (a) Prove that the fundamental group of a graph is determined by the first homology group. (See Theorem 5.1 of Chapter VI of *Algebraic Topology: An Introduction*.)
   (b) Prove that the fundamental group of a noncompact surface is determined by the first homology group. (See Exercise 5.6 of Chapter VI of *Algebraic Topology: An Introduction*.)

Bibliography for Chapter III

[1] E. Artin and R. H. Fox, Some wild cells and spheres in three-dimensional space, *Ann. Math.* **49** (1948), 979–990.
[2] S. Eilenberg and N. E. Steenrod, *Foundations of Algebraic Topology*, Princeton University Press, Princeton, 1952.
[3] J. G. Hocking and G. S. Young, *Topology*, Addison-Wesley, Reading 1961.
[4] E. Moise, *Geometric Topology in Dimensions 2 and 3*. Springer-Verlag, New York, 1977.

CHAPTER IV
Homology of CW-complexes

§1. Introduction

The purpose of this chapter is to develop a systematic procedure for determining the homology groups of a certain class of topological spaces. The class of topological spaces chosen consists of the CW-complexes of J. H. C. Whitehead. The procedure developed is a natural generalization and extension of the method used in the preceding chapter to determine the homology groups of graphs and compact 2-manifolds.

§2. Adjoining Cells to a Space

The reader may have noticed that there was an analogy in the way the exact homology sequence and Excision property were applied in §III.3 to determine the homology groups of a graph, and the way they were applied in §III.4 to determine the homology groups of a compact surface. The reason behind this analogy may be stated as follows: A graph may be obtained by adjoining the edges to the vertices, and each edge is homeomorphic to $\mathbf{R}^1$. A compact surface may be obtained by adjoining an open disc (which is homeomorphic to $\mathbf{R}^2$) to a certain graph (which is a union of one or more circles with a single point in common).

It is natural to expect there would be a higher-dimensional analogy of these two cases, in which one considers spaces which are obtained by adjoining higher-dimensional open "discs" or "open solid balls" to a given space, and then uses the Excision property, etc. in an analogous way to compute the homology groups of the resulting space. In this section we will study such a higher-dimensional analogue. We will even consider the case where an infinite number of n-dimensional "discs" or "balls" are all attached at once.

In the next section, we will consider spaces that are built up one dimension at a time by first attaching open 2-dimensional discs to a graph (as in the case of a surface), then open 3-dimensional balls to the resulting space, etc.

In this and the following sections, we will use the following terminology and notation for any integer $n \geq 1$:

$$E^n = \{x \in \mathbf{R}^n \,|\, |x| \leq 1\} \quad \text{(closed } n\text{-dimensional disc or ball),}$$
$$U^n = \{x \in \mathbf{R}^n \,|\, |x| < 1\} \quad \text{(open } n\text{-dimensional disc or ball),}$$
$$S^{n-1} = \{x \in \mathbf{R}^n \,|\, |x| = 1\} \quad ((n-1)\text{-dimensional sphere).}$$

The sphere S^{n-1} is called the "boundary" of E^n. Note that U^n is homeomorphic to $\mathbf{R}^n$, and that it is contractible.

In this section we assume that X^* is a Hausdorff space, and that X is a closed subset of X^* such that $X^* - X$ is the disjoint union of open subsets e^n_λ, $\lambda \in \Lambda$; each e^n_λ is assumed to be homeomorphic to U^n, and is called an n-cell or open n-cell. Finally, it is assumed that each n-cell e^n_λ is "attached" to X by means of a so-called characteristic map. This means that for each index $\lambda \in \Lambda$ there exists a continuous map

$$f_\lambda : E^n \to \overline{e}^n_\lambda$$

such that f_λ maps U^n homeomorphically onto e^n_λ and $f_\lambda(S^{n-1}) \subset X$.

If there are only a finite number of n-cells, then we need impose no other conditions. However, if the number of n-cells is infinite, then we must impose the following further condition in order to avoid various pathological situations: It is assumed that a subset A of X^* is closed if and only if $A \cap X$ and $f_\lambda^{-1}(A)$ are closed for all $\lambda \in \Lambda$. This last condition is often expressed by saying that "X^* has the weak topology determined by the maps f_λ and the inclusion map $X \to X^*$." Note that this condition is automatically satisfied in case the number of cells is finite (since the finite union of closed sets is closed in any topological space and a compact subset of a Hausdorff space is closed).

Intuitively speaking, we can think of the space X^* as obtained from X by the "pasting on" of the n-cells e^n_λ. The characteristic map f_λ describes precisely how the cell e^n_λ is pasted onto X. In Chapter III there were examples of the cases where $n = 1$ or 2 and the number of cells attached is finite. The student should construct other examples to illustrate some of the various possibilities inherent in this definition.

In this section, we wish to consider the following problem. Suppose X is a space whose homology groups are known. Let X^* be a space obtained from X by adjoining n-cells so that the above conditions hold. How are the homology groups of X^* related to those of X? The obvious way to attack this problem is to consider the exact sequence of the pair (X^*,X). This requires that we determine the homology groups of the pair (X^*,X). This we can do by application of the techniques of the last section. The result may be stated as follows:

Theorem 2.1. *Let X^* be a space obtained by attaching a collection of n-cells $(n > 0)$ $\{e_\lambda^n | \lambda \in \Lambda\}$ to X so that the hypotheses listed above hold. Then $H_q(X^*,X) = 0$ for all $q \neq n$. For each index $\lambda \in \Lambda$, the characteristic map f_λ induces a monomorphism of relative homology groups $f_{\lambda*}: H_n(E^n, S^{n-1}) \to H_n(X^*,X)$ and $H_n(X^*,X)$ is the direct sum of the image subgroups. Thus $H_n(X^*,X)$ is a free abelian group with basis in 1-1 correspondence with the set of cells $\{e_\lambda^n | \lambda \in \Lambda\}$.*

Corollary 2.2. *The homomorphism $i_*: H_q(X) \to H_q(X^*)$ is an isomorphism except possibly for $q = n$ and $q = n - 1$; the only nontrivial part of the homology sequence of the pair (X^*,X) is the following:*

$$0 \to H_n(X) \xrightarrow{i_*} H_n(X^*) \to H_n(X^*,X) \to H_{n-1}(X) \xrightarrow{i_*} H_{n-1}(X^*) \to 0.$$

PROOF OF THEOREM 2.1. The closed ball E^n and the sphere S^{n-1} both have center at the origin, 0, and radius 1. We also need to consider the closed ball of radius $\frac{1}{2}$ with center at the origin:

$$D^n = \{x \in \mathbf{R}^n | |x| \leq \tfrac{1}{2}\}.$$

Let

$$D_\lambda = f_\lambda(D^n),$$
$$a_\lambda = f_\lambda(0),$$
$$\mathscr{D} = \bigcup_{\lambda \in \Lambda} D_\lambda,$$
$$A = \{a_\lambda | \lambda \in \Lambda\},$$
$$X' = X^* - A.$$

Note that f_λ maps the pair $(D^n, D^n - \{0\})$ homeomorphically onto $(D_\lambda, D_\lambda - \{a_\lambda\})$, and that the subsets D_λ, $\lambda \in \Lambda$, are pairwise disjoint. Consider the following diagram,

$$H_q(\mathscr{D},\mathscr{D} - A) \xrightarrow{1} H_q(X^*,X') \xleftarrow{2} H_q(X^*,X),$$

where both arrows denote homomorphisms induced by inclusion maps. We assert that both homomorphisms in this diagram are isomorphisms for all q. For the homomorphism represented by arrow 2, this follows from the fact that X is a deformation retract of X', and by using the five-lemma. For the homomorphism represented by arrow 1, it is a consequence of the excision property.

Next, note that the arcwise connected components of $\mathscr{D}$ are obviously the sets D_λ. Hence $H_q(\mathscr{D}, \mathscr{D} - A)$ is the direct sum of the groups $H_q(D_\lambda, D_\lambda - \{a_\lambda\})$ for all $\lambda \in \Lambda$. Moreover,

$$H_q(D_\lambda, D_\lambda - \{a_\lambda\}) = \begin{cases} 0 & \text{for } q \neq n, \\ \mathbf{Z} & \text{for } q = n. \end{cases}$$

From this it follows that $H_q(X^*,X) = 0$ for $q \neq n$, and that $H_n(X^*,X)$ is a free abelian group with basis in 1-1 correspondence with the set of n-cells $\{e_\lambda^n\}$. To complete the proof, consider the following commutative diagram:

$$
\begin{array}{ccccc}
H_n(D, D - A) & \xrightarrow{\ 1\ } & H_n(X^*,X') & \xleftarrow{\ 2\ } & H_n(X^*,X) \\
\uparrow{\scriptstyle f'_{\lambda*}} & & \uparrow{\scriptstyle f''_{\lambda*}} & & \uparrow{\scriptstyle f_{\lambda*}} \\
H_n(D^n, D^n - \{0\}) & \xrightarrow{\ 3\ } & H_n(E^n, E^n - \{0\}) & \xleftarrow{\ 4\ } & H_n(E^n,S^{n-1}).
\end{array}
$$

The vertical arrows denote homomorphisms induced by f_λ. Since f_λ maps $(D^n, D^n - \{0\})$ homeomorphically onto $(D_\lambda, D_\lambda - \{a_\lambda\})$, it follows that $f'_{\lambda*}$ maps $H_n(D^n, D^n - \{0\})$ isomorphically onto the direct summand $H_n(D_\lambda, D_\lambda - \{a_\lambda\})$ of $H_n(\mathcal{D}, \mathcal{D} - A)$. We have already proved that arrows 1 and 2 are isomorphisms; by exactly the same method, one can prove that arrows 3 and 4 are isomorphisms. Putting all these facts together suffices to prove that $f_{\lambda*}: H_n(E^n, S^{n-1}) \to H_n(X^*,X)$ has the desired properties. Q.E.D

To close this section, we call the reader's attention to the naturality of the exact sequence of the pair (X^*,X). Thus if X^* is obtained from X by the adjunction of n-cells, and Y^* is similarly obtained from Y by the adjunction of n-cells, and $\varphi:(X^*,X) \to (Y^*,Y)$ is a continuous map of pairs, then we get a ladderlike commutative diagram of maps of the homology sequence of (X^*,X) into that of (Y^*,Y). Of course this is a special case of naturality of the exact sequence of a pair, but it is important and we will make use of it.

§3. CW-complexes

One of the problems encountered in a systematic exposition of algebraic topology is deciding on a suitable category of spaces to be studied. If the category chosen is too narrow and restricted, the theorems are not likely to be applicable in other parts of mathematics. On the other hand, if the category chosen is too broad and inclusive, many of the theorems one desires to prove will become very difficult or false (algebraic topology is mainly concerned with topological spaces which are sufficiently nice locally so as to be non-pathological). The category of CW-complexes (introduced by J. H. C. Whitehead in 1949) has proven to be a reasonable compromise between the various extremes. Roughly speaking, a CW-complex is built up by the successive adjunction of cells of dimensions 1, 2, 3, . . . , etc., as described in the preceding section. Our treatment of this topic is rather brief; hence it may be advisable for the student to read further on this topic. The original paper on the subject is by J. H. C. Whitehead [10]. The book by Lundell and Weingram [5] is rather complete. Other references are Cooke and Finney [2, Chapter I], Hilton [3], Hu [4], and Massey [6].

The original reason for the term "CW-complex" may be explained as follows: The letter C stands for *closure-finite* and W stands for *weak topology*.

Definition 3.1. A structure of CW-complex is prescribed on a space X (which is *always* assumed to be Hausdorff) by the prescription of an ascending sequence of *closed* subspaces

$$X^\circ \subset X^1 \subset X^2 \subset \cdots$$

which satisfy the following conditions:

  (i) X° has the discrete topology.
 (ii) For $n > 0$, X^n is obtained from X^{n-1} by adjoining a collection of n-cells so that the conditions explained in §2 hold.
(iii) X is the union of the subspaces X^i for $i \geq 0$.
 (iv) The space X and the subspaces X^q all have the weak topology: A subset A is closed if and only if $A \cap \bar{e}^n$ is closed for all n-cells, e^n, $n = 0, 1, 2, \ldots$.

The subset X^n is called the *n-skeleton*. The points of X^0 are called *vertices* or 0-cells. A CW-complex is *finite* or *infinite* according as the number of cells is finite or infinite. If $X = X^n$ for some integer n, the CW-complex is called *finite-dimensional*, and the least such integer n is called the *dimension*.

Note that for finite CW-complexes, Condition (iv) is superfluous. This fact greatly simplifies the theory in the finite case, which will be our main interest.

EXAMPLE 3.1. The n-sphere, S^n, can be given a CW-complex structure such that there are only two cells, a 0-cell and an n-cell. In other words, the k-skeleton is a single point for $0 \leq k < n$, and the n-skeleton is S^n. The characteristic map, by which the n-cell is attached, maps the boundary of E^n to a single point.

EXAMPLE 3.2. A finite graph, as defined in III.3, is a finite, 1-dimensional CW-complex with an additional condition imposed on the characteristic maps by which the 1-cells are attached.

EXAMPLE 3.3. In §III.3 we determined the homology groups of a compact, orientable surface of genus $g > 0$ (i.e., the connected sum of g tori). This amounted to prescribing a finite, 2-dimensional CW-complex structure on each of these surfaces, such that there is a single 0-cell, $2g$ 1-cells, and a single 2-cell. In the case of a nonorientable surface of genus g (i.e., the connected sum of g projective planes) we used a CW-complex having a single 0-cell, g 1-cells, and a single 2-cell.

EXAMPLE 3.4. To *triangulate* a compact 2-manifold, as explained in Chapter I of *Algebraic Topology: An Introduction*, gives it the structure of a finite,

2-dimensional CW-complex. The vertices are the 0-cells, the edges are the 1-cells, and the triangles are the 2-cells. Similarly, the more general subdivision of a compact 2-manifold discussed in Section 8 of Chapter I, *loc. cit.*, also gives rise to a CW-complex.

EXAMPLE 3.5. Suppose that X and Y are finite CW-complexes with skeletons $\{X^k\}$ and $\{Y^k\}$ respectively. Then one can specify a CW-complex on the product space $X \times Y$ such that the n-skeleton is the union of the subspaces $X^0 \times Y^n$, $X^1 \times Y^{n-1}$, $X^2 \times Y^{n-2}, \ldots, X^n \times Y^0$. The product of a p-cell of X and a q-cell of Y is a $p + q$-cell of $X \times Y$; the attaching map of such a product cell is the product of the attaching maps. The details of this construction will be described in §VI.2.

EXAMPLE 3.6. A more subtle and interesting example is a real, complex, or quaternionic projective space. Given any field F, an n-dimensional projective space over F is defined to be the set of all 1-dimensional subspaces in an $(n + 1)$-dimensional vector space over F. This definition is valid even if the field F is noncommutative (although then one should distinguish between right and left vector spaces over F). Since any $(n + 1)$-dimensional vector space over F is isomorphic to the space F^{n+1} of all $(n + 1)$-tuples of elements of F, we may as well restrict ourselves to consideration of F^{n+1}. Any point $(x_1, \ldots, x_{n+1})$ of F^{n+1} different from $(0, \ldots, 0)$ determines a unique 1-dimensional subspace, and hence a unique point of the corresponding projective space. Two such $(n + 1)$-tuples, $(x_1, \ldots, x_{n+1})$ and $(y_1, \ldots, y_{n+1})$ determine the same point of projective space if and only if there exists a nonzero element λ of F such that $y_i = \lambda x_i$ for $1 \le i \le n + 1$. In books on projective geometry, such an $(n + 1)$-tuple is referred to as a set of *homogeneous coordinates* for the corresponding point in projective space.

We will only be interested in the cases where F is the field of real numbers, complex numbers, or quaternions. In each of these cases the field F has a standard topology, and the vector space F^{n+1} is given the product topology. The corresponding projective space can be looked on as a quotient space of $F^{n+1} - \{0\}$, and it is customary to give it the quotient space topology. Alternatively, the projective space can be topologized as a quotient space of the unit sphere with center at the origin in F^{n+1}.

There is an obvious imbedding of F^n in F^{n+1}, defined by $(x_1, \ldots, x_n) \to (x_1, \ldots, x_n, 0)$. This leads to a corresponding imbedding of the $(n - 1)$-dimensional projective space into the n-dimensional projective space over F. This kind of imbedding will define the skeletons of a CW-complex on these projective spaces. We will now discuss in more detail each of the cases:

Case 1: F = real numbers. The n-dimensional real projective space, denoted by RP^n, is the set of all 1-dimensional subspaces of $\mathbf{R}^{n+1}$. It may be topologized as a quotient space of $\mathbf{R}^{n+1} - \{0\}$, or of the unit sphere, S^n. Each 1-dimensional subspace of $\mathbf{R}^{n+1}$ intersects S^n in a pair of antipodal points. Hence S^n is a 2-sheeted covering space of RP^n (see *Algebraic Topology: An Introduction*, Example 8.2 on p. 166). The inclusions $\mathbf{R}^1 \subset \mathbf{R}^2 \subset \cdots \subset \mathbf{R}^{n+1}$ give rise to corresponding inclusions of real projective spaces:

$$RP^0 \subset RP^1 \subset RP^2 \subset \cdots \subset RP^n.$$

It is clear that RP^0 is a single point, and easy to verify that RP^1 is a circle. We will take these subspaces as the skeletons of a CW-complex. We assert that RP^k is obtained from RP^{k-1} by the adjunction of a single cell of dimension k. Using homogeneous coordinates in RP^k, the characteristic map

$$f_k : E^k \to RP^k$$

is defined by the formula

$$f_k(x_1, \ldots, x_k) = (x_1, \ldots, x_k, \sqrt{1 - |x|^2}),$$

where $x = (x_1, \ldots, x_k)$. We leave it to the reader to verify that f_k maps $E^k - S^{k-1}$ homeomorphically onto $RP^k - RP^{k-1}$, and S^{k-1} onto RP^{k-1} (but not homeomorphically).

Case 2: $F =$ complex numbers. The n-dimensional complex projective space, denoted by CP^n, is the set of all 1-dimensional subspaces of the complex vector space $\mathbf{C}^{n+1}$. The inclusions

$$\mathbf{C}^1 \subset \mathbf{C}^2 \subset \cdots \subset \mathbf{C}^{n+1}$$

give rise to corresponding inclusions of complex projective spaces,

$$CP^0 \subset CP^1 \subset \cdots \subset CP^n.$$

Once again, CP^0 is a single point, and it may be shown without too much difficulty that CP^1 is homeomorphic to S^2. In this case, CP^k is obtained from CP^{k-1} by the adjunction of a single cell of dimension $2k$. The adjunction map

$$f_k : E^{2k} \to CP^k$$

is defined by the formula

$$f_k(z_1, \ldots, z_k) = (z_1, \ldots, z_k, \sqrt{1 - |z|^2}).$$

Here we are using the following notational conventions: $z = (z_1, \ldots, z_k)$ is a point of $\mathbf{C}^k = \mathbf{R}^{2k}$. On the right-hand side of this formula we are using homogeneous coordinates in CP^k. The norm of z is defined by

$$|z| = (|z_1|^2 + |z_2|^2 + \cdots + |z_k|^2)^{1/2}.$$

E^{2k} is the unit ball in $\mathbf{C}^k = \mathbf{R}^{2k}$. Once again, it can be verified that f_k maps S^{2k-1} onto CP^{k-1}, and $E^{2k} - S^{2k-1}$ homeomorphically onto $CP^k - CP^{k-1}$. Hence we can take CP^k as the $2k$-skeleton of CP^n for $k = 0, 1, \ldots, n$. The $2k + 1$-dimensional skeleton is the same as the $2k$-dimensional skeleton. There are cells of dimensions $0, 2, 4, \ldots, 2n$.

Case 3: $F =$ quaternions. This case is very similar to the preceding. The n-dimensional quaternionic projective space is denoted by QP^n. We have inclusions,

$$QP^0 \subset QP^1 \subset \cdots \subset QP^n.$$

QP^0 is a single point and QP^1 is homeomorphic to S^4. QP^k is obtained from QP^{k-1} by adjunction of a single cell of dimension $4k$. The formula for the characteristic map is the same as in the two preceding cases, using quaternions in place of real or complex numbers. QP^n is a CW-complex having a single cell in each of the dimensions $0, 4, 8, \ldots, 4n$.

For further details about these projective spaces, the reader is referred to Bourbaki [1] or Porteous [8].

Not every Hausdorff space admits a CW-complex structure. If it does admit such a structure, then usually it admits infinitely many different such structures (e.g., consider a finite regular graph as a CW-complex, and consider all its subdivisions).

Among the nice properties of a CW-complex, we list the following without proof:

(i) A CW-complex is paracompact, and hence normal.
(ii) A CW-complex is *locally contractible*, i.e., every point has a basic family of contractible neighborhoods.
(iii) A compact subset of a CW-complex meets only a finite number of cells. A CW-complex is compact if and only if it is finite.
(iv) A function f defined on a CW-complex is continuous if and only if the restriction of f to the closure $\bar{e}^n$ of every n-cell is continuous ($n = 0, 1, 2, \ldots$).

A subset A of a CW-complex is called a *subcomplex* if A is a union of cells of X, and if for any cell e^n,

$$e^n \subset A \Longrightarrow \bar{e}^n \subset A.$$

If this is the case, it may be shown that the sets

$$A^n = A \cap X^n, \qquad n = 0, 1, 2, \ldots,$$

define a CW-complex structure on A.

For example, the skeletons X^n are subcomplexes.

Definition 3.2. A continuous map of $f : X \to Y$ of one CW-complex into another is called *cellular* if $f(X^n) \subset Y^n$ for $n = 0, 1, 2, \ldots$ (here X^n and Y^n denote the n-skeletons of X and Y).

In [10] J. H. C. Whitehead proves that any continuous map $X \to Y$ is homotopic to a cellular map.

§4. The Homology Groups of a CW-complex

The purpose of this section is to apply the results of §2 to CW-complexes in a systematic way.

Let $K = \{K^n \mid n = 0,1,2,\ldots\}$ denote a structure of CW-complex on the topological space X (each K^n is a closed subset of X). We will define $K^n = \varnothing$ for $n < 0$. Since K^n is obtained from K^{n-1} by the adjunction of n-cells (by definition), we can apply the results of Theorem 2.1 to conclude that

$$H_q(K^n, K^{n-1}) = 0$$

for $q \neq n$ and that $H_n(K^n, K^{n-1})$ is a free abelian group with basis in 1-1 correspondence with the n-cells of K.

Lemma 4.1. $H_q(K^n) = 0$ for all $q > n$.

The proof is by induction on n. For $n = 0$, the lemma is trivial, since K^0 is a discrete space (by definition). The inductive step is proved by using the homology sequence of the pair (K^n, K^{n-1}).

We will now associate with the CW-complex K certain "chain groups" $C_n(K)$, $n = 0, 1, 2, \ldots$ and then we will prove that the nth homology group obtained from these chain groups is naturally isomorphic to $H_n(X)$. The definitions are as follows:

$$C_n(K) = H_n(K^n, K^{n-1}),$$

and

$$d_n : C_n(K) \to C_{n-1}(K)$$

is defined to be the composition of homomorphisms,

$$H_n(K^n, K^{n-1}) \xrightarrow{\partial_*} H_{n-1}(K^{n-1}) \xrightarrow{j_{n-1}} H_{n-1}(K^{n-1}, K^{n-2}),$$

where ∂_* is the boundary operator of the pair (K^n, K^{n-1}) and j_{n-1} is the homomorphism induced by the inclusion map. Of course one must verify that $d_{n-1} d_n = 0$, but this is easy. We will find it convenient to denote the n-dimensional groups of cycles, bounding cycles, and homology classes derived from these chain groups by the notations

$$Z_n(K), \quad B_n(K), \quad \text{and} \quad H_n(K)$$

respectively; here $Z_n(K) = \text{kernel } d_n$, $B_n(K) = \text{image } d_{n+1}$, and $H_n(K) = Z_n(K)/B_n(K)$.

For the statement of the main theorem, consider the following diagram:

$$H_n(X) \xleftarrow{k_n} H_n(K^n) \xrightarrow{j_n} H_n(K^n, K^{n-1}) = C_n(K).$$

Here j_n and k_n are homomorphisms induced by inclusion maps.

Theorem 4.2. *In the above diagram*:

k_n *is an epimorphism.*
j_n *is a monomorphism.*
image $j_n = Z_n(K)$.
kernel $k_n = j_n^{-1}(B_n(K))$.

Thus $j_n \circ k_n^{-1}$ *defines an isomorphism*

$$\theta_n : H_n(X) \to H_n(K).$$

This theorem asserts that $H_n(X) \approx H_n(K)$; however, it says even more, in that a certain composition of maps is asserted to be an isomorphism. This additional information is important in certain cases.

PROOF OF THEOREM 4.2. First of all, note that for $n \geq 1$ the only nontrivial part of the homology sequence of the pair (K^n, K^{n-1}) is the following:

$$0 \to H_n(K^n) \xrightarrow{j_n} H_n(K^n, K^{n-1}) \xrightarrow{\partial_*} H_{n-1}(K^{n-1}) \xrightarrow{i_n} H_{n-1}(K^n) \to 0. \quad (4.1)$$

This is a consequence of Theorem 2.1 and Lemma 4.1. It follows that the homomorphism

$$i_n : H_q(K^{n-1}) \to H_q(K^n)$$

is an isomorphism except for $q = n$ and $q = n - 1$; in particular it is an isomorphism for $q < n - 1$, i.e., for $n > q + 1$. Thus we have the following commutative diagram for each integer $q \geq 0$:

$$(4.2)$$

The horizontal arrows are all isomorphisms from what we have just said.

In case X is finite dimensional, $K^m = X$ for some sufficiently large integer m, and it follows from this diagram that

$$k_\alpha : H_q(K^\alpha) \to H_q(X)$$

is an isomorphism for any integer $\alpha > q$. We wish to derive this same conclusion in case X is infinite dimensional. For this purpose, recall Property (iii) of CW-complexes mentioned in the preceding section: Any compact subset of a CW-complex meets only a finite number of cells. It follows that any compact subset C of X is contained in some skeleton K^m. If one now applies Proposition III.6.1, the desired conclusion follows quite easily. The details are left to the reader. Note the particular case $\alpha = q + 1$: the homomorphism

$$k_{q+1} : H_q(K^{q+1}) \to H_q(X)$$

is an isomorphism.

Next, we consider the exact sequence (4.1). It follows from exactness that

$$j_n: H_n(K^n) \to H_n(K^n, K^{n-1})$$

is a monomorphism for all integers n, and

$$i_n: H_{n-1}(K^{n-1}) \to H_{n-1}(K^n)$$

is an epimorphism for all n. In view of the commutativity of the diagram

$$H_n(K^n) \xrightarrow{i_{n+1}} H_n(K^{n+1})$$

$$k_n \searrow \qquad \swarrow k_{n+1}$$

$$H_n(X)$$

and the fact that k_{n+1} is an isomorphism, it follows that k_n is onto, and kernel $k_n = $ kernel i_{n+1}. Thus we may replace exact sequence (4.1) by the following:

$$0 \to H_n(K^n) \xrightarrow{j_n} H_n(K^n, K^{n-1}) \xrightarrow{\partial_*} H_{n-1}(K^{n-1}) \xrightarrow{k_{n-1}} H_{n-1}(X) \to 0. \quad (4.3)$$

Since $d_n = j_{n-1}\partial_*$, and j_{n-1} is a monomorphism, we see that

$$Z_n(K) = \text{kernel } d_n = \text{kernel } \partial_*$$
$$= \text{image } j_n.$$

Next, we see that

$$\text{kernel } k_{n-1} = \text{image } \partial_*$$
$$= j_{n-1}^{-1} (\text{image } j_{n-1}\partial_*)$$
$$= j_{n-1}^{-1} (\text{image } d_n)$$
$$= j_{n-1}^{-1} (B_{n-1}(K))$$

as required.

This completes the proof of Theorem 4.2.   Q.E.D.

We will now consider some applications of this theorem:

(1) Suppose X is a CW-complex which is n-dimensional. Then

$$H_q(X) = 0 \quad \text{for } q > n.$$

(2) Suppose X is a CW-complex with only a finite number of n-dimensional cells. Then $H_n(X)$ is a finitely generated abelian group (hence it is a direct sum of cyclic groups).

(3) Suppose X is a CW-complex with no n-dimensional cells. Then $H_n(X) = 0$.

(4) *The Euler characteristic.* Let $K = \{K^n\}$ be a structure of *finite* CW-complex on the space X (hence X is compact). Denote the number of n-cells of K by α_n. The *Euler characteristic* of K is defined to be the integer

$$\chi(K) = \sum_{n \geq 0} (-1)^n \alpha_n.$$

We will now outline a proof that $\chi(K)$ is actually a homotopy type invariant of the space X; it does not depend on K.

Define a subset of an abelian group to be *linearly independent* if it satisfies the usual condition with *integer* coefficients. Then define the *rank* of an abelian group to be the cardinal number of a maximal linearly independent subset. Earlier, we defined the rank of a free abelian group to be the cardinal number of a basis; it is an exercise in matrix theory to prove that the two definitions are equivalent in the case of free abelian groups.

For any abelian group A, let $r(A)$ denote the rank of A. One can now prove the following facts about the rank of abelian groups:

(a) If B is a subgroup or quotient group of A, then $r(B) \le r(A)$. Hence any finitely generated abelian group has finite rank.

(b) Let $0 \to A \to B \to C \to 0$ be a short exact sequence abelian groups with B of finite rank. Then

$$r(B) = r(A) + r(C).$$

The proofs are left to the reader.

The proof of invariance of the Euler characteristic of a finite CW-complex depends on the following lemma:

Lemma 4.3. *Let K be a finite CW-complex on the space X. Then*

$$\sum_n (-1)^n r(C_n(K)) = \sum_n (-1)^n r(H_n(K)).$$

We leave the proof, which depends on Statements (a) and (b) above, to the reader.

Corollary 4.4. *Let $K = \{K^n\}$ be a finite CW-complex on the space X. Then the Euler characteristic satisfies the following equation:*

$$\chi(K) = \sum_n (-1)^n r(H_n(X)).$$

Hence $\chi(K)$ is independent of the choice of the CW-complex K on the space X.

(5) *The homology groups of n-dimensional projective space.* Using the CW-complexes on CP^n and QP^n described in the previous section, the following results are immediate:

$$H_q(CP^n) = \begin{cases} \mathbf{Z} & \text{for } q \text{ even and } 0 \le q \le 2n, \\ 0 & \text{otherwise,} \end{cases}$$

$$H_q(QP^n) = \begin{cases} \mathbf{Z} & \text{for } q \equiv 0 \bmod 4 \text{ and } 0 \le q \le 4n, \\ 0 & \text{otherwise.} \end{cases}$$

On the other hand, the methods we have developed do not suffice to determine the homology groups of RP^n. All one can prove using these methods is that $H_q(RP^n)$ is a cyclic group for $0 \leq q \leq n$ and is 0 otherwise (of course $H_0(RP^n)$ is infinite cyclic).

Next, we will discuss the homomorphism induced by a cellular map of one CW-complex into another. Let $K = \{K^n\}$ be a CW-complex on the space X, and let $L = \{L^n\}$ be a CW-complex on the space Y, and let $f: X \rightarrow Y$ be a cellular map, i.e., $f(K^n) \subset L^n$ for all n. Then for each integer n, f induces a homomorphism of the homology sequence of the pair (K^n, K^{n-1}) into the homology sequence of the pair (L^n, L^{n-1}). Thus we have the following commutative diagram:

$$0 \longrightarrow H_n(K^n) \xrightarrow{j_n} H_n(K^n, K^{n-1}) \xrightarrow{\partial_*} H_{n-1}(K^{n-1}) \xrightarrow{i_n} H_{n-1}(K^n) \longrightarrow 0$$
$$\downarrow{f_n} \qquad\qquad \downarrow{\phi_n} \qquad\qquad \downarrow{f_{n-1}} \qquad\qquad \downarrow{f_n}$$
$$0 \longrightarrow H_n(L^n) \xrightarrow{j_n} H_n(L^n, L^{n-1}) \xrightarrow{\partial_*} H_{n-1}(L^{n-1}) \xrightarrow{i_n} H_{n-1}(L^n) \longrightarrow 0.$$

Here $f_n: K^n \rightarrow L^n$ is the map induced by f, as is $\varphi_n: (K^n, K^{n-1}) \rightarrow (L^n, L^{n-1})$. In view of the definition of the boundary operator $d_n: C_n(K) \rightarrow C_{n-1}(K)$ above, it follows that the following diagram is commutative for all n:

$$\begin{array}{ccc} C_n(K) & \xrightarrow{\varphi_n} & C_n(L) \\ \downarrow{d_n} & & \downarrow{d_n} \\ C_{n-1}(K) & \xrightarrow{\varphi_{n-1}} & C_{n-1}(L). \end{array}$$

Hence by exactly the same reasoning used in §II.3, we conclude that the collection of homomorphisms $\{\varphi_n\}$ induce homomorphisms

$$\varphi_*: H_n(K) \rightarrow H_n(L), \qquad n = 0, 1, 2, \ldots.$$

Theorem 4.5. *The induced homomorphisms* $f_*: H_n(X) \rightarrow H_n(Y)$ *and* $\varphi_*: H_n(K) \rightarrow H_n(L)$ *correspond under the isomorphisms* θ_n *of Theorem 4.2; i.e., the following diagram is commutative for all* n:

$$\begin{array}{ccc} H_n(X) & \xrightarrow{\theta_n} & H_n(K) \\ \downarrow{f_*} & & \downarrow{\varphi_*} \\ H_n(Y) & \xrightarrow{\theta_n} & H_n(L). \end{array}$$

PROOF. This follows immediately from the fact that the following diagram is commutative for all n, together with the definition of θ_n contained in Theorem 4.2:

$$\begin{array}{ccccc} H_n(X) & \xleftarrow{k_n} & H_n(K^n) & \xrightarrow{j_n} & H_n(K^n, K^{n-1}) \\ \downarrow{f} & & \downarrow{f_n} & & \downarrow{\varphi_n} \\ H_n(Y) & \xleftarrow{k_n} & H_n(L_n) & \xrightarrow{j_n} & H_n(L^n, L^{n-1}). \end{array}$$

We will conclude this section with a discussion of the effective computability of the various concepts introduced in this section. First of all, the groups $C_n(K)$, $n = 0, 1, 2, \ldots$ are free groups with basis in 1-1 correspondence with the set of n-cells of K, hence they may be considered to be well determined. To compute the homology groups $H_n(K) \approx H_n(X)$, we must determine the homomorphisms

$$d_n: C_n(K) \to C_{n-1}(K), \qquad n = 0, 1, 2, \ldots.$$

In general, these homomorphisms will depend on the choice of the characteristic maps by which the various cells are attached, and there seems to be no universal, simple, method for their determination. The following simple example illustrates this point. Let X be a torus and Y a Klein bottle. We may choose CW-complexes K and L on X and Y respectively each of which has one vertex, two 1-cells, and one 2-cell. Thus $C_n(K) \approx C_n(L)$ for all n. However, since $H_n(K) \not\approx H_n(L)$ for $n = 1$ or 2, it follows that the boundary homomorphisms d_n for K and L must be essentially different (compare §III.4). The reason, of course, lies in the fact that the 2-cell is attached by different maps in the two cases.

The situation is even worse as regards the computation of the homomorphisms $\varphi_n: C_n(K) \to C_n(L)$ mentioned above. Here an example is furnished by the case $X = Y = S^n$, the n-sphere. We proved earlier (cf. Exercise III.3.2) that there exist continuous maps $S^n \to S^n$ of every possible degree. If we take $K = L$ to be a CW-complex with one vertex and one n-cell, then a map $S^n \to S^n$ will be cellular if and only if the vertex is mapped onto the vertex; and this can always be arranged by an appropriate homotopic deformation of any given map. Thus it is clear that in such cases Theorem 4.6 is of no help in determining the homomorphism induced by a continuous map.

One of our objectives will be to introduce a more restricted class of CW-complexes and cellular maps such that the boundary operator and the induced homomorphism are actually computable.

§5. Incidence Numbers and Orientations of Cells

This section is devoted to some material of a more or less technical nature which will be used in the computation of homology groups of CW-complexes.

As in the preceding section, let $K = \{K^n\}$ be a CW-complex on the space X. For each n-cell, e_λ^n, there is a characteristic map,

$$f_\lambda: (E^n, S^{n-1}) \to (K^n, K^{n-1})$$

and according to Theorem 2.1 the induced homomorphism on the n-dimensional relative homology groups is a monomorphism, and $H_n(K^n, K^{n-1})$ is the direct sum of the image subgroups. The characteristic map f_λ corresponding to the cell e_λ^n is by no means unique, and it is conceivable that this direct sum decomposition of the group $H_n(K^n, K^{n-1})$ depends on the choices of the characteristic maps. Before proceeding further, it is important to point out that this is not the case; the direct sum decomposition of $H_n(K^n, K^{n-1})$ is *canonical*, and independent of the choices of the characteristic maps. This may be proved as follows. For any n-cell e_λ^n, $n > 0$, let

$$\dot{e}_\lambda^n = \bar{e}_\lambda^n - e_\lambda^n.$$

We will call $\dot{e}_\lambda^n$ the *boundary* of e_λ^n, even though it need not coincide with the boundary in the sense of point set topology. We can factor the characteristic map f_λ through the pair $(\bar{e}_\lambda^n, \dot{e}_\lambda^n)$, as follows:

$$
\begin{array}{ccc}
(E^n, S^{n-1}) & \xrightarrow{\;g_\lambda\;} & (\bar{e}^n, \dot{e}^n) \\
& {\scriptstyle f_\lambda} \searrow & \downarrow {\scriptstyle l_\lambda} \\
& & (K^n, K^{n-1}).
\end{array}
$$

Here l_λ is an inclusion map. Passing to homology, we obtain the following commutative diagram:

$$
\begin{array}{ccc}
H_n(E^n, S^{n-1}) & \xrightarrow{\;g_{\lambda*}\;} & H_n(\bar{e}^n, \dot{e}^n) \\
& {\scriptstyle f_{\lambda*}} \searrow & \downarrow {\scriptstyle l_{\lambda*}} \\
& & H_n(K^n, K^{n-1}).
\end{array}
$$

We can apply Theorem 2.1 with $(X^*, X) = (\bar{e}_\lambda^n, \dot{e}_\lambda^n)$ to conclude that $g_{\lambda*}$ is an isomorphism. Hence

$$\text{image } f_{\lambda*} = \text{image } l_{\lambda*}$$

and therefore image $f_{\lambda*}$ is independent of the choice of the characteristic map f_λ, as was to be proved. Note that this also proves that $l_{\lambda*}$ is a monomorphism, and $H_n(K^n, K^{n-1})$ is the direct sum of the images for all $\lambda \in \Lambda$.

Since the group $H_n(\bar{e}_\lambda^n, \dot{e}_\lambda^n)$ is infinite cyclic for $n > 0$, there are two ways to choose a generator and the choices are negatives of each other. We will call a generator of the group $H_n(\bar{e}_\lambda^n, \dot{e}_\lambda^n)$ an *orientation* of the cell e_λ^n.

Assume we have chosen an orientation $a_\lambda^n \in H_n(\bar{e}_\lambda^n, \dot{e}_\lambda^n)$ for each n-cell e_λ^n; let

$$b_\lambda^n = l_{\lambda*}(a_\lambda^n) \in C_n(K).$$

Then the set $\{b_\lambda^n\}$ is a basis for the chain group $C_n(K)$.

The foregoing remarks are only valid if $n > 0$; the case $n = 0$ must be modified, as follows. By definition, $C_0(K) = H_0(K^0)$, and

$$H_0(K^0) = \sum_{\lambda \in \Lambda} H_0(e_\lambda^0),$$

where $\{e_\lambda^0 | \lambda \in \Lambda\}$ denotes the set of 0-cells (or vertices) of K. For each λ, the augmentation homomorphism

$$\varepsilon_* : H_0(e_\lambda^0) \to Z$$

is a *natural* isomorphism. We will *always* choose $a_\lambda^0 \in H_0(e_\lambda^0)$ to be the unique element such that $\varepsilon_*(a_\lambda^0) = 1$, and let $b_\lambda^0 \in H_0(K^0) = C_0(K)$ be the element corresponding to a_λ^0. Thus $\{b_\lambda^0 | \lambda \in \Lambda_0\}$ is a basis for $C_0(K)$.

The distinction between the cases $n = 0$ and $n > 0$ may be summarized as follows: For $n > 0$, an n-cell has two orientations, and there is no reason to prefer one orientation over the other. On the other hand, a 0-cell consists of a single point, and the question of choice of orientation does not arise in this case.

Assume, then, that the bases $\{b_\lambda^n | \lambda \in \Lambda_n\}$ have been chosen for the chain groups $C_n(K)$ for $n = 0, 1, 2, \ldots$, as described above. The boundary homomorphisms

$$d_n : C_n(K) \to C_{n-1}(K), \qquad n = 1, 2, 3, \ldots$$

are completely determined by the value of d_n on the basis elements; and we may uniquely express $d_n(b_\lambda^n)$ as a linear combination of the b_μ^{n-1}'s. It is customary to use the following notation for this purpose:

$$d_n(b_\lambda^n) = \sum_\mu [b_\lambda^n : b_\mu^{n-1}] b_\mu^{n-1}.$$

The integral coefficient $[b_\lambda^n : b_\mu^{n-1}]$ is called the *incidence number* of the cells e_λ^n and e_μ^{n-1} (with respect to the chosen orientations). Obviously, the homomorphism d_n is completely determined by the incidence numbers, and vice-versa. The most important properties of the incidence numbers are summarized in the following two lemmas.

Lemma 5.1. *The incidence numbers of a CW-complex have the following properties:*

(a) *For any n-cell e_λ^n, $[b_\lambda^n : b_\mu^{n-1}] = 0$ for all but a finite number of $(n-1)$-cells e_μ^{n-1}.*

(b) *For any n-cell e_λ^n and $(n-2)$-cell e_ν^{n-2},*

$$\sum_\mu [b_\lambda^n : b_\mu^{n-1}][b_\mu^{n-1} : b_\nu^{n-2}] = 0.$$

(c) *For any 1-cell e_λ^1, $\sum_\mu [b_\lambda^1 : b_\mu^0] = 0$.*

(d) *$[-b_\lambda^n : b_\mu^{n-1}] = [b_\lambda^n : -b_\mu^{n-1}] = -[b_\lambda^n : b_\mu^{n-1}].$*

PROOF: The proof of (a) is a direct consequence of the definition of incidence numbers, and the proof of (b) follows from the relation $d_{n-1}d_n = 0$. To prove (c), recall that $C_1(K) = H_1(K^1, K^0)$, $C_0(K) = H_0(K^0, K^{-1}) = H_0(K^0)$, and $d_1 : C_1(K) \to C_0(K)$ is the homomorphism

$$\partial_* : H_1(K^1, K^0) \to H_0(K^0)$$

in the homology sequence of the pair (K^1, K^0). Now consider the following diagram, which is commutative:

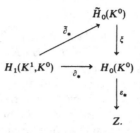

The vertical line is exact by Proposition II.2.1, hence $\varepsilon_* \xi = 0$. Therefore

$$\varepsilon_* \partial_* = \varepsilon_* \xi \tilde{\partial}_* = 0.$$

Hence we obtain

$$0 = \varepsilon_* \partial_*(b_\lambda^1) = \varepsilon_* d_1(b_\lambda^1)$$
$$= \varepsilon_* \sum_\mu [b_\lambda^1 : b_\mu^0] b_\mu^0 = \sum_\mu [b_\lambda^1 : b_\mu^0] \varepsilon_*(b_\mu^0)$$
$$= \sum_\mu [b_\lambda^1 : b_\mu^0]$$

since b_μ^0 was chosen so that $\varepsilon_*(b_\mu^0) = 1$.

The proof of (d) is trivial.   Q.E.D.

Lemma 5.2. *If the cell e_μ^{n-1} is not contained in the closure of the cell e_λ^n, then $[b_\lambda^n : b_\mu^{n-1}] = 0$.*

PROOF: Earlier in this section, it was pointed out that the canonical direct sum decomposition of the group $C_n(K) = H_n(K^n, K^{n-1})$ is determined by the monomorphisms

$$l_{\lambda*} : H_n(\bar{e}_\lambda^n, \dot{e}_\lambda^n) \to H_n(K^n, K^{n-1})$$

for all n-cells e_λ^n of K. Corresponding to this direct sum decomposition, there are projections of $C_n(K)$ onto each of the summands. We assert that these projections may be described in terms of the following commutative diagram:

$$\begin{array}{ccc} H_n(\bar{e}_\lambda^n, \dot{e}_\lambda^n) & \xrightarrow{\ l_{\lambda*}\ } & \\ \downarrow{\scriptstyle l'_{\lambda*}} & \searrow & H_n(K^n, K^{n-1}). \\ H_n(K^n, K^n - e_\lambda^n) & \nearrow{\scriptstyle m_{\lambda*}} & \end{array}$$

Here l'_λ and m_λ are inclusion maps. We assert that $l'_{\lambda*}$ is an isomorphism and $m_{\lambda*}$ composed with the inverse of $l'_{\lambda*}$ gives the projection of $C_n(K)$ onto the direct summand corresponding to the cell e_λ^n. The proof that $l'_{\lambda*}$ is an isomorphism is based on Theorem 2.1, and is exactly the same as the proof that $l_{\lambda*}$ is a monomorphism whose image is a direct summand. To prove the assertion about $m_{\lambda*}$, one must prove that if $e_\lambda^n \neq e_\mu^n$, then $m_{\mu*}l_{\lambda*} = 0$; this is an easy consequence of Lemma 5.3 below.

In view of these facts, and the definition of incidence numbers, it is clear that in order to prove $[b_\lambda^n, b_\mu^{n-1}] = 0$, we must prove that the following composition of homomorphisms is zero:

$$H_n(\overline{e}_\lambda^n, \dot{e}_\lambda^n) \xrightarrow{l_{\lambda*}} H_n(K_n, K^{n-1}) \xrightarrow{\partial_*} H_{n-1}(K^{n-1}) \xrightarrow{j_{n-1}} H_{n-1}(K^{n-1}, K^{n-2})$$

$$\downarrow{m_{\mu*}}$$

$$H_{n-1}(K^{n-1}, K^{n-1} - e_\mu^{n-1}).$$

We can imbed this sequence of homomorphisms in the following commutative diagram:

$$\begin{array}{ccccc}
H_n(K^n, K^{n-1}) & \xrightarrow{\partial_*} & H_{n-1}(K^{n-1}) & \xrightarrow{j_{n-1}} & H_{n-1}(K^{n-1}, K^{n-2}) \\
\uparrow{l_{\lambda*}} & & \uparrow & & \downarrow{m_{\mu*}} \\
H_n(\overline{e}_\lambda^n, \dot{e}_\lambda^n) & \xrightarrow{\partial'} & H_{n-1}(\dot{e}_\lambda^n) & \xrightarrow{j_*} & H_{n-1}(K^{n-1}, K^{n-1} - e_\mu^{n-1}).
\end{array}$$

By commutativity of the squares in this diagram, we see that we must prove

$$j_*\partial' = 0.$$

Since e_μ^{n-1} is not contained in $\overline{e}_\lambda^n$, the inclusion map $j: \dot{e}_\lambda^n \to (K^{n-1}, K^{n-1} - e_\mu^{n-1})$ is homotopic to a map of $\dot{e}_\lambda^n$ into $K^{n-1} - e_\mu^{n-1}$ (to see this, choose a point $x_0 \in e_\mu^{n-1}$ such that $x_0 \notin \overline{e}_\lambda^n$; the required homotopy of the map j is defined by means of a "radical projection" outward from the point x_0 to the boundary of the cell e_μ^{n-1}). It follows from Lemma 5.3 below that $j_* = 0$, and the proof is complete.   Q.E.D.

Lemma 5.3. *Let $f:(X,A) \to (Y,B)$ be a map of pairs which is homotopic to a map $g:(X,A) \to (Y,B)$ such that $g(X) \subset B$. Then the induced homomorphism*

$$f_*: H_n(X,A) \to H_n(Y,B)$$

is zero for all n.

PROOF. By the homotopy property, $f_* = g_*$, hence we must prove that $g_* = 0$. The hypotheses imply that g can be factored, as follows:

$$(X,A) \xrightarrow{g'} (B,B) \xrightarrow{i} (Y,B).$$

Passing to homology, we have

$$H_n(X,A) \xrightarrow{g'_*} H_n(B,B) \xrightarrow{i_*} H_n(Y,B).$$

Since $H_n(B,B) = 0$ for all n, the result follows.

§6. Regular CW-complexes

We will now introduce a special category of CW-complexes which have the property that their homology groups are effectively computable (at least in case the complex is finite).

Definition 6.1. A CW-complex is *regular* if for each cell e^n, $n > 0$, there exists a characteristic map $f : E^n \to \bar{e}^n$ which is a homeomorphism.

We recall that previously we have only required that the characteristic map be a homeomorphism of U^n onto e^n, and map S^{n-1} into the $(n-1)$-skeleton. We are now requiring in addition that the characteristic map be a homeomorphism of S^{n-1} into K^{n-1}.

To clarify the definition, we present in Figure 9 an example of a CW-complex on the closed 2-dimensional disc which is *not* regular. There are three vertices, three edges, and one 2-cell:

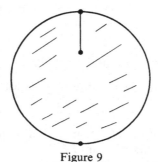

Figure 9

We now list three basic geometric properties of regular CW-complexes:

(1) If $m < n$ and e^m and e^n are cells such that $e^m \cap \bar{e}^n \neq \varnothing$, then $e^m \subset \bar{e}^n$.

(2) For any n-cell e^n, $n \geq 0$, $\bar{e}^n$ and $\dot{e}^n$ are the underlying spaces of sub-complexes. Also, $\dot{e}^n$ is the union of closures of $(n-1)$-cells.

Before stating the third property, we need a definition. We say e^m is a *face* of e^n if $e^m \subset \bar{e}^n$, and denote this by $e^m \leq e^n$. Clearly, every cell is a face of itself; we say e^m is a *proper* face of e^n if it is a face of e^n, and $e^m \neq e^n$ (notation: $e^m < e^n$). This definition makes sense in a regular cell complex mainly because of property (1).

(3) Let e^n and e^{n+2} be cells of a regular cell complex such that e^n is a face of e^{n+2}. Then there are exactly two $(n+1)$-cells e^{n+1} such that $e^n < e^{n+1} < e^{n+2}$.

It should be emphasized that (1), (2), and (3) need not be true for non-regular CW-complexes. The proofs depend on Brouwer's theorem on invariance of domain, Corollary III.6.7.

The proofs of (1), (2), and (3) are given by Cooke and Finney [2] or Massey [7]. We will not reproduce these proofs here. Actually, in any specific case it will be clear that these properties hold.

§7. Determination of Incidence Numbers for a Regular Cell Complex

Let K be a *regular* cell complex on the space X. We will denote the n-cells of K by the symbol e_λ^n, where the index λ ranges over a certain set Λ_n, $n = 0, 1, 2, \ldots$. We assume orientations b_λ^n have been chosen for each cell e_λ^n as described in §5.

Lemma 7.1. *The incidence numbers* $[b_\lambda^n : b_\mu^{n-1}]$ *in a regular cell complex* K *satisfy the following four conditions*:

(1) *If* e_μ^{n-1} *is not a face of* e_λ^n, *then* $[b_\lambda^n : b_\mu^{n-1}] = 0$.
(2) *If* e_μ^{n-1} *is a face of* e_λ^n, *then* $[b_\lambda^n : b_\mu^{n-1}] = \pm 1$.
(3) *If* e_μ^0 *and* e_ν^0 *are the two vertices which are faces of the 1-cell* e_λ^1, *then*

$$[b_\lambda^1 : b_\mu^0] + [b_\lambda^1 : b_\nu^0] = 0.$$

(4) *Let* e_λ^n *and* e_ρ^{n-2} *be cells such that* $e_\rho^{n-2} < e_\lambda^n$; *let* e_μ^{n-1} *and* e_ν^{n-1} *denote the unique* $(n-1)$-*cells* e^{n-1} *such that* $e_\rho^{n-2} < e^{n-1} < e_\lambda^n$. *Then*

$$[b_\lambda^n : b_\mu^{n-1}][b_\mu^{n-1} : b_\rho^{n-2}] + [b_\lambda^n : b_\nu^{n-1}][b_\nu^{n-1} : b_\rho^{n-2}] = 0.$$

PROOF: Condition (1) is a consequence of Lemma 5.2 and the definition of the term face.

In order to prove Statement (2), we will make use of Statement (2) of §6. According to this statement, $\bar{e}_\lambda^n$ is a subcomplex of K which contains the cell e_μ^{n-1}, and it is easy to see that it doesn't matter whether we compute the incidence number $[b_\lambda^n : b_\mu^{n-1}]$ relative to the subcomplex $\bar{e}_\lambda^n$ or to the whole complex K. Let $L = \{L^q\}$ denote this subcomplex on the space $\bar{e}_\lambda^n$. Then $L^n = \bar{e}_\lambda^n$ is a closed n-dimensional ball, and $L^{n-1} = \dot{e}_\lambda^n$ is an $(n-1)$-sphere. We will use the method of proof of Lemma 5.2 to prove the present lemma. Thus we see that in the following commutative diagram

$$H_n(L^n, L^{n-1}) \xrightarrow{\partial_*} \tilde{H}_{n-1}(L^{n-1}) \xrightarrow{j_{n-1}} H_{n-1}(L^{n-1}, L^{n-2})$$

$$\searrow^{k_*} \qquad \qquad \downarrow^{m_{\mu*}}$$

$$H_{n-1}(L^{n-1}, L^{n-1} - e_\mu^{n-1})$$

we must prove that $k_*\partial_*$ is an isomorphism. We will prove this by proving that both ∂_* and k_* are isomorphisms.

To prove that ∂_* is an isomorphism, one considers the homology sequence of the pair (L^n, L^{n-1}). Since $L^n = \bar{e}_\lambda^n$ is contractible, $\tilde{H}_q(L^n) = 0$, and the desired result follows.

To prove that k_* is an isomorphism, one considers the homology sequence of the pair $(L^{n-1}, L^{n-1} - e_\mu^{n-1})$; k_* is one of the homomorphisms in this exact sequence. We will prove that $L^{n-1} - e_\mu^{n-1}$ is contractible, from which it will follow that

$$\tilde{H}_q(L^{n-1} - e_\mu^{n-1}) = 0$$

for all q, and hence that k_* is an isomorphism. To prove that $L^{n-1} - e_\mu^{n-1}$ is contractible, recall that L^{n-1} is an $(n-1)$-sphere. Let x be a point of e_μ^{n-1}; then $L^{n-1} - e_\mu^{n-1}$ is obviously a deformation retract of $L^{n-1} - \{x\}$; and $L^{n-1} - \{x\}$ is homeomorphic to R^{n-1}, hence contractible. Therefore $L^{n-1} - e_\mu^{n-1}$ is also contractible.

Statement (3) is a consequence of Part (c) of Lemma 5.1 and Statement (1), together with the obvious fact that any 1-cell in a regular CW-complex has exactly two vertices which are faces.

Statement (4) follows from Part (b) of Lemma 5.1, Statement (1), and Statement (3) of §6.   Q.E.D.

Our main theorem now asserts that the four conditions of the lemma just proved completely characterize the incidence numbers of a regular CW-complex.

Theorem 7.2. *Let K be a regular CW-complex on the topological space X. For each pair $(e_\lambda^n, e_\mu^{n-1})$ consisting of an n-cell and an $(n-1)$-cell of K, let there be given an integer $\alpha_{\lambda\mu}^n = 0$ or ± 1 such that the following four conditions hold:*

(1) *If e_μ^{n-1} is not a face of e_λ^n, then $\alpha_{\lambda\mu}^n = 0$.*
(2) *If e_μ^{n-1} is a face of e_λ^n, then $\alpha_{\lambda\mu}^n = \pm 1$.*
(3) *If e_μ^0 and e_ν^0 are the two vertices of the 1-cell e_λ^1, then*

$$\alpha_{\lambda\mu}^1 + \alpha_{\lambda\nu}^1 = 0.$$

(4) *Let e_λ^n and e_ρ^{n-2} be cells of K such that $e_\rho^{n-2} < e_\lambda^n$; let e_μ^{n-1} and e_ν^{n-1} denote the unique $(n-1)$-cells e^{n-1} such that $e_\rho^{n-2} < e^{n-1} < e_\lambda^n$.*

Then

$$\alpha_{\lambda\mu}^n \alpha_{\mu\rho}^{n-1} + \alpha_{\lambda\nu}^n \alpha_{\nu\rho}^{n-1} = 0.$$

Under these assumptions, it is possible to choose an orientation b_λ^n for each cell e_λ^n in one and only one way such that

$$[b_\lambda^n : b_\mu^{n-1}] = \alpha_{\lambda\mu}^n$$

for all pairs $(e_\lambda^n, e_\mu^{n-1})$.

PROOF: We will prove the existence of the required orientation b_λ^n on the cell e_λ^n by induction on n. For $n = 0$ there is no choice: a 0-cell has a unique orientation, which we denote by b_λ^0.

Next, let e_λ^1 be a 1-cell, and let e_μ^0 and e_ν^0 be the two vertices which are faces of it, It is clear that one of the two possible orientations of e_λ^1, which we will denote by b_λ^1, satisfies the equation

$$[b_\lambda^1 : b_\mu^0] = \alpha_{\lambda\mu}^1.$$

Then since

$$\alpha_{\lambda\mu}^1 + \alpha_{\lambda\nu}^1 = 0$$
$$[b_\lambda^1 : b_\mu^0] + [b_\lambda^1 : b_\nu^0] = 0$$

it follows that

$$[b_\lambda^1 : b_\nu^0] = \alpha_{\lambda\nu}^1,$$

as required.

Now we make the inductive step. Assume that an orientation b_ξ^q for each cell e_ξ^q has been chosen for all $q < n$ such that the required conditions hold. Let e_λ^n be an n-cell of K, and let $e_{\mu_0}^{n-1}$ be an $(n-1)$-cell which is a face of e_λ^n. Once again, it is clear that we can choose one of the two possible orientations of e_λ^n, which we will denote by b_λ^n, so that

$$[b_\lambda^n : b_{\mu_0}^{n-1}] = \alpha_{\lambda\mu_0}^n. \tag{7.1}$$

We must prove that if e_ν^{n-1} is any other face of e_λ^n, then

$$[b_\lambda^n : b_\nu^{n-1}] = \alpha_{\lambda\nu}^n. \tag{7.2}$$

For this purpose, consider the subcomplex L of K consisting of all the cells of $\bar{e}_\lambda^n$. Then

$$z = \sum_\nu \alpha_{\lambda\nu}^n b_\nu^{n-1},$$

where the summation is over all $(n-1)$-cells of L, is a nonzero $(n-1)$-chain of L. A routine calculation using the properties of regular cell complexes and the inductive hypothesis shows that

$$d_{n-1}(z) = 0,$$

i.e., z is a cycle. A similar argument shows that

$$z' = \sum_\nu [b_\lambda^n : b_\nu^{n-1}] b_\nu^{n-1}$$

is also a nonzero cycle. Since $\bar{e}_\lambda^n = L^{n-1}$ is an $(n-1)$-sphere, it follows that

$$H_{n-1}(L) = Z_{n-1}(L)$$

is an infinite cyclic group. Therefore z and z' are both multiples of a generator of this group. Since $\{b_\nu^{n-1}\}$ is a basis for $C_{n-1}(L)$, and we are assuming that Equation (7.1) holds, it follows that z and z' must be the *same* multiple of a generator of $Z_{n-1}(L)$, i.e., $z = z'$. By comparing coefficients of z and z', we see that (7.2) holds for all ν. This completes the proof of the existence of the desired orientations.

The proof of uniqueness of orientations is also done by induction on n. For $n = 0$, orientations are unique by definition. Assume inductively that orientations have been proven unique for all cells of dimension $< n$; let e_λ^n be an n-cell. Choose an $(n-1)$-dimensional face e_μ^{n-1} of e_λ^n. By Statement (d) of Lemma 5.1, changing the orientation of e_λ^n would change the incidence number $[b_\lambda^n : b_\mu^{n-1}]$, which is not allowed.   Q.E.D.

Notational Convention. From now on, we will usually only need to consider one choice of orientation for the cells of a regular CW-complex. Therefore we will use the same symbol for a cell and its orientation. Thus

$[e_\lambda^n : e_\mu^{n-1}] = 0$ or ± 1 denotes the incidence number of the *oriented* cells e_λ^n and e_μ^{n-1}. This calculated sloppiness in notation is customary and convenient.

The uniqueness statement of Theorem 7.2 is important, because it shows that we can specify orientations for the cells of a regular CW-complex by specifying a set of incidence numbers for the complex. This is one of the most convenient ways of specifying orientations of cells. Regular CW-complexes are often more convenient than other CW-complexes, because of this simple method for specifying the orientation of cells.

The method of using this theorem is quite simple. We assume we have given a list of cells of K together with the information as to whether $e_\lambda^{n-1} < e_\mu^n$ for any two cells e_λ^{n-1} and e_μ^n. For each 1-cell e^1, choose incidence numbers between it and its two vertices so that Conditions (2) and (3) of Lemma 7.1 (or Theorem 7.2) hold. Define all other incidence numbers between vertices and 1-cells to be 0 (Condition (1)).

Now assume, inductively, that incidence numbers have been chosen between all cells of dimension $< n$. Let e^n be an 1-cell. Choose a face e_0^{n-1} of e^n, and choose $[e^n : e^{n-1}]$ to be $+1$ or -1. Using Condition (4), determine $[e^n : e_\lambda^{n-1}]$ for all $(n-1)$-cells e_λ^{n-1} which are faces of e^n and have an $(n-2)$-face in common with e^{n-1}. Spread out over the boundary e^n by repeating this process. Theorem 7.2 assures us that we will never reach a contradiction by this process. Repeat this process for each n-cell of K, and then use Condition (1) to define all other incidence numbers between $(n-1)$- and n-cells.

Here is a convenient way to indicate incidence relations between low-dimensional cells on a diagram:

(a) Between 0-cells and 1-cells. Let e^1 be a 1-cell with vertices e_0^0 and e_1^0. Consider the two incidence numbers $[e^1 : e^0]$ and $[e^1 : e_1^0]$; one of these is 1, the other is -1. Draw an arrow on e^1 indicating the direction *from* the vertex corresponding to -1 to the vertex corresponding to $+1$ as in Figure 10.

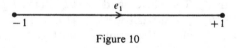

$$e_1$$
$$-1 \qquad\qquad\qquad\qquad\qquad\qquad\qquad +1$$

Figure 10

(b) Between 1-cells and 2-cells. Let e^2 be a 2-cell and let e^1 be a face of e^2, as shown in Figure 11. We assume the orientation chosen for e^1 is indicated by means of an arrow, as shown. Indicate the orientation of e^2 by indicating a direction of rotation of e^2 about its center. This direction of rotation will

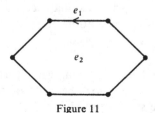

Figure 11

be the *same* as that indicated by the arrow on e^1 if $[e^2 : e^1] = +1$, otherwise it will be the opposite. Note that the resultant direction of rotation of e^2 is independent of the choice of the face e^1.

(c) Between 2-cells and 3-cells. We can indicate orientations of 3-cells by assigning to them a right- or left-handed corkscrew. We assume that all the faces of a given 3-cell e^3 have their orientations indicated as described in the preceding paragraph. Let e^2 be a face of e^3. If $[e^3 : e^2] = +1$, assign to e^3 the kind of corkscrew needed to *bore into* e^3 from the outside, through the face e^2, rotating in the direction indicated by the orientation of e^2. If $[e^3 : e^2] = -1$, assign to e^3 the kind of corkscrew needed to *bore out* of e^3 through the face e^2, rotating in the direction indicated by the orientation of e^2. Note once again that the type of corkscrew assigned to e^3 is independent of the choice of the face e^2.

EXERCISES

7.1. Divide an orientable surface of genus n into $4n$ quadrilaterals. There will be $2n + 2$ vertices and $8n$ 1-cells. Figure 12 indicates the case $n = 2$:

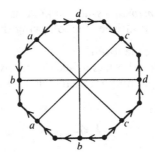

Figure 12

Compute incidence numbers.

7.2. Consider real projective 3-space as obtained by identifying diametrically opposite points on the boundary of the regular octahedron

$$\{(x,y,z) \in \mathbf{R}^3 \,|\, |x| + |y| + |z| \leq 1\}.$$

Divide the octahedron into eight tetrahedra by means of the coordinate planes (i.e., there is one tetrahedron in each octant). Compute incidence numbers. Note: This process can be generalized to define a regular CW-complex on real projective n-space.

7.3. Let K be a regular CW-complex on X. Define K to be an *almost simplicial complex* if the following conditions hold for all $n \geq 0$:

(a) Each n-cell has exactly $n + 1$ vertices.
(b) Any set of $n + 1$ vertices is the set of vertices of *at most* one n-cell (it need not be the set of vertices of any n-cell).

Prove the following two facts about almost simplicial complexes:

1. An n-cell has exactly $n + 1$ faces of dimension $n - 1$.
2. Incidence numbers for an almost simplicial complex can be described explicitly as follows: Each cell is uniquely described by listing its vertices. Linearly order all the vertices (in any order whatsoever) and agree to always list vertices in the given order. If e^n has vertices $v_0, v_1, \ldots, v_n$ in the given order, and the face e^{n-1} has only the vertex v_i omitted, then set $[e^n : e^{n-1}] = (-1)^i$.

(Note: A *simplicial complex*, as defined in most books, is an almost simplicial complex with certain additional geometric structure. This additional structure is irrelevant as far as computing homology groups is concerned.)

§8. Homology Groups of a Pseudomanifold

In this section we apply the results of §7 to determine the structure of certain homology groups of a special class of regular CW-complexes. This special class is of fairly wide occurrence.

Definition 8.1. An *n-dimensional pseudomanifold* is an n-dimensional finite, regular CW-complex which satisfies the following three conditions:

(1) Every cell is a face of some n-cell.
(2) Every $(n - 1)$-dimensional cell is a face of exactly two n-cells.
(3) Given any two n-cells, e^n and e'^n, there exists a sequence of n-cells

$$e^n_0, e^n_1, \ldots, e^n_k$$

such that $e^n_0 = e^n$, $e^n_k = e'^n$, and e^n_{i-1} and e^n_i have a common $(n - 1)$-dimensional face $(i = 1, 2, \ldots, k)$.

Some authors call an n-dimensional pseudomanifold a *simple n-circuit*.

A regular CW-complex on a compact connected 2-manifold is an example of a 2-dimensional pseudomanifold. More generally it may be shown that a regular CW-complex on a compact connected n-manifold is an n-dimensional pseudomanifold. An example of a pseudomanifold which is not a manifold may be constructed as follows: Let K be a regular CW-complex on a compact, connected 2-manifold. Form the quotient by identifying two vertices which are not both vertices of the same 2-cell. The quotient space has an obvious structure of regular CW-complex, which may be shown to be a 2-dimensional pseudomanifold.

It may be proved that the above definition is "topologically invariant" in the sense that it expresses a condition on the underlying space rather than a condition on the particular regular CW-complex chosen on the space (a proof of this fact is contained in the book by Seifert and Threlfall, [9], Chapter 5).

Let K be an n-dimensional pseudomanifold, and let e_1^n and e_2^n be n-cells of K which have a common $(n-1)$-dimensional face e^{n-1}. We define orientations for e_1^n and e_2^n to be *coherent* (with respect to the common face e^{n-1}) if the incidence numbers satisfy the following relation:

$$[e_1^n : e^{n-1}] + [e_2^n : e^{n-1}] = 0.$$

Note that this condition is independent of the choice of the orientation for the cell e^{n-1}. A set of orientations for all the n-cells of K is said to be *coherent*, if it is coherent in the above sense for any pair of n-cells which have a common face of dimension $n-1$.

In connection with the above definition, it should be pointed out that a pair of n-cells in an n-dimensional pseudomanifold may have more than one common $(n-1)$-dimensional face; in such a case it is essential to specify the common face with respect to which given orientations are asserted to be coherent. An example is the following subdivision of the projective plane with four vertices, $v_1, \ldots, v_4$, seven edges, $e_1, \ldots, e_7$, and four 2-cells, A, B, C, and D (see Figure 13). The 2-cells A and B have the edges e_1 and e_3 in common; if A and B are oriented coherently with respect to the edges e_1, the orientations are not coherent with respect to the edge e_3, and vice-versa.

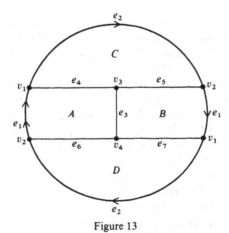

Figure 13

Given an n-dimensional pseudomanifold K, either all the n-cells of K can be simultaneously oriented so any pair having a common $(n-1)$-dimensional face are oriented coherently, or they can not be so oriented. In the former case, K is said to be *orientable*, in the latter case *nonorientable*.

Theorem 8.1. *If K is an orientable n-dimensional pseudomanifold, then $H_n(K)$ is infinite cyclic; if K is nonorientable then $H_n(K) = 0$.*

The details of the proof are left to the reader. Note that since K is an n-dimensional CW-complex, $H_n(K) = Z_n(K)$. If K is orientable, and the n-cells are oriented so that any pair having a common face of dimension $n - 1$ are coherently oriented, then the sum of all the n-cells (thus oriented) is an n-cycle; moreover, any n-cycle is an integral multiple of this sum. If K is nonorientable, then one proves that there are no nonzero n-cycles.

In view of the invariance of the homology groups of a regular CW-complex K, this theorem shows that the concepts of *orientability* and non-orientability really only depend on the underlying topological space involved, and not on the choice of the regular cell complex K.

The next theorem describes the structure of the torsion subgroup of $H_{n-1}(K)$.

Theorem 8.2. *Let K be an n-dimensional pseudomanifold. If K is orientable, then $H_{n-1}(K)$ is torsion-free. If K is nonorientable, then the torsion subgroup of K is cyclic of order two.*

PROOF. Let k denote the number of n-cells of K. We assert that it is possible to enumerate the n-cells of K in order $e_1^n, e_2^n, \ldots, e_k^n$ and to choose $(n - 1)$-cells e_i^{n-1}, $2 \le i \le k$, of K such that the following condition holds: e_i^{n-1} is a common face of e_i^n and some n-cell e_j^n with $j < i$. The proof of this assertion is left to the reader.

Assume that the n-cells have been enumerated and the $(n - 1)$-cells $e_2^{n-1}, \ldots, e_k^{n-1}$ have been chosen so the above conditions hold. Choose an arbitrary orientation for the cell e_1^n; then orient e_2^n so that its orientation is coherent to that of e_1^n with respect to the face e_2^{n-1}. Next orient e_3^n so it is coherent with respect to the face e_3^{n-1} to either e_1^n or e_2^n as is relevant. Continue in this manner, orienting all the n-cells in succession, so each e_i^n is coherently oriented with some e_j^n, $j < i$, with respect to e_i^{n-1}. Once the orientation of e_1^n is chosen, this condition uniquely determines the orientations of the rest of the n-cells. It is easy to see that if K is orientable, then the result is a coherent orientation of all the n-cells of K.

We next assert that any $(n - 1)$-cycle z of K is homologous to a cycle z' such that the coefficient of each of the cells $e_2^{n-1}, \ldots, e_k^{n-1}$ in z' is 0. The proof, which is easy, is left to the reader.

With these preparations out of the way, we can now prove the theorem. Let u be a homology class of finite order of $H_{n-1}(K)$, i.e., $q \cdot u = 0$ for some integer q. Let $z \in Z_{n-1}(K)$ be a representative cycle for u. By the above argument, we may assume that the coefficients of the cells $e_2^{n-1}, \ldots, e_k^{n-1}$ in the cycle z are all 0. Since $qu = 0$, there exists an n-chain

$$c = \sum_{i=1}^{k} \alpha_i e_i^n$$

such that

$$d(c) = q \cdot z.$$

In view of the way we have oriented the n-cells e_i^n, and the fact that the coefficients of $e_2^{n-1}, \ldots, e_k^{n-1}$ are 0 in z, we conclude that

$$\alpha_1 = \alpha_2 = \cdots = \alpha_k.$$

If K is orientable, we see that

$$d(c) = 0,$$

hence $q \cdot z = 0$, $z = 0$, and $u = 0$, as required. In the nonorientable case, consider the n-chain

$$c' = \sum_{i=1}^{k} e_i^n.$$

Then $d(c')$ is a nonzero $(n-1)$-cycle which assigns the coefficient 0 or ± 2 to every $(n-1)$-cell of K. Hence

$$y = \tfrac{1}{2} d(c')$$

is an $(n-1)$-dimensional cycle of K, and its homology class is an element of order 2 in $H_{n-1}(K)$. Note that the coefficient of any $(n-1)$-cell is 0 or ± 1 in the expression for the cycle y. Since $c = \alpha c'$ for some integer α, we see that

$$d(c) = \alpha d(c'),$$

$$q \cdot z = 2\alpha y,$$

$$z = \frac{2\alpha}{q} y,$$

hence the homology class of z is a multiple of that of y. Thus the torsion subgroup of $H_{n-1}(K)$ is the cyclic group generated by the homology class of y.   Q.E.D.

Since it may be shown that any regular CW-complex on a compact connected n-manifold is an n-dimensional pseudomanifold, the above results apply in particular to all compact n-manifolds which can be "subdivided" so as to define a regular CW-complex structure. It is known that every compact n-manifold admits such a subdivision if $n \le 3$; the question is still open for manifolds of dimension > 3.

Bibliography for Chapter IV

[1] N. Bourbaki, *Topologie Générale*, Hermann et Cie., Paris 1947, Chapters VI and VIII.

[2] G. E. Cooke and R. L. Finney, *Homology of Cell Complexes*, Princeton University Press, Princeton, 1967.

[3] P. J. Hilton, *An Introduction Homotopy Theory*, Cambridge University Press, New York, 1953, Chapter VII.

[4] S. T. Hu, *Elements of General Topology*, Holden-Day, San Francisco, 1964, Chapter IV.

[5] A. T. Lundell and S. Weingram, *The Topology of CW-complexes*, Van Nostrand Reinhold Co., New York, 1969.

[6] W. S. Massey, *Algebraic Topology: An Introduction*, Springer-Verlag, New York, 1977, pp. 215–217.

[7] W. S Massey, *Homology and Cohomology Theory: An Approach Based on Alexander–Spanier Cochains*, Marcel Dekker, Inc., New York, 1978, Chapter 5.

[8] I. R. Porteous, *Topological Geometry*, Van Nostrand Reinhold Co., New York, 1969.

[9] H. Seifert and W. Threlfall, *Lehrbuch der Topologie*, Chelsea Publishing Co., New York, 1947.

[10] J. H. C. Whitehead, Combinatorial homotopy I, *Bull Amer. Math. Soc.* **55** (1949), 213–245.

CHAPTER V

Homology with Arbitrary Coefficient Groups

§1. Introduction

This chapter is more algebraic in nature than the preceding chapters. In §2 we discuss chain complexes. This discussion mainly puts on a formal basis many facts that the reader must know by now. Nevertheless, there *is* some point to a systematic organization of the ideas involved, and certain new ideas and techniques are introduced. The remainder of the chapter is concerned with homology groups with arbitrary coefficients. These new homology groups are a generalization of those we have considered up to now. In the application of homology theory to certain problems they are often convenient and sometimes necessary.

Starting in §2, we make systematic use of tensor products. It is assumed that the reader knows the definition and basic properties of tensor products of abelian groups.

§2. Chain Complexes

Much of this section consists of terminology and definitions which it will be very convenient to use from now on.

Definition 2.1. A *chain complex* $K = \{K_n, \partial_n\}$ is a sequence of abelian groups K_n, $n = 0, \pm 1, \pm 2, \ldots$, and a sequence of homomorphisms $\partial_n : K_n \to K_{n-1}$ which are required to satisfy the condition

$$\partial_{n-1}\partial_n = 0$$

for all n.

For any such chain complex $K = \{K_n, \partial_n\}$ we define

$$Z_n(K) = \text{kernel } \partial_n,$$
$$B_n(K) = \text{image } \partial_{n+1}.$$

Then $B_n(K) \subset Z_n(K) \subset K_n$, and we can define

$$H_n(K) = Z_n(K)/B_n(K),$$

called the nth *homology group of* K.

EXAMPLE 2.1. For any space X, we have previously defined the chain complexes

$$Q(X) = \{Q_n(X), \partial_n\}$$
$$D(X) = \{D_n(X), \partial_n\}$$
$$C(X) = \{C_n(X), \partial_n\}$$

and for any pair (X, A),

$$C(X, A) = \{C_n(X, A), \partial_n\}.$$

Definition 2.2. Let $K = \{K_n, \partial_n\}$ and $K' = \{K'_n, \partial'_n\}$ be chain complexes. A *chain map* $f : K \to K'$ consists of a sequence of homomorphisms $f_n : K_n \to K'_n$ such that the commutativity condition

$$f_{n-1} \partial_n = \partial'_n f_n$$

holds for all n.

EXAMPLE 2.2. A continuous map $\varphi : X \to Y$ induces chain maps

$$\varphi_\# : Q(X) \to Q(Y)$$
$$\varphi_\# : D(X) \to D(Y)$$
$$\varphi_\# : C(X) \to C(Y)$$

etc.

If $f : K \to K'$ is a chain map, then $f_n[Z_n(K)] \subset Z_n(K')$ and $f_n[B_n(K)] \subset B_n(K')$, hence there is induced a homomorphism

$$f_* : H_n(K) \to H_n(K')$$

for all n.

Note that the set of all chain complexes and chain maps constitutes a category, and that H_n is a functor from this category to the category of abelian groups and homomorphisms. Note also that if f and $g : K \to K'$ are chain maps, their *sum*,

$$f + g = \{f_n + g_n\},$$

is also a chain map, and

$$(f + g)_* = f_* + g_* : H_n(K) \to H_n(K').$$

In other words, H_n is an *additive functor*.

Definition 2.3. Let $f, g: K \rightarrow K'$ be chain maps. A *chain homotopy* $D: K \rightarrow K'$ *between* f *and* g is a sequence of homomorphisms

$$D_n: K_n \rightarrow K'_{n+1}$$

such that

$$f_n - g_n = \partial'_{n+1} D_n + D_{n-1} \partial_n$$

for all n. Two chain maps are said to be *chain homotopic* if there exists a chain homotopy between them (notation: $f \simeq g$).

EXAMPLE 2.3. If $\varphi_0, \varphi_1: X \rightarrow Y$ are continuous maps, any homotopy between φ_0 and φ_1 gives rise to a chain homotopy between the induced chain maps $\varphi_{0\#}$ and $\varphi_{1\#}$ on cubical singular chains (see §II.4).

The reader should prove the following two facts for himself:

Proposition 2.1. *Let* $f, g: K \rightarrow K'$ *be chain maps. If* f *and* g *are chain homotopic, then*

$$f_* = g_*: H_n(K) \rightarrow H_n(K')$$

for all n.

Proposition 2.2. *Chain homotopy is an equivalence relation on the set of all chain maps from* K *to* K'.

EXERCISES

2.1. By analogy with the category of topological spaces and continuous maps, complete the following definitions:

   (a) A chain map $f: K \rightarrow K'$ is a *chain homotopy equivalence* if _____.
   (b) A chain complex K' is a *subcomplex* of the chain complex K if _____.
   (c) A subcomplex K' of the chain complex K is a *retract* of K if _____.
   (d) A subcomplex K' of the chain complex K is a *deformation retract* of K if

   _____.

   (e) If K' is a subcomplex of K, the quotient complex K/K' is _____.

   In each case, what assertions can be made about the homology groups of the various chain complexes involved, and about the homomorphisms induced by the various chain maps?

2.2. Let f, g, f', and g' be chain maps $K \rightarrow K'$. If f is chain homotopic to f', and g is chain homotopic to g', then prove that $f + g$ is chain homotopic to $f' + g'$.

2.3. Let $f, g: K \rightarrow K'$ and $f', g': K' \rightarrow K''$ be chain maps, D a chain homotopy between f and g, and D' a chain homotopy between f' and g'. Using D and D', construct an explicit chain homotopy between $f'f$ and $g'g: K \rightarrow K''$.

2.4. Let D be a chain homotopy between the maps f and $g: K \rightarrow K$ (of K into itself). Use D to construct an explicit chain homotopy between $f^n = fff \cdots f$ and $g^n = gg \cdots g$ (n-fold iterates).

Definition 2.4. A sequence of chain complexes and chain maps

$$\cdots \to K \xrightarrow{f} K' \xrightarrow{g} K' \to \cdots$$

is *exact* if for each integer n the sequence of abelian groups

$$\cdots \to K_n \xrightarrow{f_n} K'_n \xrightarrow{g_n} K''_n \to \cdots.$$

is exact in the usual sense.

We will be especially interested in *short* exact sequences of chain complexes, i.e., those of the form

$$E : 0 \to K' \xrightarrow{f} K \xrightarrow{g} K'' \to 0.$$

This means that for each n, f_n is an monomorphism, g_n is an epimorphism, and image $f_n = $ kernel g_n. Given any such short exact sequence of chain complexes, we can follow the procedure of §II.5 to define a *connecting homomorphism* or *boundary operator*

$$\partial_E : H_n(K'') \to H_{n-1}(K')$$

for all n, and then prove that the following sequence of abelian groups

$$\cdots \xrightarrow{\partial_E} H_n(K') \xrightarrow{f_*} H_n(K) \xrightarrow{g_*} H_n(K'') \xrightarrow{\partial_E} H_{n-1}(K') \to \cdots$$

is exact. One can also prove the following important naturality property of this connecting homomorphism or boundary operator: Let

$$
\begin{array}{ccccccccc}
E: & 0 & \longrightarrow & K' & \xrightarrow{f} & K & \xrightarrow{g} & K'' & \longrightarrow & 0 \\
 & & & \downarrow{\varphi} & & \downarrow{\psi} & & \downarrow{\omega} & & \\
F: & 0 & \longrightarrow & L' & \xrightarrow{h} & L & \xrightarrow{i} & L'' & \longrightarrow & 0
\end{array}
$$

be a commutative diagram of chain complexes and chain maps. It is assumed that the two rows, denoted by E and F, are short exact sequences. Then the following diagram is commutative for each n:

$$
\begin{array}{ccc}
H_n(K'') & \xrightarrow{\partial_E} & H_{n-1}(K') \\
\downarrow{\omega_*} & & \downarrow{\varphi_*} \\
H_n(L'') & \xrightarrow{\partial_F} & H_{n-1}(L').
\end{array}
$$

EXERCISES

2.5. Define the direct sum and direct product of an arbitrary family of chain complexes in the obvious way. How is the homology of such a direct sum or product related to the homology of the individual chain complexes of the family?

2.6. Let $E: 0 \to K' \xrightarrow{f} K \xrightarrow{g} K'' \to 0$ be a short exact sequence of chain complexes. By a *splitting homomorphism* for such a sequence we mean a sequence $s = \{s_n\}$ such that for each n, $s_n: K_n'' \to K_n$, and $g_n s_n =$ identity map of K_n'' onto itself. Note that we do *not* demand that s should be a chain map. Assume that such a splitting homomorphism exists.

(a) Prove that there exist unique homomorphisms $\varphi_n: K_n'' \to K_{n-1}'$ for all n such that

$$f_{n-1}\varphi_n = \partial_n s_n - s_{n-1}\partial_n''.$$

(b) Prove that $\partial_{n-1}'\varphi_n + \varphi_{n-1}\partial_n'' = 0$ for all n.

(c) Let $s' = \{s_n'\}$ be another sequence of splitting homomorphisms, and $\varphi_n': K_n'' \to K_{n-1}'$ the unique homomorphisms such that $f_{n-1}\varphi_n' = \partial_n s_n' - s_{n-1}'\partial_n''$. Prove that there exists a sequence of homomorphisms $D_n: K_n'' \to K_n'$ such that

$$\varphi_n - \varphi_n' = \partial_n' D_n - D_{n-1}\partial_n''$$

for all n.

(d) Prove that the connecting homomorphism $\partial_E: H_n(K'') \to H_{n-1}(K')$ is induced by the sequence of homomorphisms $\{\varphi_n\}$ in the same sense that a chain map induces homomorphisms of homology groups. (Note: The sequence of homomorphisms $\{\varphi_n\}$ can be thought of as a "chain map of degree -1." The sequence of homomorphisms $\{D_n\}$ in Part (c) is a chain homotopy between $\{\varphi_n\}$ and $\{\varphi_n'\}$.)

We will conclude this section on chain complexes with a discussion of a construction called the algebraic mapping cone of a chain map.

Definition 2.5. Let $K = \{K_n, \partial_n\}$ and $K' = \{K_n', \partial_n'\}$ be chain complexes and $f: K \to K'$ a chain map. The algebraic mapping cone of f, denoted by $M(f) = \{M(f)_n, d_n\}$ is a chain complex defined as follows:

$$M(f)_n = K_{n-1} \oplus K_n' \quad \text{(direct sum)}.$$

The boundary operator $d_n: M(f)_n \to M(f)_{n-1}$ is defined by

$$d_n(x, x') = (-\partial_{n-1}x, \partial_n' x' + f_{n-1}x)$$

for any $x \in K_{n-1}$ and $x' \in K_n'$. It is trivial to verify that $d_{n-1}d_n = 0$.

Next, define $i_n: K_n' \to M(f)_n$ by $i_n(x') = (0, x')$. The sequence of homomorphisms $i = \{i_n\}$ is easily seen to be a chain map $K' \to M(f)$. Similarly, the sequence of projections $j_n: M(f)_n \to K_{n-1}$ (defined by $j_n(x, x') = x$) is almost a chain map. However, it reduces degrees by one, and instead of commuting with the boundary operators, we have the relation

$$\partial_{n-1}j_n = -j_{n-1}d_n.$$

It is a "chain map of degree -1." It induces a homomorphism of homology groups which reduces degrees by one.

The chain maps i and j define a short exact sequence of chain complexes:

$$0 \to K' \overset{i}{\to} M(f) \overset{j}{\to} K \to 0.$$

As usual, this short exact sequence of chain complexes gives rise to a long exact homology sequence:

$$\cdots \to H_n(K') \overset{i_*}{\to} H_n(M(f)) \overset{j_*}{\to} H_{n-1}(K) \overset{d_*}{\to} H_{n-1}(K') \to \cdots.$$

Here d_* denotes the connecting homomorphism. It is now an easy matter to check that

$$d_* = f_* : H_n(K) \to H_n(K')$$

for all n. Thus we have imbedded the homomorphisms f_* induced by the given chain map in a long exact sequence; and this has been done in a natural way. That is the whole point of introducing the algebraic mapping cone. This long exact sequence will be called the *exact homology sequence of f*.

Remark. The topological analog of this construction is described in §X.3.

Our first application of the algebraic mapping cone is to prove the following basic theorem. We will see other applications later on.

Theorem 2.3. *Let $K = \{K_n, \partial_n\}$ and $K' = \{K'_n, \partial'_n\}$ be chain complexes such that K_n and K'_n are free abelian groups for all n. Then a chain map $f : K \to K'$ is a chain homotopy equivalence if and only if the induced homomorphism $f_* : H_n(K) \to H_n(K')$ is an isomorphism for all n.*

The *only if* part of this theorem is a triviality, hence we will be concerned only with the *if* part. First, we need a couple of lemmas.

Recall that if the identity map and the zero map of a chain complex K into itself are chain homotopic, then $H_n(K) = 0$ for all n. The first lemma is a partial converse of this statement.

Lemma 2.4. *Let K be a chain complex such that $Z_n(K)$ is a direct summand of K_n for all n, and $H_n(K) = 0$ for all n. Then the identity map and the zero map of K into itself are chain homotopic.*

PROOF: For each n, choose a direct sum decomposition

$$K_n = Z_n(K) \oplus A_n.$$

Since $H_n(K) = 0$, $B_n(K) = Z_n(K)$ for all n. It follows that ∂_n maps A_n isomorphically onto $Z_{n-1}(K)$. We now define the chain homotopy $D_n : K_n \to K_{n+1}$ as follows: D_n restricted to A_n is the zero map, and D_n restricted to $Z_n(K)$ shall map $Z_n(K)$ isomorphically onto A_{n+1} by the inverse of the isomorphism ∂_{n+1}. It is now easily verified that

$$D_{n-1}\partial_n(x) + \partial_{n+1}D_n(x) = x$$

for any $x \in K_n$.   Q.E.D.

Lemma 2.5. *Let K be a chain complex such that K_n is a free abelian group. Then $Z_{n+1}(K)$ is a direct summand of K_{n+1}.*

PROOF. Since K_n is free abelian, it follows by a standard theorem of algebra that the subgroup $B_n(K)$ is also free abelian. Because ∂_{n+1} is a homomorphism of K_{n+1} onto the free group $B_n(K)$, we can conclude that $Z_{n+1}(K) = $ kernel ∂_{n+1} is a direct summand.   Q.E.D.

PROOF OF THEOREM 2.3. We assume that the induced homomorphism $f_*: H_n(K) \to H_n(K')$ is an isomorphism for all n, and will prove that f is a chain homotopy equivalence. Let $M(f)$ denote the algebraic mapping cone of f; our assumption implies that $H_n(M(f)) = 0$ for all n. Since K and K' are both chain complexes of free abelian groups, it follows that $M(f)$ is also a chain complex of free abelian groups. Hence $Z_n(M(f))$ is a direct summand of $M(f)_n$ for all n by Lemma 2.5. Therefore we can apply Lemma 2.4 to $M(f)$ to conclude that there exists a chain homotopy $D_n: M(f)_n \to M(f)_{n+1}$ such that

$$d_{n+1}D_n(a) + D_{n-1}d_n(a) = a \qquad (2.1)$$

for any $a \in M(f)_n$. Making use of the fact that $M(f)_n$ is a direct sum for any n, we see that there exist unique homomorphisms

$$D_n^{11}: K_{n-1} \to K_n,$$
$$D_n^{12}: K_n' \to K_n,$$
$$D_n^{21}: K_{n-1} \to K_{n+1}',$$
$$D_n^{22}: K_n' \to K_{n+1}',$$

such that

$$D_n(x,x') = (D_n^{11}x + D_n^{12}x', D_n^{21}x + D_n^{22}x')$$

for any $x \in K_{n-1}$ and $x' \in K_n'$. With this notation, Equation (2.1) is equivalent to the following four equations:

$$-\partial_n D_n^{11} - D_n^{11}\partial_{n-1} + D_{n-1}^{12}f_{n-1} = 1, \qquad (2.2)$$

$$-\partial_n D_n^{12} + D_{n-1}^{12}\partial_n' = 0, \qquad (2.3)$$

$$f_n D_n^{11} + \partial_{n+1}' D_n^{21} - D_{n-1}^{21}\partial_{n-1} + D_{n-1}^{22}f_{n-1} = 0 \qquad (2.4)$$

$$f_n D_n^{12} + \partial_{n+1}' D_n^{22} + D_{n-1}^{22}\partial_n' = 1'. \qquad (2.5)$$

In these equations, the symbols 1 and 1′ denote the identity maps of the chain complexes K and K' respectively. Equation (2.3) implies that the sequence of homomorphisms $D^{12} = \{D_n^{12}\}$ is a chain map $K' \to K$. Similarly, Equation (2.2) implies that

$$D^{12}f \simeq 1,$$

while Equation (2.5) implies that

$$fD^{12} \simeq 1'.$$

This completes the proof.   Q.E.D.

EXERCISES

2.7. Assume we have given a commutative diagram

$$
\begin{array}{ccc}
K & \xrightarrow{f} & K' \\
\downarrow{\varphi} & & \downarrow{\psi} \\
L & \xrightarrow{g} & L'
\end{array}
$$

of chain complexes and chain maps. Show that the pair of chain maps (φ, ψ) induces a chain map $M(f) \to M(g)$, and gives rise to a commutative diagram involving the exact homology sequences of f and g (this is a naturality statement for the algebraic mapping cone).

2.8. Let $f, g : K \to K'$ be chain maps. Show that any chain homotopy D between f and g gives rise to a chain map $M(f) \to M(g)$ which induces isomorphisms $H_n(M(f)) \approx H_n(M(g))$ for all n. What is the relation between the exact homology sequences of f and g in this case?

2.9. Assume that

$$E : 0 \to K \xrightarrow{f} K' \xrightarrow{g} K'' \to 0$$

is a short exact sequence of chain complexes and chain maps. Prove that the exact homology sequence of f and the exact homology sequence of g are both isomorphic to the exact homology sequence of E.

§3. Definition and Basic Properties of Homology with Arbitrary Coefficients

In II.2 we defined an element of the group $Q_n(X)$ to be a finite linear combination $a_1 T_1 + a_2 T_2 + \cdots + a_k T_k$ of singular n-cubes with integral coefficients. As the reader may have already suspected, one could equally well use linear combinations of n-cubes with coefficients in an arbitrary ring, rather than the ring of integers. In fact, one can even go further, and allow the coefficients $a_1, a_2, \ldots$ above to be elements of an arbitrary abelian group (written additively). It turns out that the entire theory we have developed so far can be re-done with very little change with this added degree of generality. For certain problems the resulting homology groups with other coefficients are more convenient, or perhaps even essential. Examples to illustrate this point will be given later.

For our purposes, it will be quicker and more convenient to develop the properties of homology groups with arbitrary coefficients by using the theory of tensor products. This we will now proceed to do. The motivation for this approach is as follows: Recall that $Q_n(X)$ is a free abelian group with basis consisting of the set of singular n-cubes in X. Let G be an abelian group. It follows that any element of the group $G \otimes Q_n(X)$ has a *unique* expression of the form

$$a_1 \otimes T_1 + a_2 \otimes T_2 + \cdots + a_k \otimes T_k,$$

where $T_1, T_2, \ldots$ are singular n-cubes in X, and $a_1, a_2, \ldots$ are elements of the given group G. We can look on this expression as a linear combination of the singular n-cubes $T_1, T_2, \ldots$ with coefficients in G, as desired. This motivates the following definition.

Definition 3.1. Let $K = \{K_n, \partial_n\}$ be a chain complex and G an abelian group. Then $K \otimes G$ denotes the chain complex $\{K_n \otimes G, \partial_n \otimes 1_G\}$, where 1_G denotes the identity map of G. If $f = \{f_n\}$ is a chain map $K \to L$, then $f \otimes 1_G : K \otimes G \to L \otimes G$ denotes the chain map $\{f_n \otimes 1_G\}$. Finally, if $D : K \to L$ is a chain homotopy between f and $g : K \to L$, then $D \otimes 1_G : K \otimes G \to L \otimes G$ denotes the chain homotopy $\{D_n \otimes 1\}$ between $f \otimes 1$ and $g \otimes 1$.

Of course, in the above definition it is necessary to verify that $K \otimes G$ is actually a chain complex, that $f \otimes 1_G$ is a chain map, and that $D \otimes 1_G$ is a chain homotopy between $f \otimes 1_G$ and $g \otimes 1_G$. However, these are trivialities. A more serious problem is the following: Suppose that

$$0 \to K' \xrightarrow{f} K \xrightarrow{g} K'' \to 0$$

is a short exact sequence of chain complexes and chain maps. We would like to be able to conclude that for *any* abelian group G, the sequence

$$0 \to K' \otimes G \xrightarrow{f \otimes 1} K \otimes G \xrightarrow{g \otimes 1} K'' \otimes G \to 0$$

is also exact. Then we could define the corresponding long exact homology sequence. Unfortunately, it is generally not true that Sequence (3.2) will be exact; all we can expect is that the sequence

$$K' \otimes G \xrightarrow{f \otimes 1} K \otimes G \xrightarrow{g \otimes 1} K'' \otimes G \to 0$$

will be exact (right exactness of the tensor product). Thus we will not be able to define a long exact homology sequence without some further assumptions. Experience has shown that the following assumption suffices for most of the applications we have in mind. Define a short exact sequence of chain complexes

$$0 \to K' \xrightarrow{f} K \xrightarrow{g} K'' \to 0$$

to be *split*, or *split exact* if for each integer n, image f_n is a direct summand of K_n. Alternatively, we can require that for each integer n there exists a homomorphism $s_n \colon K_n'' \to K_n$ such that $g_n s_n =$ identity map of K_n'' (such a homomorphism is called a *splitting homomorphism*). Note that we do *not* require that the sequence of homomorphisms s_n should be a chain map; such an assumption would be far too strong for our purposes.

Lemma 3.1. *If the sequence* $0 \to K' \xrightarrow{f} K \xrightarrow{g} K'' \to 0$ *is split exact, then so is the sequence* $0 \to K' \otimes G \xrightarrow{f \otimes 1} K \otimes G \xrightarrow{g \otimes 1} K'' \otimes G \to 0$.

In fact, if $\{s_n\}$ is a sequence of splitting homomorphisms for the original short exact sequence, then $\{s_n \otimes 1\}$ is a sequence of splitting homomorphisms for the second sequence.

Lemma 3.2. *If K'' is a chain complex of free abelian groups, then any short exact sequence* $0 \to K' \to K \to K'' \to 0$ *is split exact.*

The proof is easy.

Since most of the chain complexes we will encounter are composed of free abelian groups, this lemma will find frequent application.

We will now apply these ideas to the homology groups of topological spaces.

Given any topological space X, we have the following short exact sequence of chain complexes:

$$0 \to D(X) \to Q(X) \to C(X) \to 0.$$

All three of these chain complexes consist of free abelian groups, and the sequence is split exact. Therefore if we define new chain complexes as follows:

$$D(X;G) = D(X) \otimes G,$$
$$Q(X;X) = Q(X) \otimes G,$$
$$C(X;G) = C(X) \otimes G,$$

then the resulting sequence

$$0 \to D(X;G) \to Q(X;G) \to C(X;G) \to 0$$

is also split exact. Thus we can consider $D_n(X;G) = D_n(X) \otimes G$ as a subgroup of $Q_n(X;G) = Q_n(X) \otimes G$, and $C_n(X;G)$ is the quotient group, $Q_n(X;G)/D_n(X;G)$. As was remarked above, an element of $Q_n(X;G)$ has a unique expression as a linear combination of singular n-cubes in X with coefficients in G; obviously, $D_n(X;G)$ is the subgroup consisting of linear combinations of degenerate singular cubes.

If A is any subspace of X, we have the short exact sequence of chain complexes:

$$0 \to C(A) \xrightarrow{i} C(X) \xrightarrow{j} C(X,A) \to 0.$$

Once again each of the chain complexes consists of free abelian groups and the sequence is split exact. Therefore if we define

$$C(X,A;G) = C(X,A) \otimes G,$$

then the resulting sequence

$$0 \to C(A;G) \xrightarrow{i \otimes 1} C(X;G) \xrightarrow{j \otimes 1} C(X,A;G) \to 0$$

is also split exact. Thus we can regard $C_n(A;G)$ as a subgroup of $C_n(X;G)$, and $C_n(X,A;G)$ is the quotient group $C_n(X;G)/C_n(A;G)$. It is customary to denote the group $H_n(C(X,A;G))$ by the notation $H_n(X,A;G)$ and call it the *relative homology group of* (X,A) *with coefficient group* G.

If $\varphi:(X,A) \to (Y,B)$ is a continuous map of one pair of spaces into another, then we have the induced chain map

$$\varphi_{\#}:C(X,A) \to C(Y,B).$$

Hence we get an induced chain map

$$\varphi_{\#} \otimes 1_G: C(X,A;G) \to C(Y,B;G)$$

and an induced homomorphism of homology groups, which we will denote by

$$\varphi_*:H_n(X,A;G) \to H_n(Y,B;G).$$

If two maps φ_0, $\varphi_1:(X,A) \to (Y,B)$ are homotopic (as maps of pairs), then any homotopy between them defines a chain homotopy $D:C(X,A) \to C(Y,B)$ between the chain maps

$$\varphi_{0\#}, \varphi_{1\#}:C(X,A) \to C(Y,B)$$

(see §II.4). Hence $D \otimes 1_G$ is a chain homotopy between $\varphi_{0\#} \otimes 1_G$ and $\varphi_{1\#} \otimes 1_G$. It follows that the induced homomorphisms

$$\varphi_{0*}, \varphi_{1*}:H_n(X,A;G) \to H_n(Y,B;G)$$

are the same.

It is now an easy matter to check that all the properties of homology theory which were proved in §§II.2–II.5 remain true for homology theory with coefficients in an abelian group G. In particular, given any pair (X,A), we have a *natural* exact homology sequence,

$$\cdots \xrightarrow{\partial_*} H_n(A;G) \xrightarrow{i_*} H_n(X;G) \xrightarrow{j_*} H_n(X,A;G) \xrightarrow{\partial_*} \cdots.$$

Also, one can check by direct computation that if P is a space consisting of a single point,

$$H_q(P;G) = \begin{cases} G & \text{for } q = 0, \\ \{0\} & \text{for } q \neq 0. \end{cases}$$

In order to define reduced homology groups in dimension 0, it is convenient for any space $X \neq \varnothing$ to define the *augmented* chain complex $\tilde{C}(X)$ as follows:

$$\tilde{C}_q(X) = C_q(X) \quad \text{if } q \neq -1,$$
$$\tilde{C}_{-1}(X) = Z,$$
$$\tilde{\partial}_q = \partial_q \quad \text{if } q \neq 0 \text{ or } -1,$$
$$\partial_0 = \varepsilon \quad (\text{see §II.2}),$$
$$\partial_{-1} = 0.$$

Then $\tilde{H}_q(X) = H_q(\tilde{C}(X))$. We next define

$$\tilde{C}(X;G) = \tilde{C}(X) \otimes G$$
$$\tilde{H}_q(X;G) = H_q(\tilde{C}(X;G)).$$

One readily verifies that

$$\tilde{H}_q(X;G) = H_q(X;G) \quad \text{if } q \neq 0,$$

while for $q = 0$ there is a split exact sequence

$$0 \to \tilde{H}_0(X;G) \to H_0(X,G) \xrightarrow{\varepsilon_*} G \to 0$$

relating the reduced and unreduced 0-dimensional homology groups.

In order to prove the excision property for homology with arbitrary coefficients, it is convenient to have the following lemma.

Lemma 3.3. *Let K and K' be chain complexes of free abelian groups, and let $f:K \to K'$ be a chain map such that the induced homomorphism $f_*:H_n(K) \to H_n(K')$ is an isomorphism for all n. Then for any coefficient group G, the chain map $f \otimes 1_G:K \otimes G \to K' \otimes G$ also induces isomorphisms*

$$(f \otimes 1_G)_*:H_n(K \otimes G) \approx H_n(K' \otimes G)$$

for all n.

PROOF: By Theorem 2.3, f is a chain homotopy equivalence. It follows readily that $f \otimes 1_G:K \otimes G \to K' \otimes G$ is also a chain homotopy equivalence. Hence $(f \otimes 1_G)_*$ is an isomorphism, as required,   Q.E.D

Now suppose that the hypotheses of the excision property hold as stated in Theorem II.6.2, i.e., (X,A) is a pair and W is a subset of A such that $\bar{W}$ is contained in the interior of A. Then it should be clear how to apply the lemma we have just proved in order to conclude that the inclusion map $(X - W, A - W) \to (X,A)$ induces an isomorphism $H_n(X - W, A - W; G) \approx H_n(X,A;G)$ for any n. Thus the excision property also holds true for homology with coefficients in any group G.

In a similar way, one can use Lemma 3.3 above to prove that Theorem II.6.3 holds true for homology with coefficients in an arbitrary group G: If $\mathcal{U}$ is a generalized open covering of C, then the chain map

$$\sigma \otimes 1_G:C(X,A,\mathcal{U}) \otimes G \to C(X,A) \otimes G$$

induces isomorphisms on homology groups. This result can then be used to prove the exactness of the Mayer–Vietoris sequence (Theorem III.5.1) for homology with coefficient group G. The details are left to the reader.

Later on in this chapter, we will indicate an alternative method of proving the excision property and exactness of the Mayer–Vietoris sequence without using Theorem 2.3.

§4. Intuitive Geometric Picture of a Cycle with Coefficients in G

In Chapter I we emphasized the intuitive picture of a 1-cycle as a collection of oriented closed curves with integral "multiplicities" attached to each, a 2-cycle as a collection of oriented closed surfaces, etc. The intuitive picture of a cycle with coefficients in a group G is basically similar, except now the multiplicity assigned to each closed curve or closed surface must be an element of G rather than an integer.

If the group G has elements of finite order, then certain new possibilities arise. For example, suppose G is a cyclic group of order n generated by an element of $g \in G$. Let x and y be distinct points in the space X, and suppose we have n distinct oriented curves in X, starting at x and ending at y. If the element g is assigned as the multiplicity of each curve, then the "sum" of all these oriented curves is a 1-cycle, because $n \cdot g = 0$.

If the group G is infinitely divisible, certain other new phenomena occur. Consider, for example, the case where G is the additive group of rational numbers. Suppose that z is an n-dimensional cycle in X with coefficient group G, and that qz is homologous to 0 for some integer $q \neq 0$. Since we can divide by q in this case, we can conclude that z is homologous to 0.

The above are just examples of two of the many things that can occur. The reader will undoubtedly encounter other examples as he proceeds in the study of this subject.

§5. Coefficient Homomorphisms and Coefficient Exact Sequences

Let $h: G_1 \to G_2$ be a homomorphism of abelian groups. Then we get an obvious homomorphism

$$1 \otimes h: C_n(X,A;G_1) \to C_n(X,A;G_2)$$

for any pair (X,A) and all integers n. These homomorphisms fit together to define a chain map $C(X,A;G_1) \to C(X,A;G_2)$ which we may as well continue to denote by the same symbol, $1 \otimes h$, and hence there is an induced homomorphism

$$h_{\#}: H_n(X,A;G_1) \to H_n(X,A;G_2).$$

The reader should verify the following two naturality properties of this induced homomorphism:

(a) For any continuous map $f:(X,A) \to (Y,B)$, the following diagram is commutative:

$$
\begin{array}{ccc}
H_n(X,A;G_1) & \xrightarrow{\ f_*\ } & H_n(Y,B;G_1) \\
\Big\downarrow h_* & & \Big\downarrow h_* \\
H_n(X,A;G_2) & \xrightarrow{\ f_*\ } & H_n(Y,B;G_2).
\end{array}
$$

(b) For any pair (X,A), the following diagram is commutative:

$$
\begin{array}{ccc}
H_n(X,A;G_1) & \xrightarrow{\ \partial_*\ } & H_{n-1}(A;G_1) \\
\Big\downarrow h_* & & \Big\downarrow h_* \\
H_n(Y,B;G_2) & \xrightarrow{\ \partial_*\ } & H_{n-1}(B;G_2).
\end{array}
$$

The induced homomorphism $h_{\#}$ is important in the further development of homology theory. As an example, we give the following application. Let R be an arbitrary ring, and assume that the abelian group G is also a left R-module, i.e., R operates on the left on G as a set of endomorphisms, satisfying the usual conditions. Any element $r \in R$ defines an endomorphism $G \to G$ by the rule $x \to rx$ for $x \in G$. There is an induced endomorphism of $H_n(X,A;G)$ according to the procedure developed in the preceding paragraphs. Thus for each element $r \in R$ we have defined an endomorphism of $H_n(X,A;G)$. We leave it to the reader to verify that these induced endomorphisms define on $H_n(X,A;G)$ a structure of left R-module. The naturality properties (a) and (b) above show that f_* and ∂_* respectively are homomorphisms of left R-modules.

An especially important case occurs when R is a commutative field and G is a vector space over R. Then $H_n(X,A;G)$ is also a vector space over R, and the induced homomorphisms f_* and ∂_* are R-linear. In this case all the machinery of vector space theory and linear algebra can be applied to problems arising in homology theory, which is often a substantial advantage.

Next, suppose that

$$
0 \to G' \xrightarrow{h} G \xrightarrow{k} G'' \to 0
$$

is a short exact sequence of abelian groups. This gives rise to the following sequence of chain maps and chain complexes for any pair, (X,A):

$$
0 \to C(X,A;G') \xrightarrow{1 \otimes h} C(X,A;G) \xrightarrow{1 \otimes k} C(X,A;G'') \to 0.
$$

We assert that *this sequence of chain complexes is exact*. This assertion is an easy consequence of the fact that $C(X,A)$ is a chain complex of free abelian groups. As a consequence, we get a corresponding long exact homology sequence:

$$\cdots \xrightarrow{\beta} H_n(X,A;G') \xrightarrow{h_\#} H_n(X,A;G) \xrightarrow{k_\#} H_n(X,A;G'') \xrightarrow{\beta} H_{n-1}(X,A;G')$$
$$\xrightarrow{h_\#} \cdots.$$

The connecting homomorphism β of this exact sequence is called the *Bockstein operator* corresponding to the given short exact sequence of coefficient groups. As is so often the case, this label is a misnomer, because this homomorphism was introduced by other mathematicians before Bockstein.

The reader should formulate and prove the naturality properties of the Bockstein operator *vis-a-vis* homomorphisms induced by continuous maps and the boundary homomorphism of the exact sequence of a pair (X,A):

$$
\begin{array}{ccc}
H_n(X,A;G'') & \xrightarrow{\ \beta\ } & H_{n-1}(X,A;G') \\
\downarrow{\scriptstyle \partial_*} & & \downarrow{\scriptstyle \partial_*} \\
H_{n-1}(A;G'') & \xrightarrow{\ \beta\ } & H_{n-2}(A;G').
\end{array}
$$

CAUTION. The question as to whether or not this diagram is commutative is a bit subtle.)

EXERCISE

5.1. Using the methods of §III.4, determine the homology groups of the real projective plane for the case where the coefficient group G is cyclic of order 2. Then determine the long exact homology sequence corresponding to the following short exact sequence of coefficient groups:

$$0 \to \mathbf{Z} \xrightarrow{h} \mathbf{Z} \xrightarrow{k} \mathbf{Z}_2 \to 0.$$

Here $h(n) = 2n$ for any $n \in \mathbf{Z}$.

The coefficient homomorphism and Bockstein operator are additional elements of structure on the homology groups of a space. The fact that homomorphisms induced by continuous maps must commute with them places a definite limitation on such induced homomorphisms.

§6. The Universal Coefficient Theorem

We will next take up the relation between integral homology groups and homology groups with various coefficients.

Let $K = \{K_n, \partial_n\}$ be an arbitrary chain complex. There is a natural homomorphism

$$\alpha : H_n(K) \otimes G \to H_n(K \otimes G)$$

defined as follows. Let $u \in H_n(K)$ and $x \in G$. Choose a representative cycle $u' \in Z_n(K)$ for u. Then it is immediate that $u' \otimes x \in K_n \otimes G$ is a cycle; define $\alpha(u \otimes x)$ to be the homology class of $u' \otimes x$. Of course it must be verified that this definition is independent of the choice of u', and that α is a homomorphism.

As usual, α is natural in several different senses:

(a) If $f: K \to K'$ is a chain map, then the following diagram is commutative:

$$
\begin{array}{ccc}
H_n(K) \otimes G & \xrightarrow{\alpha} & H_n(K \otimes G) \\
\downarrow{\scriptstyle f_* \otimes 1_G} & & \downarrow{\scriptstyle (f \otimes 1_G)_*} \\
H_n(K') \otimes G & \xrightarrow{\alpha'} & H_n(K' \otimes G).
\end{array}
$$

(b) If $E: 0 \to K' \to K \to K'' \to 0$ is a split exact sequence of chain complexes, then the following diagram is commutative:

$$
\begin{array}{ccc}
H_n(K'') \otimes G & \xrightarrow{\alpha''} & H_n(K'' \otimes G) \\
\downarrow{\scriptstyle \partial_E \otimes 1_G} & & \downarrow{\scriptstyle \partial_{E \otimes G}} \\
H_{n-1}(K') \otimes G & \xrightarrow{\alpha'} & H_{n-1}(K' \otimes G).
\end{array}
$$

(Note: The fact that E is split exact assures exactness on tensoring with G.)

(c) If $h: G_1 \to G_2$ is a homomorphism of coefficient groups, then the following diagram is commutative:

$$
\begin{array}{ccc}
H_n(K) \otimes G_1 & \xrightarrow{\alpha_1} & H_n(K \otimes G_1) \\
\downarrow{\scriptstyle 1 \otimes h} & & \downarrow{\scriptstyle h_*} \\
H_n(K) \otimes G_2 & \xrightarrow{\alpha_2} & H_n(K \otimes G_2).
\end{array}
$$

If $0 \to G' \to G \to G'' \to 0$ is a short exact sequence of abelian groups, and K is a chain complex of *free* abelian groups, then we might expect a commutative diagram involving the Bockstein operator, but such does not exist.

For our purposes, the most important case of the homomorphism is where $K = C(X, A)$; then we obtain a homomorphism

$$
\alpha: H_n(X, A) \otimes G \to H_n(X, A; G)
$$

with all the above naturality properties.

Lemma 6.1. *If G is a free abelian group, then the homomorphism $\alpha: H_n(K) \otimes G \to H_n(K \otimes G)$ is an isomorphism.*

PROOF: First, one considers the case where $G = \mathbf{Z}$, which is trivial. In the general case, G is a direct sum of infinite cyclic groups, and α obviously "respects" such direct sum decompositions (because of Property (c) above).

Q.E.D.

In order to make further progress, we must make use of the Tor functor (the first derived functor of the tensor product). For any two abelian groups A and B, we will use the notation $\mathrm{Tor}(A,B)$ to denote $\mathrm{Tor}_1^{\mathbf{Z}}(A,B)$. The definition and properties of this functor are given in most books on homological algebra, e.g., Cartan and Eilenberg [1], Hilton and Stammbach [2] or Mac Lane [3]. Here is a list of some of the principal properties of this functor:

(1) (Symmetry) $\mathrm{Tor}(A,B)$ and $\mathrm{Tor}(B,A)$ are naturally isomorphic.

(2) If either A or B is torsion-free, then $\mathrm{Tor}(A,B) = 0$.

(3) Let $0 \to F_1 \overset{h}{\to} F_0 \overset{k}{\to} A \to 0$ be a short exact sequence with F_0 a *free* abelian group; it follows that F_1 is also free. Then there is an exact sequence, as follows:

$$0 \to \mathrm{Tor}(A,B) \to F_1 \otimes B \xrightarrow{h \otimes 1} F_0 \otimes B \xrightarrow{k \otimes 1} A \otimes B \to 0.$$

Since any abelian group A is the homomorphic image of some free abelian group F_0, we can use this property to define $\mathrm{Tor}(A,B)$, or to determine it in specific cases.

(4) For any abelian group G, $\mathrm{Tor}(\mathbf{Z}_n,G)$ is isomorphic to the subgroup of G consisting of all $x \in G$ such that $nx = 0$ (this may be proved by use of (3)). In particular, $\mathrm{Tor}(\mathbf{Z}_n,\mathbf{Z}_m)$ is a cyclic group whose order is the g.c.d. of m and n.

(5) Tor is an additive functor in each variable, i.e., for direct sums

$$\mathrm{Tor}\left(\sum_i A_i, B\right) \approx \sum_i \mathrm{Tor}(A_i,B).$$

(6) Let $0 \to A' \underset{h}{\to} A \underset{k}{\to} A'' \to 0$ be a short exact sequence of abelian groups; then we have the following long exact sequence:

$$0 \to \mathrm{Tor}(A',B) \xrightarrow{\mathrm{Tor}(h,1)} \mathrm{Tor}(A,B) \xrightarrow{\mathrm{Tor}(k,1)} \mathrm{Tor}(A'',B)$$

$$\to A' \otimes B \xrightarrow{h \otimes 1} A \otimes B \xrightarrow{k \otimes 1} A'' \otimes B \to 0$$

(this is a generalization of (3)).

With these preliminaries out of the way, we can state and prove the universal coefficient theorem:

Theorem 6.2. *Let K be a chain complex of free abelian groups, and let G be an arbitrary abelian group. Then there exists a split short exact sequence*

$$0 \to H_n(K) \otimes G \overset{\alpha}{\to} H_n(K \otimes G) \overset{\beta}{\to} \mathrm{Tor}(H_{n-1}(K),G) \to 0.$$

The homomorphism β is natural vis-a-vis chain maps and coefficient homo-morphisms. The splitting is natural vis-a-vis coefficient homomorphisms, but it is not natural with respect to chain mappings.

PROOF. As mentioned in (3) above, we may choose a free abelian group F_0 such that there is an epimorphism $k : F_0 \to G$; let F_1 denote the kernel of k. Then F_1 is free, and we have the following short exact sequence:

$$0 \to F_1 \overset{h}{\to} F_0 \overset{k}{\to} G \to 0.$$

Now consider the following commutative diagram:

$$
\begin{array}{ccccccc}
H_n(K \otimes F_1) & \overset{h_\#}{\longrightarrow} & H_n(K \otimes F_0) & \overset{k_\#}{\longrightarrow} & H_n(K \otimes G) & \overset{\beta_0}{\longrightarrow} & H_{n-1}(K \otimes F_1) \\
\uparrow {\scriptstyle \alpha_1} & & \uparrow {\scriptstyle \alpha_0} & & \uparrow {\scriptstyle \alpha} & & \\
H_n(K) \otimes F_1 & \overset{1 \otimes h}{\longrightarrow} & H_n(K) \otimes F_0 & \overset{1 \otimes k}{\longrightarrow} & H_n(K) \otimes G & \longrightarrow & 0.
\end{array}
$$

The top line is part of the long exact sequence corresponding to the given short exact sequence of coefficients, with Bockstein operator β_0. The bottom line is exact, and both α_0 and α_1 are isomorphisms by Lemma 6.1. From this diagram it readily follows that α is a monomorphism, and image $\alpha =$ image $k_\# = $ kernel β_0.

Next, consider the following somewhat analogous diagram:

$$
\begin{array}{ccccccccc}
0 & \longrightarrow & \mathrm{Tor}(H_{n-1}(K),G) & \longrightarrow & H_{n-1}(K) \otimes F_1 & \overset{1 \otimes h}{\longrightarrow} & H_{n-1}(K) \otimes F_0 & \overset{1 \otimes k}{\longrightarrow} & H_{n-1}(K) \otimes G & \longrightarrow & 0 \\
& & \boxed{1} \downarrow & & \downarrow {\scriptstyle \alpha_1} & & \downarrow {\scriptstyle \alpha_0} & & \downarrow {\scriptstyle \alpha} & & \\
& & H_n(K \otimes G) & \overset{}{\underset{\beta_0}{\to}} & H_{n-1}(K \otimes F_1) & \overset{}{\underset{h_\#}{\to}} & H_{n-1}(K \otimes F_0) & \overset{}{\underset{k_\#}{\to}} & H_{n-1}(K \otimes G).
\end{array}
$$

The top line of this diagram is the exact sequence mentioned in Property (3) above. Once again, α_1 and α_0 are isomorphisms, and the diagram is commutative. It follows easily from this diagram that there exists a unique homomorphism

$$\beta : H_n(K \otimes G) \to \mathrm{Tor}(H_{n-1}(K),G)$$

which makes the left-hand square (labelled 1) of this diagram commutative. Furthermore, β is an epimorphism, and kernel $\beta = $ kernel β_0.

Thus we have defined the homomorphism β, and proved the exactness of the sequence mentioned in the theorem. We leave it to the reader to prove that β is natural *vis-a-vis* chain maps and coefficient homomorphisms. It remains to prove that the sequence splits. For this purpose, we will use the following trick. We may consider the sequence of abelian groups $\{H_n(K)\}$ as a chain complex with $\partial_n = 0$ for all n; we will denote this chain complex by $H(K)$. With this notation, it is clear that $H_n(H(K)) = H_n(K)$. We assert *that there exists a chain map $f : K \to H(K)$ such that the induced homomorphism $f_* : H_n(K) \to H_n(H(K))$ is the identity map of $H_n(K)$ onto itself.* To prove this assertion, note that our hypothesis that K_n is a free abelian group for each n implies that $Z_n(K)$ is a direct summand of K_n. Hence we may *choose* a direct sum decomposition

$$K_n = Z_n(K) \oplus L_n$$

for each n. Define $f_n : K_n \to H_n(K)$ by $f_n | Z_n(K) =$ natural homomorphisms of $Z_n(K)$ onto $H_n(K)$, and $f_n | L_n = 0$. It is readily verified that the sequence of homomorphisms $f = \{f_n\}$ is a chain map with the required properties. The definition of f obviously depends on the choice of the direct sum decomposition.

Next, by the naturality of α, we have the following commutative diagram:

$$
\begin{array}{ccc}
H_n(K) \otimes G & \xrightarrow{\ \alpha\ } & H_n(K \otimes G) \\
\downarrow{\scriptstyle f_* \otimes 1_G} & & \downarrow{\scriptstyle (f \otimes 1_G)_*} \\
H_n(H(K)) \otimes G & \xrightarrow{\ \alpha'\ } & H_n(H(K) \otimes G).
\end{array}
$$

However, it is readily checked that $H_n(H(K)) \otimes G = H_n(K) \otimes G = H_n(H(K) \otimes G)$ and that $f_* \otimes 1_G$ and α' are both the identity maps. Hence it follows from the commutativity of the diagram that image α is a direct summand of $H_n(K \otimes G)$, as required. Incidentally, this furnishes an alternative proof that α is a monomorphism.

Using this procedure, it is easy to prove that the direct sum decomposition is natural vis-a-vis coefficient homomorphisms.   Q.E.D

We will give an example later to prove that it is impossible to choose the direct sum decomposition so it is natural with respect to chain maps.

Corollary 6.3. *For any pair (X,A) and any abelian group G there exists a split short exact sequence*:

$$0 \to H_n(X,A) \otimes G \xrightarrow{\alpha} H_n(X,A;G) \xrightarrow{\beta} \mathrm{Tor}(H_{n-1}(X,A),G) \to 0.$$

The homomorphisms α and β are natural with respect to homomorphisms induced by continuous maps of pairs and coefficient homomorphisms. The splitting can be chosen to be natural with respect to coefficient homomorphisms, but not with respect to homomorphisms induced by continuous maps.

These results show that the structure of the homology group $H_n(X,A;G)$ is *completely* determined by the structure of the integral homology groups $H_n(X,A)$ and $H_{n-1}(X,A)$. However, this does *not* imply that the homomorphism $f_* : H_n(X,A;G) \to H_n(Y,B;G)$ is determined by the homomorphisms $f_* : H_n(X,A) \to H_n(Y,B)$ and $f_* : H_{n-1}(X,A) \to H_{n-1}(Y,B)$ (here $f : (X,A) \to (Y,B)$ denotes a continuous maps of pairs). A convincing example will be given later.

6.1. Decide whether or not the following diagram is commutative for any pair (X,A) and any abelian group G:

$$
\begin{array}{ccc}
H_n(X,A;G) & \xrightarrow{\ \beta\ } & \mathrm{Tor}(H_{n-1}(X,A),G) \\
\Big\downarrow{\scriptstyle \partial_*} & & \Big\downarrow{\scriptstyle \mathrm{Tor}(\partial_*,1_G)} \\
H_{n-1}(A;G) & \xrightarrow{\ \beta\ } & \mathrm{Tor}(H_{n-2}(A),G).
\end{array}
$$

6.2. Prove that $\alpha : H_i(X,A) \otimes G \to H_i(X,A;G)$ is an isomorphism for any pair (X,A) and any group G for $i = 0$ or 1.

6.3. Let X be a finite regular graph. Express the structure of the homology groups $H_q(X;G)$ in terms of the Euler characteristic and number of components of X.

6.4. Describe the structure of the homology groups $H_q(X;G)$ for any group G in the following cases:
   (a) $X = S^n$.
   (b) X is a compact orientable 2-manifold.
   (c) X is a compact nonorientable 2-manifold.

6.5. Let X be an n-dimensional pseudomanifold in the sense of §IV.8. Determine the structure of $H_n(X,G)$ in case X is (a) orientable and (b) nonorientable.

We will conclude this section by giving another proof of the excision property for homology with arbitrary coefficient groups. Let (X,A) be a pair, and W a subset of A such that $\overline{W}$ is contained in the interior of A. Then the inclusion map $i : (X - W, A - W) \to (X,A)$ induces a chain map

$$
i_\# : C(X - W, A - W) \to C(X,A).
$$

It is easy to verify that $i_\#$ is a monomorphism; thus we can consider $C(X - W, A - W)$ as a subcomplex of $C(X,A)$. Hence we have the following short exact sequence of chain complexes:

$$
0 \to C(X - W, A - W) \xrightarrow{i_\#} C(X,A) \to \frac{C(X,A)}{C(X - W, A - W)} \to 0.
$$

This short exact sequence of chain complexes gives rise to a long exact homology sequence, as usual. Because the excision property is true for integral homology, we can conclude that

$$
H_n\left(\frac{C(X,A)}{C(X - W, A - W)} \right) = 0
$$

for all n. Next, one must verify that the quotient complex

$$
\frac{C(X,A)}{C(X - W, A - W)}
$$

is a chain complex of free abelian groups, and that the short exact sequence above is split exact. This is not difficult, and is left to the reader. One now completes the proof by tensoring the short exact sequence with G, and considering the resulting long exact homology sequence. By using Theorem 6.2, one proves that

$$H_n\left(\frac{C(X,A)}{C(X-W,A-W)}\otimes G\right)=0$$

for all n. It follows from exactness that

$$i_*:H_n(X-W,A-W;G)\to H_n(X,A;G)$$

is an isomorphism for all n, as desired.

This technique can also be used to prove that the chain map

$$\sigma\otimes 1_G:C(X,A;\mathcal{U})\otimes G\to C(X,A)\otimes G$$

induces isomorphisms in homology in all dimensions (here $\mathcal{U}$ is a generalized open covering of X).

§7. Further Properties of Homology with Arbitrary Coefficients

Practically all the properties we have proved for integral homology have analogs for homology with arbitrary coefficients. For example, the reader should have no difficulty verifying that Proposition III.6.1 is true for homology with coefficients in any group G.

The material in Chapter IV on the homology of CW-complexes readily generalizes to the case of an arbitrary coefficient group. We will quickly indicate how this goes.

Let $K=\{K^n\}$ be a CW-complex on the space X. Using the universal coefficient theorem (Corollary 6.3) it is readily shown that $H_q(K^n,K^{n-1};G)=0$ for $q\neq n$, and that

$$\alpha:H_n(K^n,K^{n-1})\otimes G\to H_n(K^n,K^{n-1};G)$$

is an isomorphism. Thus if we define

$$C_n(K;G)=H_n(K^n,K^{n-1};G),$$

then

$$C_n(K;G)=C_n(K)\otimes G.$$

Next, we define a boundary operator $C_n(K;G)\to C_{n-1}(K;G)$ as the composition of the homomorphisms

$$H_n(K^n,K^{n-1};G)\overset{\partial_*}{\to}H_{n-1}(K^{n-1};G)\overset{j_{n-1}}{\longrightarrow}H_{n-1}(K^{n-1},K^{n-2};G)$$

by analogy with that defined in §IV.4. It is then true that this boundary operator is $d_n\otimes 1_G$, where $d_n:C_n(K)\to C_{n-1}(K)$ is defined in §IV.4. In other words,

$$C(K;G)=C(K)\otimes G.$$

One can now prove an analog of Theorem IV.4.2 for the case of an arbitrary coefficient group G. Essentially this analog says that $H_n(X;G)$ is naturally isomorphic to $H_n(C(K;G))$. Similarly, there is an analog of Theorem IV.4.6 for homomorphisms induced by cellular maps: Assume $K = \{K^n\}$ is a CW-complex on X, $L = \{L^n\}$ is a CW-complex on Y, and $f: X \to Y$ is a cellular map, i.e., $f(K^n) \subset L^n$. Then f induces a chain map $\varphi: C(K) \to C(L)$, and we have a commutative diagram:

$$H_n(X;G) \approx H_n(C(K) \otimes G)$$

$$\Big\downarrow f_* \qquad\qquad \Big\downarrow (\varphi \otimes 1_G)_*$$

$$H_n(Y;G) \approx H_n(C(L) \otimes G).$$

These results can be summarized as follows: To extend the results of §IV.4 from integral homology to homology with arbitrary coefficient group G, simply tensor all chain complexes and chain maps with G. In particular, this applies to the computation of the homology of regular CW-complexes as described in §IV.7.

There is one case where the computation of the homology of regular cell complexes becomes *greatly* simplified, namely, the case where $G = \mathbf{Z}_2$. In this case every incidence number must be 0 or 1, and we see that $[e^n : e^{n-1}] = 1$ or 0 according as e^{n-1} is or is not a face of e^n. Thus the four rules given in Theorem IV.7.2 for determining incidence numbers reduce to two rules, and it is not necessary to use an inductive procedure. Of course mod 2 homology ignores much of the structure of integral homology, but for some problems it is more appropriate than integral homology.

EXAMPLE 7.1. Let P^2 denote the real projective plane. In III.4 we found that the only nonzero homology groups of P^2 were

$$H_0(P^2) = \mathbf{Z},$$
$$H_1(P^2) = \mathbf{Z}_2.$$

Thus if $f: P^2 \to S^2$ is any continuous map, then

$$f_*: H_0(P^2) \to H_0(S^2)$$

is an isomorphism (both are connected spaces), while for $q \neq 0$,

$$f_*: H_q(P^2) \to H_q(S^2)$$

must be the zero map. Hence there is no possibility of distinguishing between different homotopy classes of such maps using integral homology. We will now show that one *can* distinguish two different homotopy classes using mod 2 homology. To prove this, recall that there is a CW-complex, K, on P^2 having a single cell in dimensions 0, 1, and 2; this was used to compute the homology of P^2 in Example III.4.3, although it was not called a CW-complex at that time. Thus $C_0(K)$, $C_1(K)$, and $C_2(K)$ are infinite cyclic groups,

$$d_2:C_2(K) \rightarrow C_1(K)$$

has degree ± 2, and

$$d_1:C_1(K) \rightarrow C_0(K)$$

has degree 0. Analogously, there is a CW-complex L on S^2 having a single vertex, a single 2-cell, and no 1-cells. There is an obvious cellular map

$$f:P^2 \rightarrow S^2$$

defined by shrinking the 1-skeleton, K^1, to a point, namely, L^0. The open 2-cell of K is mapped homomorphically onto the open 2-cell of L. We wish to compute the induced homomorphism on mod 2 homology. To this end, we determine the chain transformation

$$f':C(K) \rightarrow C(L)$$

induced by the cellular map f. The only nontrivial problem is to determine the homomorphism $C_2(K) \rightarrow C_2(L)$. But this is easily settled by Theorem IV.2.1. Let $g:(E^2,S^1) \rightarrow (K^2,K^1)$ be the characteristic map for the unique 2-cell of K. In view of the way the map $f:P^2 \rightarrow S^2$ was defined, it is clear that

$$h = fg:(E^2,S^1) \rightarrow (L^2,L^1)$$

is a characteristic map for the only 2-cell of L. Thus we have the following commutative diagram:

$$(E^2,S^1) \xrightarrow{\ g\ } (K^2,K^1)$$

$$\searrow^{h} \qquad \downarrow^{f}$$

$$(L^2,L1).$$

Hence we have the following commutative diagram:

$$H_2(E^2,S^1) \xrightarrow{\ g_*\ } H_2(K^2,K^1)$$

$$\searrow^{h_*} \qquad \downarrow^{f_*}$$

$$H_2(L^2,L^1).$$

By Theorem IV.2.1, g_* and h_* are isomorphisms; it follows that f_* is also an isomorphism. Therefore the chain map $f':C(K) \rightarrow C(L)$ is completely determined. All that remains is to tensor with $\mathbf{Z}_2$ and then pass to homology. The end result is that

$$f_*:H_2(P^2;\mathbf{Z}_2) \rightarrow H_2(S^2;\mathbf{Z}_2)$$

is an isomorphism. On the other hand, if $\varphi:P^2 \rightarrow S^2$ is the constant map, then

$$\varphi_*:H_2(P^2,\mathbf{Z}_2) \rightarrow H_2(S^2,\mathbf{Z}_2)$$

is the 0 homomorphism. Thus f and φ are not homotopic.

Note that f and φ must (of necessity) induce the same homomorphism on integral homology groups. This proves our earlier assertion that the induced homomorphisms on integral homology groups do not suffice to determine the induced homomorphisms on homology groups with other coefficients.

Finally, this example also show that the splitting of the short exact sequence of the universal coefficient theorem (Corollary 6.3) can *not* be chosen to be natural. Consider the following commutative diagram involving the universal coefficient theorems for $H_2(P^2,\mathbf{Z}_2)$ and $H_2(S^2,\mathbf{Z}_2)$ and the homomorphism induced by the map $f\colon P^2 \to S^2$ described above:

$$
\begin{array}{ccccccccc}
0 & \longrightarrow & H_2(P^2) \otimes Z_2 & \xrightarrow{\ \alpha_1\ } & H_2(P^2,Z_2) & \xrightarrow{\ \beta_1\ } & \mathrm{Tor}(H_1(P^2),Z_2) & \longrightarrow & 0 \\
 & & \Big\downarrow{\scriptstyle f_*\otimes 1} & & \Big\downarrow{\scriptstyle f_*} & & \Big\downarrow{\scriptstyle \mathrm{Tor}(f_*,1)} & & \\
0 & \longrightarrow & H_2(S^2) \otimes Z_2 & \xrightarrow{\ \alpha_2\ } & H_2(S^2,Z_2) & \xrightarrow{\ \beta_2\ } & \mathrm{Tor}(H_1(S^2),Z_2) & \longrightarrow & 0.
\end{array}
$$

In the top line, $H_2(P^2) \otimes \mathbf{Z}_2 = 0$ and β_1 is an isomorphism. In the bottom line, $\mathrm{Tor}(H_1(S^2),\mathbf{Z}_2) = 0$ and α_2 is an isomorphism. As we have just proved, the vertical arrow labelled f_* is an isomorphism; however, this fact contradicts the possibility of any splitting of these two short exact sequences which is natural with respect to homomorphisms induced by continuous maps.

We will conclude this section with a brief consideration of the mod 2 homology of a pseudomanifold.

Let K be an n-dimensional nonorientable pseudomanifold; by Theorem IV.8.1, $H_n(K) = 0$. For some purposes, this is a defect in the theory; we need a nonzero homology class in the top dimension. This matter is partially remedied by using mod 2 homology. Indeed we find that $H_2(K,\mathbf{Z}_2) = \mathbf{Z}_2$ (use the universal coefficient theorem and Theorems IV.8.1 and IV.8.2). A representative cycle for the nonzero element of $H_n(K,\mathbf{Z}_2)$ is obtained by taking the sum of all the n-cells of K. Since we are using $\mathbf{Z}_2$ as coefficient group, we do not need to worry about orientations.

Bibliography for Chapter V

[1] H. Cartan and S. Eilenberg, *Homological Algebra*, Princeton University Press, Princeton, 1956, Chapter VI, §1.
[2] P. Hilton and U. Stammbach, *A Course in Homological Algebra* Springer-Verlag New York, 1971, pp. 112–115.
[3] S. Mac Lane, *Homology*, Springer-Verlag, New York, 1963, Chapter V, §6.

The Homology of Product Spaces

§1. Introduction

If two or more spaces are related to each other in some way, we would naturally expect that their homology groups should also be related in some way. Some of the most important theorems in the preceding chapters bear out this expectation: If A is a subspace of X, the exact homology sequence of the pair (X,A) describes the relations between the homology groups of A and the homology groups of X. If the space X is the union of two subspaces U and V, then the Mayer–Vietoris sequence gives relations between the homology groups of $U,V,\ U \cap V$, and X.

The main theorems of this chapter are of this same general nature. The Eilenberg–Zilber theorem asserts that the singular chain complex of the product of two spaces, $C(X \times Y)$, is chain homotopy equivalent to the tensor product of the chain complexes of the two factors denoted by $C(X) \otimes C(Y)$. The Künneth theorem expresses the homology groups of the product space $X \times Y$ in terms of the homology groups of X and the homology groups of Y. The derivation of the Künneth theorem from the Eilenberg–Zilber theorem is purely algebraic.

These theorems are somewhat more complicated than most of our previous theorems, such as the exactness of the Mayer–Vietoris sequence. Nevertheless, they are of basic importance in homology theory.

The material on CW-complexes in §2 is not essential for most of the rest of the chapter. It is introduced mainly to motivate the definition of the tensor product of chain complexes.

§2. The Product of CW-complexes and the Tensor Product of Chain Complexes

Let $K = \{K^n\}$ be a CW-complex on the space X, and $L = \{L^n\}$ a CW-complex on the space Y. We wish to prove that $X \times Y$ is a CW-complex in a natural way. In order to understand this situation better, we need the following basic facts about open and closed cells; our notation is that of §IV.2.

(a) $E^m \times E^n$ is homeomorphic to E^{m+n}; under any such homeomorphism, $(E^m \times S^{n-1}) \cup (S^{m-1} \times E^n)$ corresponds to the boundary S^{m+n-1}.

(b) $U^m \times U^n$ is homeomorphic to U^{m+n}.

In view of Statement (b), it is natural to demand that an open cell of a CW-complex on $X \times Y$ should be the product of an open cell of K with an open cell of L. Therefore we *define* the n-skeleton of a CW-structure on $X \times Y$ by

$$M^n = \bigcup_{p+q=n} K^p \times L^q$$

for $n = 0, 1, 2, \ldots$. Then the subsets $M^n \subset X \times Y$ are closed, $M^n \subset M^{n+1}$ for all $n \geq 0$, M^0 is discrete (because it is the product of discrete spaces) and

$$X \times Y = \bigcup_{n=0}^{\infty} M^n.$$

If e^m is an m-cell of K with characteristic map $f : E^m \to \bar{e}^m$, and if e^n is an n-cell of L with characteristic map $g : E^n \to \bar{e}^n$, then $f \times g : E^m \times E^n \to \bar{e}^m \times \bar{e}^n$ has all the required properties for a characteristic map of the product cell $e^m \times e^n$. Thus it only remains to check that the product topology on $X \times Y$ is the same as the weak topology determined by the closed cells. If both K and L are *finite* CW-complexes, then M will also have only a finite number of cells, and there is nothing to prove. J. H. C. Whitehead proved that if one of the factors is locally compact then the product topology agrees with the weak topology. However, Dowker gave an example to show that in the general case, the two topologies on $X \times Y$ do *not* agree. See Lundell and Weingram [5] for details. Fortunately, there is an easy way out of this difficulty; one can agree to give $X \times Y$ the weak topology, so that it *is* a CW-complex. The weak topology will be *larger* than the product topology in general (i.e., it will have more open (or closed) sets), but the *compact* sets will be the same in both topologies. Therefore the identity map,

$$X \times Y_{\text{(weak top.)}} \to X \times Y_{\text{(prod. top.)}}$$

is a continuous map, and induces an isomorphism on singular homology groups. See N. E. Steenrod [8] for details.

However, we do not want to get involved with these fine points now. The reader can restrict his attention to finite CW-complexes, knowing full well that the generalization to infinite CW-complexes is not difficult.

Next, let us assume that K and L are regular CW-complexes. Then it is readily seen that $M = \{M^n\}$ (as defined above) is a *regular* CW-complex on $X \times Y$ (provided $X \times Y$ is given the weak topology). As usual, there are many choices for orientations of the cells of $X \times Y$, and hence of incidence numbers. Let us assume that orientations (and hence incidence numbers) have been chosen for the cells of K and L. It seems plausible to expect that there should be a way to use these chosen orientations of the cells of K and L to define *canonical* orientations of the cells of M. The following theorem shows that this expectation is justified. The actual result is stated in terms of incidence numbers rather than orientations. However, this does not matter from a logical point of view, since there is a 1-1 correspondence between incidence numbers and orientations of cells in any regular CW-complex.

Theorem 2.1. *Let K be a regular CW-complex on X with cells e_1^m, and let L be a regular CW-complex on Y with cells σ_j^n. Assume that the incidence numbers have been chosen for both K and L. Then incidence numbers are defined for the product cells on $X \times Y$ by the following rules:*

$$[e^m \times \sigma^n : e^{m-1} \times \sigma^n] = [e^m : e^{m-1}]$$
$$[e^m \times \sigma^n : e^m \times \sigma^{n-1}] = (-1)^m[\sigma^n : \sigma^{n-1}]$$
$$[e_i^m \times \sigma_j^n : e_k^p \times \sigma_l^q] = 0 \quad \text{if } e_i^m \neq e_k^p \text{ and } \sigma_j^n \neq \sigma_l^q.$$

To prove this theorem, we must verify that Statements (1)–(4) of Theorem IV.7.2 are true with the stated choices of incidence numbers. This we leave to the reader as a nontrivial exercise.

Obviously one could establish other conventions for the incidence numbers of a product complex, but the one given by this theorem is universally accepted.

Now let us consider the group of n-chains, $C_n(M)$ of the regular CW-complex M on $X \times Y$. It has as basis the oriented product cells $e_i^p \times \sigma_j^q$, $p + q = n$. This suggests that we should identify $C_n(M)$ with the direct sum of tensor products,

$$\sum_{p+q=n} C_p(K) \otimes C_q(L).$$

Using the formulas for incidence numbers in the theorem, we see that

$$\partial(e_i^p \times \sigma_j^q) = (\partial e_i^p) \times \sigma_j^q + (-1)^p e_i^p \times (\partial \sigma_j^q),$$

where the right-hand side of this equation is to be interpreted in an obvious way. Since this formula holds true for the basis elements, we can extend it to linear combinations of the basis elements, obtaining the formula

$$\partial(u \otimes v) = (\partial u) \otimes v + (-1)^p u \otimes (\partial v)$$

for any $u \in C_p(K)$ and $v \in C_q(L)$.

This suggests the following definition.

Definition 2.1. Let $C' = \{C'_n, \partial'_n\}$ and $C'' = \{C''_n, \partial''_n\}$ be chain complexes. Their *tensor product* $C = C' \otimes C''$ is the chain complex defined as follows: The groups are

$$C_n = \sum_{p+q=n} C'_p \otimes C''_q,$$

and the homomorphisms $\partial_n : C_n \to C_{n-1}$ are defined by

$$\partial_n(u \otimes v) = (\partial'_p u) \otimes v + (-1)^p u \otimes (\partial''_q v)$$

for any $u \in C'_p$, $v \in C''_q$, $p + q = n$.

Of course, one must verify that $\partial_{n-1}\partial_n = 0$.

In view of this definition, we can assert that *the chain complex $C(M)$ is isomorphic to $C(K) \otimes C(L)$, where K, L, and M are regular CW-complexes on X, Y, and $X \times Y$, as described above.* Of course it remains to determine the relation between the homology groups of the tensor product of two chain complexes and the homology of each of the factors. We will describe the solution to this problem in §4. This result shows that the algebraic operation of taking the tensor product of chain complexes corresponds to the topological operation of taking the cartesian product of two spaces. Soon we will see another example of this process.

EXERCISES

2.1. Let K and L be *finite* CW-complexes on X and Y respectively. What is the relation between the Euler characteristics of X, Y, and $X \times Y$?

2.2. Let K and L be *regular* CW-complexes on X and Y respectively. Assume that orientations and incidence numbers have been chosen for the cells of K and L. Consider the canonical homeomorphism $f : X \times Y \to Y \times X$ defined by $f(x,y) = (y,x)$. Then f maps the oriented cell $e^m \times \sigma^n$ homeomorphically onto $\sigma^n \times e^m$ and induces an isomorphism

$$f_* : H_{m+n}[\overline{e^m \times \sigma^n}, (e^m \times \sigma^n)'] \to H_{m+n}[\overline{\sigma^n \times e^m}, (\sigma^n \times e^m)'].$$

Show that $f_*(e^m \times \sigma^n) = (-1)^{mn}\sigma^n \times e^m$. (*Hint*: Use induction on the dimension of the cell.)

§3. The Singular Chain Complex of a Product Space

Our immediate objective is to define a natural chain map

$$\zeta : C(X) \otimes C(Y) \to C(X \times Y)$$

for any topological spaces X and Y. Then later on we will show that ζ induces an isomorphism of homology groups. The definition of ζ is very simple; however, the proof that the induced homomorphism on homology groups is an isomorphism will be somewhat more involved.

It will be convenient to use the following notation, which is now standard: If $f: A \to B$ and $g: C \to D$ are continuous maps, then $f \times g: A \times C \to B \times D$ denotes the map defined by $(f \times g)(a,c) = (fa, gc)$. We will also find it convenient to identify $I^m \times I^n$ with I^{m+n} in the obvious way. With these conventions, if $S: I^m \to X$ and $T: I^n \to Y$ are singular cubes in X and Y respectively, then $S \times T: I^{m+n} \to X \times Y$ is a singular cube in the product space. Thus we can define a homomorphism

$$\zeta_{m,n}: Q_m(X) \otimes Q_n(Y) \to Q_{m+n}(X \otimes Y)$$

by the formula

$$\zeta_{m,n}(S \otimes T) = S \times T.$$

We assert that the homomorphisms $\zeta_{m,n}$ for all values of m and n define a chain map

$$\zeta: Q(X) \otimes Q(Y) \to Q(X \times Y).$$

To verify that ζ is a chain map, one must compute $\partial(S \times T)$. This is not difficult, if one uses the following formulas:

$$
\begin{aligned}
(A_i S) \times T &= A_i(S \times T) & 1 \leq i \leq m, \\
(B_i S) \times T &= B_i(S \times T) & 1 \leq i \leq m, \\
S \times (A_j T) &= A_{m+j}(S \times T) & 1 \leq j \leq n, \\
S \times (B_j T) &= B_{m+j}(S \times T) & 1 \leq j \leq n.
\end{aligned}
$$

It is clear that if S or T is a degenerate singular cube, then so is $S \times T$. Hence

$$\zeta_{m,n}(Q_m(X) \otimes D_n(Y)) \subset D_{m+n}(X \times Y)$$
$$\zeta_{m,n}(D_m(X) \otimes Q_n(Y)) \subset D_{m+n}(X \times Y)$$

and therefore $\zeta_{m,n}$ induces a homomorphism of quotient groups,

$$\zeta_{m,n}: C_m(X) \otimes C_n(Y) \to C_{m+n}(X \times Y)$$

and the homomorphisms $\zeta_{m,n}$ for all m and n obviously define a chain map

$$\zeta: C(X) \otimes C(Y) \to C(X \times Y).$$

Next, we point out that the chain map ζ has the following very important naturality property: Let $f: X \to X'$ and $g: Y \to Y'$ be continuous maps. Then the following diagram is obviously commutative:

$$
\begin{array}{ccc}
Q_m(X) \otimes Q_n(Y) & \xrightarrow{\zeta_{m,n}} & Q_{m+n}(X \times Y) \\
\downarrow{\scriptstyle f_* \otimes g_*} & & \downarrow{\scriptstyle (f \times g)_*} \\
Q_m(X') \otimes Q_n(Y') & \xrightarrow{\zeta_{m,n}} & Q_{m+n}(X' \times Y').
\end{array}
$$

Hence on passing to quotient groups, etc., we obtain the following commutative diagram:

$$
\begin{array}{ccc}
C(X) \otimes C(Y) & \xrightarrow{\ \zeta\ } & C(X \times Y) \\
\downarrow{\scriptstyle f_* \otimes g_*} & & \downarrow{\scriptstyle (f \times g)_*} \\
C(X') \otimes C(Y') & \xrightarrow{\ \zeta\ } & C(X' \times Y').
\end{array}
$$

Theorem 3.1 (Eilenberg–Zilber theorem). *The chain map* $\zeta : C(X) \otimes C(X) \to C(X \times Y)$ *is a chain homotopy equivalence, and hence induces isomorphisms*

$$\zeta_* : H_q(C(X) \otimes C(Y)) \rightleftarrows H_q(X \times Y).$$

We will postpone the proof of this theorem until later. For the time being, we point out that

$$\zeta_{0,0} : C_0(X) \otimes C_0(Y) \to C_0(X \times Y)$$

is an isomorphism for any spaces X and Y. However, in higher degrees, ζ is only a monomorphism, not an isomorphism.

§4. The Homology of the Tensor Product of Chain Complexes (The Künneth Theorem)

The preceding paragraphs should convince the reader of the importance of the following problem: *Let K and L be chain complexes. Is the homology of the tensor product, $K \otimes L$, determined by the homology of K and L? If so, how?* The answer to the first question is affirmative. We will now proceed to describe the details. First of all, there is a natural homomorphism

$$\alpha : H_m(K) \otimes H_n(L) \to H_{m+n}(K \otimes L)$$

which is defined as follows. Let $u \in H_m(K)$ and $v \in H_n(L)$; choose representative cycles $u' \in Z_m(K)$ for u and $v' \in Z_n(L)$ for v. It is immediate that $u' \otimes v' \in K_m \otimes L_n$ is a cycle; its homology class is, by definition, $\alpha(u \otimes v)$. Of course it is necessary to check that this definition is independent of the choices of the cycles u' and v', and that α is actually a homomorphism. The reader will note that this definition is a slight generalization of that given in §V.6.

The homomorphism α has various naturality properties. For example:

(a) If $f : K \to K'$ and $g : L \to L'$ are chain maps, then the following diagram is commutative:

$$
\begin{array}{ccc}
H_m(K) \otimes H_n(L) & \xrightarrow{\ \alpha\ } & H_{m+n}(K \otimes L) \\
\downarrow{\scriptstyle f_* \otimes g_*} & & \downarrow{\scriptstyle (f \otimes g)_*} \\
H_m(K') \otimes H_n(L') & \xrightarrow{\ \alpha'\ } & H_{m+n}(K' \otimes L').
\end{array}
$$

(b) Assume that

$$E: 0 \to K' \overset{i}{\to} K \overset{j}{\to} K'' \to 0$$

is a short exact sequence of chain complexes, and that L is a chain complex such that the following sequence is exact:

$$E \otimes L: 0 \to K' \otimes L \xrightarrow{i \otimes 1} K \otimes L \xrightarrow{j \otimes 1} K'' \otimes L \to 0$$

(sufficient conditions for $E \otimes L$ to be exact are that E be *split* exact, or that L be a chain complex of torsion-free abelian groups). Then the following diagram is commutative:

$$
\begin{array}{ccc}
H_m(K'') \otimes H_n(L) & \overset{\alpha}{\longrightarrow} & H_{m+n}(K'' \otimes L) \\
\downarrow{\scriptstyle \partial_E \otimes 1} & & \downarrow{\scriptstyle \partial_{E \otimes L}} \\
H_{m-1}(K') \otimes H_n(L) & \overset{\alpha}{\longrightarrow} & H_{m+n-1}(K' \otimes L).
\end{array}
$$

(c) There is an obvious symmetric situation: Assume that

$$E: 0 \to L' \overset{i}{\to} L \overset{j}{\to} L'' \to 0$$

is a short exact sequence of chain complexes, and that K is a chain complex such that the sequence

$$K \otimes E: 0 \to K \otimes L' \xrightarrow{1 \otimes i} K \otimes L \xrightarrow{1 \otimes j} K \otimes L'' \to 0$$

is exact. The reader should investigate the question of the commutativity of the following diagram:

$$
\begin{array}{ccc}
H_m(K) \otimes H_n(L'') & \overset{\alpha}{\longrightarrow} & H_{m+n}(K \otimes L'') \\
\downarrow{\scriptstyle 1 \otimes \partial_E} & & \downarrow{\scriptstyle \partial_{K \otimes E}} \\
H(K) \otimes H_{n-1}(L') & \overset{\alpha}{\longrightarrow} & H_{m+n-1}(K \otimes L').
\end{array}
$$

With these preliminaries taken care of, we can now state our main theorem, the so-called Künneth theorem:

Theorem 4.1. *Let K and L be chain complexes, at least one of which consists of free abelian groups. Then there exists a split exact sequence:*

$$0 \to \sum_{i+j=n} H_i(K) \otimes H_j(L) \overset{\alpha}{\to} H_n(K \otimes L) \overset{\beta}{\to} \sum_{i+j=n-1} \mathrm{Tor}(H_i(K), H_j(L)) \to 0.$$

The homomorphisms α and β are natural with respect to chain maps but the splitting is not natural.

The proof of this important theorem is not difficult; it may be found in various books on homological algebra and algebraic topology, e.g., Vick [9], Hilton and Stammbach [4], Mac Lane [6], Cartan and Eilenberg [1], or Dold [2]. Actually, the theorem can be proved under slightly more general hypotheses than we have stated it, but we will have no use for such greater generality.

This theorem can be combined with our previous results on the product of regular CW-complexes and the singular chain complex of a product space to obtain significant results on the homology of product spaces. We will state the precise results later. In the meantime, we note the following corollary for future reference:

Corollary 4.2. *Suppose that* K *and* L *are chain complexes of free abelian groups which have the homology of a point, i.e.,*

$$H_q(K) = H_q(L) = 0 \quad \text{for } q \neq 0$$
$$H_0(K) = H_0(L) = \mathbf{Z}.$$

Then $K \otimes L$ *also has the homology of a point, and*

$$\alpha : H_0(K) \otimes H_0(L) \to H_0(K \otimes L)$$

is an isomorphism.

§5. Proof of the Eilenberg–Zilber Theorem

We must define a chain map $\eta : C(X \times Y) \to C(X) \otimes C(Y)$ such that $\eta\zeta$ is chain homotopic to the identity map of $C(X) \otimes C(Y)$, and $\zeta\eta$ is chain homotopic to the identity map of $C(X \times Y)$. One way to proceed is by brute force, relying on our geometric intuition to lead us to the correct formulas. We will indicate the first few steps in such a procedure, by defining homomorphisms

$$\eta_q : Q_q(X \times Y) \to \sum_{i+j=q} Q_i(X) \otimes Q_j(Y)$$

such that on passing to the quotient groups modulo degenerate singular chains we obtain the desired chain map η.

Note that a singular n-cube $I^n \to X \times Y$ in the product space corresponds in an obvious way to a pair of singular n-cubes $S : I^n \to X$ and $T : I^n \to Y$ in each of the factors. It will be convenient to let the notation (S, T) denote the corresponding singular n-cube in the product space $X \times Y$.

It is obvious that we should define $\eta_0 : Q_0(X \times Y) \to Q_0(X) \otimes Q_0(Y)$ by the formula

$$\eta_0(S, T) = S \otimes T$$

for any singular 0-cubes $S : I^0 \to X$ and $T : I^0 \to Y$. This makes η_0 the inverse of ζ_0 (which is an isomorphism).

Next, one defines $\eta_1 : Q_1(X \times Y) \to Q_0(X) \otimes Q_1(Y) + Q_1(X) \otimes Q_0(Y)$ by the formula

$$\eta_1(S,T) = (A_1 S) \otimes T + S \otimes (B_1 T)$$

for any singular 1-cubes $S:I^1 \to X$ and $T:I^1 \to Y$.

To define η_q in general, we need a generalization of the face operators A_i and B_i. Let H be any subset of $\{1,2,\ldots,n\}$, and let K denote the complementary subset. If H has p elements and K has q elements, $p + q = n$, we will let

$$\varphi_K : K \to \{1,2,\ldots,q\}$$

denote the unique bijective, order-preserving map. If $T:I^n \to X$ is any singular n-cube, let

$$A_H T, B_H T : I^q \to X$$

denote the following maps:

$$(A_H T)(x_1,\ldots,x_q) = T(y_1,\ldots,y_n),$$

where

$$y_i = \begin{cases} 0 & \text{if } i \in H, \\ x_{\varphi_K(i)} & \text{if } i \in K, \end{cases}$$

and

$$(B_H T)(x_1,\ldots,x_q) = T(y_1,\ldots,y_n),$$

where

$$y_i = \begin{cases} 1 & \text{if } i \in H, \\ x_{\varphi_K(i)} & \text{if } i \in K. \end{cases}$$

EXAMPLE 5.1. If $H = \varnothing$, then $A_H T = B_H T = T$.

EXAMPLE 5.2. If $H = \{i\}$, then $A_H T = A_i T$ and $B_H T = B_i T$.

EXAMPLE 5.3. If $H = \{1,2,\ldots,n\}$, then $A_H T$ and $B_H T$ are singular 0-cubes represented by $T(0,\ldots,0)$ and $T(1,\ldots,1)$ respectively.

We can now define $\eta_q : Q_q(X \times Y) \to \sum_{i+j=q} Q_i(X) \otimes Q_j(Y)$ by the magic formula

$$\eta_q(S,T) = \sum \rho_{H,K} A_H(S) \otimes B_K(T),$$

where $S:I^q \to X$ and $T:I^q \to Y$ are singular q-cubes, H ranges over all subsets of $\{1,2,\ldots,q\}$ and K denotes the complementary set, and $\rho_{H,K} = \pm 1$ denotes the signature of the permutation HK of $\{1,\ldots,q\}$. (If H or K is empty, then $\rho_{H,K} = +1$.)

The student who has sufficient stamina and enthusiasm for calculating can now verify the following assertions:

(a) If (S,T) is a degenerate singular cube, then $\eta_q(S,T)$ belongs to

$\sum_{i+j=q} [D_i(X) \otimes Q_j(Y) + Q_i(X) \otimes D_j(Y)]$. Hence η_q induces a homomorphism

$$\eta_q : C_q(X \times Y) \to \sum_{i+j=q} C_i(X) \otimes C_j(Y).$$

(b) The sequence of homomorphisms $\eta = \{\eta_q\}$ is a chain map $C(X \times Y) \to C(X) \otimes C(Y)$.

(c) $\eta\zeta = $ identity map of $C(X) \otimes C(Y)$.

(d) It is possible to define a chain homotopy between $\zeta\eta$ and the identity map of $C(X \times Y)$, but the formulas are somewhat complicated.

Rather than go through the details of these lengthy calculations, it seems preferable to use a more conceptual method due to Eilenberg and Mac Lane, called the *method of acyclic models*. This method makes strong use of the naturality of the chain maps ζ and η which we have defined. By making full use of this naturality, it is possible to avoid the necessity of having explicit formulas. First, however, we have to make two brief digressions in preparation for this proof.

DIGRESSION 1 : *Some more generalities on chain complexes*

Definition 5.1. A chain complex $K = \{K_q\}$ is *positive* if $K_q = \{0\}$ for $q < 0$.

Most of the chain complexes we have considered so far have been positive. Note that the tensor product of two positive chain complexes is again a positive chain complex.

Definition 5.2. An *augmentation* of a positive chain complex $\{K_q, \partial_q\} = K$ is a homomorphism $\varepsilon : K_0 \to \mathbf{Z}$ such that $\varepsilon\partial_1 = 0$.

Observe that an augmentation ε induces a homomorphism $\varepsilon_* : H_0(K) \to \mathbf{Z}$.

Definition 5.3. A positive chain complex K with augmentation is *acyclic* if $H_q(K) = 0$ for $q \neq 0$ and $\varepsilon_* : H_0(K) \to \mathbf{Z}$ is an isomorphism.

For example, if X is a contractible space, then the chain complex $C(X)$ is acyclic.

Let K and L be positive chain complexes with augmentations. It should be clear what we mean when we say a chain map $f : K \to L$ "preserves the augmentation." For example, if $\varphi : X \to Y$ is a continuous map, the induced chain map $\varphi_\# : C(X) \to C(Y)$ obviously preserves the augmentation. In the rest of this chapter, we will be mainly concerned with chain complexes which have an augmentation, and chain maps which preserve the augmentation.

Let $K' = \{K_q', \partial_q'\}$ and $K'' = \{K_q'', \partial_q''\}$ be positive chain complexes with augmentation $\varepsilon' : K_0' \to \mathbf{Z}$ and $\varepsilon'' : K_0'' \to \mathbf{Z}$ respectively. It is customary to define an augmentation ε on the tensor product $K = K' \otimes K''$ by the simple formula

$$\varepsilon(u \otimes v) = \varepsilon'(u) \cdot \varepsilon''(v)$$

for any $u \in K'_0$ and $v \in K''_0$. With this definition, the following diagram is obviously commutative:

$$
\begin{array}{ccc}
H_0(K') \otimes H_0(K'') & \xrightarrow{\ \alpha\ } & H_0(K' \otimes K'') \\
& \searrow{\scriptstyle e'_* \otimes e''_*} \qquad \swarrow{\scriptstyle \varepsilon_*} & \\
& \mathbf{Z} \otimes \mathbf{Z} \xrightarrow[\approx]{} \mathbf{Z} &
\end{array}
$$

Proposition 5.1. *The tensor product of two free acyclic chain complexes (with augmentations) is again acyclic.*

This follows from the Corollary 4.2 to the Künneth theorem and the commutative diagram above.

DIGRESSION 2. Let us denote by $\mu_n : Q_n(X) \to C_n(X)$ the natural homomorphism of $Q_n(X)$ onto its quotient group. It is obvious that $D_n(X)$ is a direct summand of $Q_n(X)$, hence we can choose (for each space X) a homomorphism $v_n : C_n(X) \to Q_n(X)$ such that $\mu_n v_n = $ identity map of $C_n(X)$. What is surprising is that we can do this in a *natural* way. To be precise:

Lemma 5.2. *There exist homomorphisms $v_n^X : C_n(X) \to Q_n(X)$, defined for each space X and each integer $n \geq 0$ such that $\mu_n v_n^X = $ identity, and for any continuous map $f : X \to Y$, the following diagram is commutative:*

$$
\begin{array}{ccc}
C_n(X) & \xrightarrow{\ v_n^X\ } & Q_n(X) \\
\downarrow{\scriptstyle f_\#} & & \downarrow{\scriptstyle f_\#} \\
C_n(Y) & \xrightarrow{\ v_n^Y\ } & Q_n(Y).
\end{array}
$$

PROOF: In order to save words, in the rest of this section we will call a homomorphism, such as v_n^X or μ_n^X, which is defined for each space X and commutes with the homomorphism $f_\#$ induced by any continuous map f, a *natural* homomorphism. As examples, we have the *face operators*

$$ A_i, B_i : Q_n(X) \to Q_{n-1}(X) \qquad 1 \leq i \leq n, $$

which were (almost) defined in §II.2; they satisfy the identities listed in II.2. Another important example is the family of *degeneracy operators*

$$ E_i : Q_n(X) \to Q_{n+1}(X), \qquad 1 \leq i \leq n+1 $$

defined by

$$ (E_i T)(x_1, \ldots, x_{n+1}) = T(x_1, \ldots, \hat{x}_i, \ldots, x_{n+1}) $$

for any singular n-cube $T : I^n \to X$; the circumflex over x_i means that it is to be omitted. Note that image $E_i \subset D_{n+1}(X)$, and every degenerate singular

$(n + 1)$-cube is of the form $E_i T$ for some i and some n-cube T. It is a routine matter to verify the following list of identities:

$$E_i E_j = E_{j+1} E_i \qquad i \le j,$$
$$A_i E_j = E_{j-1} A_i, \qquad B_i E_j = E_{j-1} B_i, \qquad i < j,$$
$$A_j E_j = B_j E_j = 1$$
$$A_i E_j = E_j A_{i-1}, \qquad B_i E_j = E_j B_{i-1}, \qquad i > j.$$

Now consider the natural homomorphism

$$(1 - E_1 A_1)(1 - E_2 A_2) \cdots (1 - E_n A_n) \colon Q_n(X) \to Q_n(X).$$

We assert that this homomorphism annihilates $D_n(X)$, and hence defines a natural homomorphism

$$v_n \colon C_n(X) \to Q_n(X).$$

To prove this assertion, it helps to first prove the following identities:

$$(1 - E_j A_j) E_j = 0$$
$$(1 - E_j A_j) E_i = E_i (1 - E_{j-1} A_{j-1}) \quad \text{if } i < j.$$

It remains to verify that $\mu_n v_n = $ identity. This follows from the fact that for any $u \in Q_n(X)$,

$$(1 - E_1 A_1)(1 - E_1 A_2) \cdots (1 - E_n A_n)(u)$$

belongs to the same coset modulo $D_n(X)$ as u. 　　　　　　　　　　Q.E.D.

   With these digressions out of the way, we can proceed with our proof that $\zeta \colon C(X) \otimes C(Y) \to C(X \times Y)$ is a chain homotopy equivalence. The proof depends on the following three lemmas:

Lemma 5.3. *For every ordered pair of spaces (X, Y) we can choose a chain map*

$$\xi^{X,Y} \colon C(X \times Y) \to C(X) \otimes C(Y)$$

(which commutes with augmentations) such that the following naturality condition holds: For any continuous maps $f \colon X \to X'$ and $g \colon Y \to Y'$, the following diagram is commutative:

$$
\begin{array}{ccc}
C(X \times Y) & \xrightarrow{\;\xi^{X,Y}\;} & C(X) \otimes C(Y) \\
{\scriptstyle (f \times g)_\#} \downarrow & & \downarrow {\scriptstyle f_\# \otimes g_\#} \\
C(X' \times Y') & \xrightarrow[\;\xi^{X',Y'}\;]{} & C(X') \otimes C(Y').
\end{array}
$$

Lemma 5.4. *Let $\varphi^{X,Y}, \psi^{X,Y} \colon C(X) \otimes C(Y) \to C(X) \otimes C(Y)$ be a natural collection of chain maps. Then there exists a natural collection of chain homotopies*

$$D^{X,Y} \colon C(X) \otimes C(Y) \to C(X) \otimes C(Y)$$

such that

$$\varphi^{X,Y} - \psi^{X,Y} = \partial D^{X,Y} + D^{X,Y}\partial$$

for every ordered pair (X,Y) of spaces.

Lemma 5.5 *Let $\varphi^{X,Y}$, $\psi^{X,Y}:C(X \times Y) \to C(X \times Y)$ be a natural collection of chain maps. Then there exists a natural collection of chain homotopies $D^{X,Y}:C(X \times Y) \to C(X \times Y)$ such that*

$$\varphi^{X,Y} - \psi^{X,Y} = \partial D^{X,Y} + D^{X,Y}\partial.$$

In regard to the statements of these lemmas, the following points should be emphasized:

(a) All chain maps are assumed to preserve the augmentation.

(b) In each case, the adjective "natural" has the following technical meaning: Any pair of continuous maps $f:X \to X'$ and $g:Y \to Y'$ gives rise to a certain square diagram, which is required to be commutative.

It should be clear that these three lemmas imply the truth of the assertion that $\zeta:C(X) \otimes C(Y) \to C(X \times Y)$ is a chain homotopy equivalence. For, let $\xi:C(X \times Y) \to C(X) \otimes C(Y)$ be the chain map whose existence is guaranteed by Lemma 5.3. Then $\xi\zeta$ and the identity are natural chain maps $C(X) \otimes C(Y) \to C(X) \otimes C(Y)$, and hence they are chain homotopic by Lemma 5.4. Similarly, $\zeta\xi$ and the identity $C(X \times Y) \to C(X \times Y)$ are chain homotopic by Lemma 5.5. This result is known as the Eilenberg–Zilber theorem.

PROOF OF LEMMA 5.3. We will use induction on n to define homomorphisms

$$\xi_n^{X,Y}:C_n(X \times Y) \to [C(X) \otimes C(Y)]_n$$

for all spaces X and Y, which will be *natural*, and will define the required chain map (i.e., will commute with the boundary operator).

Case $n = 0$. Define

$$\xi_n^{X,Y}:C_0(X \times Y) \to C_0(X) \otimes C_0(Y)$$

by

$$\xi_0(S,T) = S \otimes T$$

for any singular 0-cubes $S:I^0 \to X$ and $T:I^0 \to Y$ (recall that $Q_0(W) = C_0(W)$ for any space W). Then it is trivial to check that ξ_0 is natural, and that it preserves the augmentation.

Case $n = 1$. Let $\imath:I^1 \to I^1$ denote the identity map. Then $(\imath,\imath):I^1 \to (I^1 \times I^1)$ is a singular 1-cube, i.e., $(\imath,\imath) \in Q_1(I^1 \times I^1)$, and

$$\partial_1(\imath,\imath) \in Q_0(I^1 \times I^1) = C_0(I^1 \times I^1)$$
$$\xi_0^{I^1,I^1}\partial_1(\imath,\imath) \in C_0(I^1) \otimes C_0(I^1)$$

and

$$\varepsilon\xi_0^{I^1,I^1}\partial_1(\imath,\imath) = \varepsilon\partial_1(\imath,\imath) = 0$$

since ξ_0 preserves augmentation. By Proposition 5.1, the chain complex $C(I^1) \otimes C(I^1)$ is acyclic. Hence we can choose an element

$$e^1 \in [C(I^1) \otimes C(I^1)]_1$$

such that

$$\partial_1(e_1) = \xi_0 \partial_1(\iota,\iota).$$

Define a homomorphism

$$\overline{\xi}_1^{X,Y} : Q_1(X \times Y) \to [C(X) \otimes C(Y)]_1$$

for any spaces X and Y by the formula

$$\overline{\xi}_1(S,T) = (S_\# \otimes T_\#)(e_1),$$

where $S: I^1 \to X$ and $T: I^1 \to Y$ are arbitrary singular 1-cubes. We now have to check two things:

(a) Naturality. If $f: X \to X'$ and $g: Y \to Y'$ are continuous maps, the following diagram is commutative:

$$
\begin{array}{ccc}
Q_1(X \times Y) & \xrightarrow{\overline{\xi}_1^{X,Y}} & [C(X) \otimes C(Y)]_1 \\
\downarrow{\scriptstyle (f \times g)_*} & & \downarrow{\scriptstyle f_* \otimes g_*} \\
Q_1(X' \times Y') & \xrightarrow[\overline{\xi}_1^{X',Y'}]{} & [C(X') \otimes C(Y')]_1.
\end{array}
$$

This is an easy calculation:

$$(f_\# \otimes g_\#)\overline{\xi}_1(S,T) = (f_\# \otimes g_\#)(S_\# \otimes T_\#)(e_1)$$
$$= ((fS)_\# \otimes (gT)_\#)(e_1)$$
$$\overline{\xi}_1(f \times g)_\#(S,T) = \overline{\xi}_1(fS,gT)$$
$$= ((fS)_\# \otimes (gT)_\#)(e_1).$$

(b) Commutativity with ∂_1, i.e., the following diagram is commutative:

$$
\begin{array}{ccc}
Q_1(X \times Y) & \xrightarrow{\overline{\xi}_1} & [C(X) \otimes C(Y)]_1 \\
\downarrow{\scriptstyle \partial_1} & & \downarrow{\scriptstyle \partial_1} \\
Q_0(X \times Y) & \xrightarrow{\xi_0} & [C(X) \otimes C(Y)]_0.
\end{array}
$$

Here the computation proceeds as follows:

$$\partial \overline{\xi}_1(S,T) = \partial(S_\# \otimes T_\#)(e_1)$$
$$= (S_\# \otimes T_\#)\partial(e_1)$$
$$= (S_\# \otimes T_\#)\xi_0\partial(\iota,\iota)$$
$$= \xi_0(S \times T)_\#\partial(\iota,\iota)$$
$$= \xi_0\partial(S \times T)_\#(\iota,\iota)$$
$$= \xi_0\partial(S,T).$$

Now define $\xi_1^{X,Y} : C_1(X \times Y) \to [C(X) \otimes C(Y)]_1$ by

$$\xi_1^{X,Y} = \bar{\xi}_1^{X,Y} v_1^{X,Y}.$$

Then ξ_1 is natural, because it is the composition of natural homomorphisms. It remains to check that ξ_1 commutes with ∂_1. Consider the following diagram:

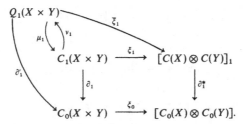

We wish to prove that

$$\partial_1^* \xi_1 = \xi_0 \partial_1.$$

We have

$$\partial_1^* \xi_1 = \partial_1^* \bar{\xi}_1 v_1 = \xi_0 \partial_1' v_1,$$
$$\xi_0 \partial_1 = \xi_0 \partial_1 \mu_1 v_1 = \xi_0 \partial_1' v_1,$$

as desired.

Inductive step. Assume that $n > 1$ and homomorphisms

$$\xi_q^{X,Y} : C_q(X \times Y) \to [C(X) \otimes C(Y)]_q$$

have been defined for all spaces X and Y and all $q < n$ so that naturality holds and the homomorphisms commute with the boundary operator. Define homomorphisms

$$\bar{\xi}_q^{X,Y} : Q_q(X \times Y) \to [C(X) \otimes C(Y)]_q$$

for all X, Y and $q < n$ by the formula

$$\bar{\xi}_q^{X,Y} = \xi_q^{X,Y} \mu_q^{X,Y}.$$

Note that $\bar{\xi}_q$ is a natural homomorphism and that the various $\bar{\xi}_q$'s commute with the boundary operators ∂_q, since they are the composition of homomorphisms having these two properties. Let $\iota : I^n \to I^n$ denote the identity map. Then $(\iota,\iota) : I^n \to I^n \times I^n$ is a singular n-cube, and $(\iota,\iota) \in Q_n(I^n \times I^n)$,

$$\bar{\xi}_{n-1}^{I^n,I^n} \partial_n(\iota,\iota) \in [C(I^n) \otimes C(I^n)]_{n-1}$$
$$\partial_{n-1} \bar{\xi}_{n-1} \partial_n(\iota,\iota) = \bar{\xi}_{n-2} \partial_{n-1} \partial_n(\iota,\iota) = 0.$$

Therefore $\bar{\xi}_{n-1} \partial_n(\iota,\iota)$ is a cycle, and since $C(I^n) \otimes C(I^n)$ is acyclic, we can choose $e_n \in [C(I^n) \otimes C(I^n)]_n$ such that

$$\partial(e_n) = \bar{\xi}_{n-1} \partial_n(\iota,\iota).$$

For any $S: I^n \to X$ and $T: I^n \to Y$, define

$$\bar{\xi}_n(S,T) = (S_\# \otimes T_\#)(e_n).$$

Then $\bar{\xi}_n$ defines a homomorphism $Q_n(X \times Y) \to [C(X) \otimes C(Y)]_n$. Exactly as for the case $n = 1$, we can prove that $\bar{\xi}_n$ is *natural*; Also, the following diagram is commutative:

$$
\begin{array}{ccc}
Q_n(X \times Y) & \xrightarrow{\;\bar{\xi}_n\;} & [C(X) \otimes C(Y)]_n \\
\downarrow{\scriptstyle \partial_n} & & \downarrow{\scriptstyle \partial_n} \\
Q_{n-1}(X \times Y) & \xrightarrow{\;\bar{\xi}_{n-1}\;} & [C(X) \otimes C(Y)]_{n-1}.
\end{array}
$$

In fact, we have

$$
\begin{aligned}
\partial_n \bar{\xi}_n(S,T) &= \partial_n(S_\# \otimes T_\#)(e_n) \\
&= (S_\# \otimes T_\#)\partial_n(e_n) \\
&= (S_\# \otimes T_\#)\bar{\xi}_{n-1}\partial_n(\iota,\iota) \\
\bar{\xi}_{n-1}\partial_n(S,T) &= \bar{\xi}_{n-1}\partial_n(S \times T)_\#(\iota,\iota) \\
&= \bar{\xi}_{n-1}(S \times T)_\#\partial_n(\iota,\iota) \\
&= (S_\# \otimes T_\#)\bar{\xi}_{n-1}\partial_n(\iota,\iota).
\end{aligned}
$$

Next, define

$$\xi_n^{X,Y}: C_n(X \times Y) \to [C(X) \otimes C(Y)]_n$$

by

$$\xi_n^{X,Y} = \bar{\xi}_n^{X,Y} \nu_n^{X,Y}.$$

Then ξ_n is natural, since it is the composition of natural homomorphisms. Also,

$$\partial_n \xi_n = \xi_{n-1}\partial_n;$$

for, we have,

$$
\begin{aligned}
\partial_n \xi_n &= \partial_n \bar{\xi}_n \nu_n = \bar{\xi}_{n-1}\partial_n \nu_n \\
&= \xi_{n-1}\mu_{n-1}\partial_n \nu_n \\
&= \xi_{n-1}\partial_n \mu_n \nu_n = \xi_{n-1}\partial_n
\end{aligned}
$$

as desired.   Q.E.D.

PROOF OF LEMMA 5.4. Once again we will use induction on n to difine homomorphisms

$$D_n^{X,Y}: [C(X) \otimes C(Y)]_n \to [C(X) \otimes C(Y)]_{n+1}$$

for all integers n and all spaces X, Y, such that

$$\varphi_n^{X,Y} - \psi_n^{X,Y} = \partial_{n+1}D_n^{X,Y} + D_{n-1}^{X,Y}\partial_n$$

and such that naturality holds.

   Case $n = 0$. We assert that the condition on φ and ψ imply that $\varphi_0^{X,Y} = \psi_0^{X,Y}$ for any spaces X and Y. The assertion is true for $X = Y = I^0$ (a single point) because $\varepsilon: C_0(I^0) \otimes C_0(I^0) \to \mathbf{Z}$ is an isomorphism, and φ and ψ are assumed to be augmentation preserving. For arbitrary spaces X and Y, let $S: I^0 \to X$ and $T: I^0 \to Y$ be singular 0-cubes. Then

$$S \otimes T = (S_\# \otimes T_\#)(\iota \otimes \iota) \in C_0(X) \otimes C_0(Y),$$

where $\iota: I^0 \to I^0$ is the identity map. Hence it follows by naturality that $\varphi_0^{X,Y} = \psi_0^{X,Y}$.

Since $\varphi_0^{X,Y} = \psi_0^{X,Y}$, we may define $D_0^{X,Y} = 0$ and all conditions will be satisfied.

For the remainder of the proof, it will be convenient to define homomorphisms

$$\bar{\varphi}_n^{X,Y}, \bar{\psi}_n^{X,Y}: [Q(X) \otimes Q(Y)]_n \to [C(X) \otimes C(Y)]_n$$

by the formulas

$$\bar{\varphi}_n^{X,Y} = \varphi_n^{X,Y}(\mu^X \otimes \mu^Y)$$
$$\bar{\psi}_n^{X,Y} = \psi_n^{X,Y}(\mu^X \otimes \mu^Y).$$

Then $\bar{\varphi}$ and $\bar{\psi}$ are natural chain mappings, since they are the composition of natural chain mappings. Also, for any integer $q \geq 0$ we let

$$\iota_q: I^q \to I^q$$

denote the identity map.

Case $n = 1$. Consider

$$\iota_0 \otimes \iota_1 \in Q_0(I^0) \otimes Q_1(I^1)$$

and

$$\iota_1 \otimes \iota_0 \in Q_1(I^1) \otimes Q_0(I^0).$$

We now compute

$$\partial_1(\bar{\varphi}_1 - \bar{\psi}_1)(\iota_0 \otimes \iota_1) = (\bar{\varphi}_0 - \bar{\psi}_0)\partial_1(\iota_0 \otimes \iota_1)$$
$$= 0$$

since $\bar{\varphi}_0 = \varphi_0 = \psi_0 = \bar{\psi}_0$. Similarly, $\partial_1(\bar{\varphi}_1 - \bar{\psi}_1)(\iota_1 \otimes \iota_0) = 0$. Since the chain complexes $C(I^0) \otimes C(I^1)$ and $C(I^1) \otimes C(I^0)$ are acyclic, we can choose elements

$$e_{01} \in [C(I^0) \otimes C(I^1)]_2$$
$$e_{10} \in [C(I^1) \otimes C(I^0)]_2$$

such that

$$\partial_2(e_{01}) = (\bar{\varphi}_1 - \bar{\psi}_1)(\iota_0 \otimes \iota_1),$$
$$\partial_2(e_{10}) = (\bar{\varphi}_1 - \bar{\psi}_1)(\iota_1 \otimes \iota_0).$$

Define $\bar{D}_1: [Q(X) \otimes Q(Y)]_1 \to [C(X) \otimes C(Y)]_2$ by

$$\bar{D}_1(S \otimes T) = \begin{cases} (S_\# \otimes T_\#)(e_{01}) & \text{if } S: I^0 \to X, \ T: I^1 \to Y, \\ (S_\# \otimes T_\#)(e_{10}) & \text{if } S: I^1 \to X, \ T: I^0 \to Y. \end{cases}$$

Then $\bar{D}_1$ is natural in the sense that the following diagram is commutative:

$$
\begin{array}{ccc}
[Q(X) \otimes Q(Y)]_1 & \xrightarrow{\ \bar{D}_1\ } & [C(X) \otimes C(Y)]_2 \\
\downarrow{\scriptstyle f_* \otimes g_*} & & \downarrow{\scriptstyle f_* \otimes g_*} \\
[Q(X') \otimes Q(Y')]_1 & \xrightarrow{\ \bar{D}_1\ } & [C(X') \otimes C(Y')]_2 .
\end{array}
$$

The verification of naturality should present no difficulty to the reader who has already gone through the details of such verifications earlier.

Next, if $S : I^0 \to X$ and $T : I^1 \to Y$ are singular cubes, we compute

$$
\begin{aligned}
\partial_2 \bar{D}_1 (S \otimes T) &= \partial_2 (S_\# \otimes T_\#)(e_{01}) \\
&= (S_\# \otimes T_\#) \partial_2 (e_{01}) \\
&= (S_\# \otimes T_\#)(\bar{\varphi}_1 - \bar{\psi}_1)(\iota_0 \otimes \iota_1) \\
&= (\bar{\varphi}_1 - \bar{\psi}_1)(S_\# \otimes T_\#)(\iota_0 \otimes \iota_1) \\
&= (\bar{\varphi}_1 - \bar{\psi}_1)(S \otimes T).
\end{aligned}
$$

Similarly, we can prove that $\partial_2 \bar{D}_1 (S \otimes T) = (\bar{\varphi}_1 - \bar{\psi}_1)(S \otimes T)$ in case $S : I^1 \to X$ and $T : I^0 \to Y$. Thus we see that

$$
\partial \bar{D}_1 = \bar{\varphi}_1 - \bar{\psi}_1,
$$

in all cases.

Now define

$$
D_1^{X,Y} : [C(X) \otimes C(Y)]_1 \to [C(X) \otimes C(Y)]_2
$$

by

$$
D_1^{X,Y} = \bar{D}_1^{X,Y}(\nu^X \otimes \nu^Y).
$$

Then D_1 is natural because it is the composition of natural homomorphisms, and

$$
\begin{aligned}
\partial_2 D_1^{X,Y} &= \partial_2 \bar{D}_1^{X,Y}(\nu^X \otimes \nu^Y) \\
&= (\bar{\varphi}_1^{X,Y} - \bar{\psi}_1^{X,Y})(\nu^X \otimes \nu^Y) \\
&= (\varphi_1^{X,Y} - \psi_1^{X,Y})(\mu^X \otimes \mu^Y)(\nu^X \otimes \nu^Y) \\
&= (\varphi_1^{X,Y} - \psi_1^{X,Y})[(\mu^X \nu^X) \otimes (\mu^Y \nu^Y)] \\
&= \varphi_1^{X,Y} - \psi_1^{X,Y}
\end{aligned}
$$

as required.

Inductive step. Assume that $n > 1$ and

$$
D_r : [C(X) \otimes C(Y)]_r \to [C(X) \otimes C(Y)]_{r+1}
$$

is defined for all $r < n$, that D_r is natural, and

$$
\varphi_r - \psi_r = \partial_{r+1} D_r + D_{r-1} \partial_r .
$$

Define $\bar{D}_r : [Q(X) \otimes Q(Y)]_r \to [C(X) \otimes C(Y)]_{r+1}$ for all $r < n$ by

$$
\bar{D}_r^{X,Y} = D_r^{X,Y}(\mu^X \otimes \mu^Y).
$$

Then $\bar{D}_r$ is natural, and

$$
\begin{aligned}
\partial_{r+1}\bar{D}_r &= \partial_{r+1}D_r(\mu^X \otimes \mu^Y) \\
&= (\varphi_r - \psi_r - D_{r-1}\partial_r)(\mu^X \otimes \mu^Y) \\
&= (\varphi_r - \psi_r)(\mu^X \otimes \mu^Y) - D_{r-1}(\mu^X \otimes \mu^Y)\partial_r \\
&= \bar{\varphi}_r - \bar{\psi}_r - \bar{D}_r\partial_r.
\end{aligned}
$$

Next, we define $\bar{D}_n$. Let (p,q) range over all pairs of nonnegative integers such that $p + q = n$, and for each such pair consider $\iota_p \otimes \iota_q \in [Q(I^p) \otimes Q(I^q)]_n$ and

$$
[\bar{\varphi}_n - \bar{\psi}_n - \bar{D}_{n-1}\partial_n](\iota_p \otimes \iota_q) \in [C(I^p) \otimes C(I^q)]_n.
$$

We now compute as follows:

$$
\begin{aligned}
\partial_n[\bar{\varphi}_n &- \bar{\psi}_n - \bar{D}_{n-1}\partial_n](\iota_p \otimes \iota_q) \\
&= [\bar{\varphi}_{n-1}\partial_n - \bar{\psi}_{n-1}\partial_n - \partial_n\bar{D}_{n-1}\partial_n](\iota_p \otimes \iota_q) \\
&= [\partial_n\bar{D}_{n-1}\partial_n - \partial_n\bar{D}_{n-1}\partial_n](\iota_p \otimes \iota_q) = 0.
\end{aligned}
$$

Since $C(I^p) \otimes C(I^q)$ is acyclic, there exists $e_{p,q} \in [C(I^p) \otimes C(I^q)]_{n+1}$ such that

$$
\partial(e_{p,q}) = [\bar{\varphi}_n - \bar{\psi}_n - \bar{D}_{n-1}\partial_n](\iota_p \otimes \iota_q).
$$

If $S:I^p \to X$ and $T:I^q \to Y$ are singular cubes, define

$$
\bar{D}_n(S \otimes T) = (S_\# \otimes T_\#)(e_{p,q}).
$$

Then $\bar{D}_n^{X,Y}$ is a homomorphism of $[Q(X) \otimes Q(Y)]_n$ into $[C(X) \otimes C(Y)]_{n+1}$. As before, we can easily prove that $\bar{D}_n$ is a *natural* homomorphism, and

$$
\begin{aligned}
\partial_{n+1}\bar{D}_n(S \otimes T) &= \partial_{n+1}(S_\# \otimes T_\#)(e_{p,q}) \\
&= (S_\# \otimes T_\#)\partial_{n+1}(e_{p,q}) \\
&= (S_\# \otimes T_\#)(\bar{\varphi}_n - \bar{\psi}_n - \bar{D}_{n-1}\partial_n)(\iota_p \otimes \iota_q) \\
&= (\bar{\varphi}_n - \bar{\psi}_n - \bar{D}_{n-1}\partial_n)(S_\# \otimes T_\#)(\iota_p \otimes \iota_q) \\
&= (\bar{\varphi}_n - \bar{\psi}_n - \bar{D}_{n-1}\partial_n)(S \otimes T).
\end{aligned}
$$

We now define

$$
D_n^{X,Y}:[C(X) \otimes C(Y)]_n \to [C(X) \otimes C(Y)]_{n+1}
$$

by

$$
D_n^{X,Y} = \bar{D}_n^{X,Y}(v^X \otimes v^Y).
$$

Then D_n is natural, and

$$
\begin{aligned}
\partial_{n+1}D_n^{X,Y} &= \partial_{n+1}\bar{D}_n^{X,Y}(v^X \otimes v^Y) \\
&= (\bar{\varphi}_n - \bar{\psi}_n - \bar{D}_{n-1}\partial_n)(v^X \otimes v^Y) \\
&= (\varphi_n - \psi_n - D_{n-1}\partial_n)(\mu^X \otimes \mu^Y)(v^X \otimes v^Y) \\
&= (\varphi_n^{X,Y} - \psi_n^{X,Y} - D_{n-1}^{X,Y}\partial_n)
\end{aligned}
$$

as required.

The reader should have no trouble by now proving Lemma 5.5 for himself, hence we will not go through the details.

The reader should reflect on the essentials of these proofs. They were concerned with certain chain complexes defined on ordered pairs (X,Y) of topological spaces, namely, $Q(X \times Y)$, $C(X \times Y)$, $Q(X) \otimes Q(Y)$ and $C(X) \otimes C(Y)$. There were certain *models* at hand: for $Q(X \times Y)$ they were the pairs (I^n,I^n) and the singular cube $(\iota,\iota):I^n \to I^n \times I^n$, $(\iota,\iota) \in Q_n(I^n \times I^n)$, while for $Q(X) \otimes Q(Y)$ they were the pairs (I^p,I^q) and the elements $\iota_p \otimes \iota_q \in Q_p(I^p) \otimes Q_q(I^q)$. The important thing about the models is that $Q_n(X \times Y)$ has a basis composed of the elements $(S,T) = (S \times T)_\#(\iota,\iota)$, while $Q_p(X) \otimes Q_q(Y)$ has a basis composed of the elements $S \otimes T = (S_\# \otimes T_\#)(\iota_p \otimes \iota_q)$. Finally these models are *acyclic*, in the sense that the chain complexes $C(I^n \times I^n)$ and $C(I^p) \otimes C(I^q)$ are acyclic. The whole procedure is explained is complete generality in the original paper of Eilenberg and Mac Lane [3]. However, such a general treatment is so abstract that it is difficult to follow; moreover, the reader who has used the method in a few specific cases should have no difficulty in applying it to new cases.

This method of acyclic models is applicable to many problems involving singular homology groups. For example, singular homology groups can be defined using singular simplexes rather than singular cubes. Then one can use the method of acyclic models to define a natural chain homotopy equivalence between cubical singular chains and simplicial singular chains. This is explained in detail in the paper of Eilenberg and Mac Lane [3] mentioned above.

EXERCISES

5.1. Prove that any two natural chain maps

$$\varphi^{X,Y}, \psi^{X,Y} : C(X) \otimes C(Y) \to C(X \times Y)$$

are chain homotopic (by a natural chain homotopy). [Note: This applies, in particular, to the natural chain map ζ defined in §3.]

5.2. Prove that there exist natural chain maps

$$\Delta^X : C(X) \to C(X) \otimes C(X)$$

and that any two such natural chain maps are chain homotopic (by a natural chain homotopy). Such a natural chain map is sometimes called a *diagonal map*.

5.3. Prove that there is a 1-1 correspondence between diagonal maps

$$\Delta^X : C(X) \to C(X) \otimes C(X)$$

as defined in the preceding exercise and natural chain maps

$$\xi^{X,Y} : C(X \times Y) \to C(X) \otimes C(Y)$$

as described in Lemma 5.3. This 1-1 correspondence is defined as follows:

(a) Given any such diagonal map $\varDelta$, define a natural chain map

$$\bar{\varDelta}^{X,Y}:C(X \times Y) \to C(X) \otimes C(Y)$$

by the formula

$$\bar{\varDelta}^{X,Y} = (\pi_{1\#} \otimes \pi_{2\#})\varDelta^{X \times Y},$$

where $\pi_1:X \times Y \to X$ and $\pi_2:X \times Y \to Y$ denote projections on the first and second factors respectively.

(b) Given a natural chain map $\xi:C(X \times Y) \to C(X) \otimes C(Y)$, define a diagonal map

$$\bar{\xi}^X:C(X) \to C(X) \otimes C(X)$$

by the formula

$$\bar{\xi}^X = \xi^{X,X}d_\#,$$

where $d:X \to X \times X$ denotes the "diagonal" map defined by $d(x) = (x,x)$ for any $x \in X$. [Note: When we study cohomology theory, we will see the real importance of such diagonal maps.]

5.4. Let $\varDelta^X:C(X) \to C(X) \otimes C(X)$ be a diagonal map, as defined in Exercise 5.2, and let $\tau^X:C(X) \otimes C(X) \to C(X) \otimes C(X)$ be the natural chain map defined by

$$\tau(u \otimes v) = (-1)^{pq}v \otimes u,$$

where $u \in C_p(X)$, and $v \in C_q(X)$. By Exercise 5.2, there exists a natural chain homotopy $D^X:C(X) \to C(X) \otimes C(X)$ between $\varDelta$ and $\tau\varDelta$, i.e.,

$$\varDelta - \tau\varDelta = \partial D + D\partial.$$

Prove by the method of acyclic models that there exist natural homomorphisms

$$D'_n:C_n(X) \to [C(X) \otimes C(X)]_{n+2}$$

such that

$$D + \tau D = \partial D' - D'\partial.$$

(Note that $\tau^2 = $ identity. One can think of D' as a "second-order chain homotopy" between the "first-order" chain homotopies D and $-\tau D'$. One could then consider third-order chain homotopies between D' and $\tau D'$, etc. This procedure leads to one method of constructing the Steenrod squaring operations in cohomology theory; see E. Spanier [7], pp. 271–276.)

§6. Formulas for the Homology Groups of Product Spaces

Our objective is to combine the Künneth theorem for chain complexes (Theorem 4.1) with the existence of the natural chain homotopy equivalences (Eilenberg–Zilber theorem)

$$C(X) \otimes C(Y) \underset{\xi}{\overset{\zeta}{\rightleftarrows}} C(X \times Y)$$

to express the homology groups of $X \times Y$ in terms of those of X and Y.

By this method, we obviously obtain a split exact sequence:

$$0 \to \sum_{p+q=n} H_p(X) \otimes H_q(Y) \xrightarrow{\alpha} H_n(X \times Y) \xrightarrow{\beta} \sum_{p+q=n-1} \mathrm{Tor}(H_p(X),H_q(Y)) \to 0.$$

The homomorphisms α and β are natural, but the splitting is not natural.

One way to generalize this theorem is the following: Let G be an abelian group. Then we have the following natural chain homotopy equivalences:

$$C(X) \otimes C(Y) \otimes G \xrightleftharpoons[\xi \otimes 1]{\zeta \otimes 1} C(X \times Y) \otimes G.$$

In the Künneth theorem we only needed to assume that one of the chain complexes K and L was free. Hence we obtain the following split exact sequence:

$$0 \to \sum_{p+q=n} H_p(X) \otimes H_q(Y;G) \xrightarrow{\alpha} H_n(X \times Y;G)$$

$$\xrightarrow{\beta} \sum_{p+q=n-1} \mathrm{Tor}(H_p(X),H_q(Y;G)) \to 0.$$

Once again, the homomorphisms α and β are natural, but the splitting is not natural.

We will now generalize these theorems to include relative homology groups. If (X,A) is a pair, then

$$C(X,A) = C(X)/C(A)$$

by definition; also, the sequence

$$0 \to C(A) \to C(X) \to C(X,A) \to 0$$

is split exact. Using these facts, plus basic properties of tensor products, it is easy to see that there is a natural isomorphism of chain complexes

$$\frac{C(X)}{C(A)} \otimes \frac{C(Y)}{C(B)} \approx \frac{C(X) \otimes C(Y)}{C(X) \otimes C(B) + C(A) \otimes C(Y)}$$

for any pairs (X,A) and (Y,B). In the denominator on the right-hand side of this equation, the plus sign does *not* mean direct sum; it refers to the least subgroup containing the two terms.

Due to the naturality of the chain maps ζ and ξ with respect to the chain maps $i_\# : C(A) \to C(X)$ and $j_\# : C(B) \to C(Y)$ induced by inclusion maps i and j, we conclude that we have chain homotopy equivalences

$$\frac{C(X) \otimes C(Y)}{C(X) \otimes C(B) + C(A) \otimes C(Y)} \xrightleftharpoons[\xi]{\zeta} \frac{C(X \times Y)}{C(X \times B) + C(A \times Y)}.$$

The inclusion maps $X \times B \to X \times B \cup A \times Y$ and $A \times Y \to X \times B \cup A \times Y$ induce an obvious chain map

$$C(X \times B) + C(A \times Y) \to C(X \times B \cup A \times Y).$$

Under certain circumstances, this chain map will be a chain homotopy equivalence; for example, this will be the case if either A or B is empty (trivially). More generally, it will be the case if the interiors of $X \times B$ and $A \times Y$ cover $X \times B \cup A \times Y$ (cf. Theorem II.6.3) in the relative topology of $(X \times B) \cup (A \times Y)$.

Definition 6.1. Let X_1 and X_2 be subspaces of some topological space X. We say $\{X_1, X_2\}$ is an *excisive couple* if the obvious chain map $C(X_1) + C(X_2) \to C(X_1 \cup X_2)$ (induced by inclusion) induces isomorphisms on homology groups.

The term "excisive" is used here because of its obvious connection with the excision property.

We note the following two sufficient conditions for $\{X_1, X_2\}$ to be an excisive couple:

(a) If $X_1 \cup X_2 = (\text{Interior } X_1) \cup (\text{Interior } X_2)$ in the relative topology of $X_1 \cup X_2$, then $\{X_1, X_2\}$ is an excisive couple. This is a consequence of Theorem II.6.3.

(b) If X is a CW-complex and X_1 and X_2 are subcomplexes, then $\{X_1, X_2\}$ is an excisive couple. This is a consequence of the theorems of §IV.4.

EXERCISES

6.1. Prove that the following conditions are equivalent to $\{X_1, X_2\}$ being an excisive couple in X:

(a) $H_q(C(X_1 \cup X_2)/(C(X_1) + C(X_2))) = 0$ for all q.
(b) The obvious chain map $C(X)/(C(X_1) + C(X_2)) \to C(X)/C(X_1 \cup X_2)$ induces isomorphisms on homology groups.
(c) The inclusion map $(X_1, X_1 \cap X_2) \to (X_1 \cup X_2, X_2)$ induces isomorphisms on homology groups.

6.2. If $\{X_1, X_2\}$ is an excisive couple, prove that the chain map $C(X_1; G) + C(X_2; G) \to C(X_1 \cup X_2; G)$ induces an isomorphism on homology groups for any coefficient group G. Then deduce that the analogues of Conditions (a), (b), and (c) of Exercise 6.1 hold for homology with coefficient group G.

In view of the above discussion, we see that if $\{A \times Y, X \times B\}$ is an excisive couple in $X \times Y$, then the composition of the Eilenberg–Zilber chain homotopy equivalence

$$C(X,A) \otimes C(Y,B) \xrightarrow{\;\simeq\;} \frac{C(X \times Y)}{C(X \times B) + C(A \times Y)}$$

and the chain map

$$\frac{C(X \times Y)}{C(X \times B) + C(A \times Y)} \to \frac{C(X \times Y)}{C(X \times B \cup A \times Y)}$$

induces an isomorphism

$$H_q(C(X,A) \otimes C(Y,B)) \approx H_q(X \times Y, X \times B \cup A \times Y)$$

for all q. Hence we have the following:

Theorem 6.1. *Let (X,A) and (Y,B) be pairs such that $\{A \times Y, X \times B\}$ is an excisive couple in $X \times Y$. Then there exists a split exact sequence*

$$0 \to \sum_{p+q=n} H_p(X,A) \otimes H_q(Y,B,;G) \overset{\alpha}{\to} H_n(X \times Y, A \times Y \cup X \times B;G)$$

$$\overset{\beta}{\to} \sum_{p+q=n-1} \text{Tor}(H_p(X,A),H_q(Y,B;G))$$

$$\to 0$$

The homomorphisms α and β are natural, but the splitting is not.

EXAMPLE 6.1. Let $K = \{K^n\}$ and $L = \{L^n\}$ be finite CW-complexes on the spaces X and Y respectively. Then

$$\alpha: H_p(K^p,K^{p-1}) \otimes H_q(L^q,L^{q-1}) \to H_{p+q}(K^p \times L^q, K^p \times L^{q-1} \cup K^{p-1} \times L^q)$$

is an isomorphism by the above theorem. Let

$$M^n = \bigcup_{p+q=n} K^p \times L^q, \qquad n = 0, 1, 2, \ldots$$

denote the CW-complex on $X \times Y$. Then composing the isomorphism α with the homomorphism induced by the inclusion map

$$(K^p \times L^q, K^{p-1} \times L^q \cup K^p \times L^{q-1}) \to (M^n,M^{n-1})$$

gives rise to a natural homomorphism

$$C_p(K) \otimes C_q(L) \to C_n(M).$$

It may be shown that this agrees with the identification

$$C_n(M) = \sum_{p+q=n} C_p(K) \otimes C_q(L)$$

we made in §2 for the case where K and L are *regular* CW-complexes.

EXERCISES

6.3. Let K and L be pseudomanifolds of dimensions m and n respectively. (a) Prove that $K \times L$ is a pseudomanifold of dimension $m + n$. (b) Prove that $K \times L$ is orientable if and only if both K and L are orientable.

6.4. Let P^2 denote the real projective plane. Compute the integral and mod 2 homology groups of $P^2 \times P^2$.

6.5. Let R be a ring, and let $K = \{K_n,\partial_n\}$ and $L = \{L_n,d_n\}$ be chain complexes such that each K_n is a right R-module, each ∂_n is a homomorphism of right R-modules, each

L_n is a left R-module, and each d_n is a homomorphism of left R-modules (we can express these conditions more briefly by saying that K *is a chain complex of right R-modules* and L *is a chain complex of left R-modules.*) The definition of $K \otimes_R L$ should be obvious; it is a chain complex of abelian groups. Define a natural homomorphism $\alpha: H_p(K) \otimes_R H_q(L) \to H_{p+q}(K \otimes_R L)$ by analogy with our earlier definition.

6.6. Let F be a commutative field, and let K and L be chain complexes of vector spaces over F. Prove that

$$\alpha: \sum_{p+q=n} H_p(K) \otimes_F H_q(L) \to H_n(K \otimes_F L)$$

is an isomorphism.

6.7. Let (X,A) and (Y,B) be pairs such that $\{A \times Y, X \times B\}$ is an excisive couple in $X \times Y$, and let F be a commutative field. Prove that there exists a natural isomorphism

$$\alpha: \sum_{p+q=n} H_p(X,A;F) \otimes_F H_q(Y,B;F) \to H_n(X \times Y, A \times Y \cup X \times B;F).$$

6.8. Let F be a commutative field and (X,A) a pair such that for all q, $H_q(X,A;F)$ has finite rank r_q over F. Define the *Poincare series* of (X,A) (over F) to be the formal power series

$$P(X,A;t) = \sum_{q \geq 0} r_q t^q.$$

Give a formula for $P(X \times Y; X \times B \cup A \times Y;t)$ in terms of $P(X,A;t)$ and $P(Y,B;t)$, assuming that $\{A \times Y, X \times B\}$ is an excisive couple.

Bibliography for Chapter VI

[1] H. Cartan and S. Eilenberg, *Homological Algebra*, Princeton University Press, Princeton, 1956, p. 113.
[2] A. Dold, *Lectures on Algebraic Topology*, Springer-Verlag, New York, 1972, p. 164.
[3] S. Eilenberg and S. Mac Lane, Acyclic nodels, *Amer. J. Math.*, **75** (1953), 189–199.
[4] P. Hilton and U. Stammbach, *A Course in Homological Algebra*, Springer-Verlag, New York, 1971, pp. 172 ff.
[5] A. T. Lundell and S. Weingram, *The Topology of CW-complexes*, Van Nostrand Reinhold Co., Princeton, 1969, 56–59.
[6] S. Mac Lane, *Homology*, Springer-Verlag, New York, 1963, Chapter V, §10.
[7] E. Spanier, *Algebraic Topology*, McGraw-Hill, New York, 1966.
[8] N. E. Steenrod, A convenient category of topological spaces, *Mich. Math J.*, **14** (1967), 133–152.
[9] J. Vick, *Homology Theory*, Academic Press, New York, 1973, Chapter 4.

CHAPTER VII

Cohomology Theory

§1. Introduction

Recall that one obtains homology groups with coefficient group G by the following process:

(a) Start with the chain complex $C(X,A) = \{C_q(X,A), \partial_q\}$.

(b) Apply the functor $\otimes G$ to obtain the new chain complex

$$C(X,A) \otimes G = C(X,A;G).$$

(c) Take the homology groups of the resulting chain complex:

$$H_q(X,A;G) = H_q(C(X,A;G)).$$

We could go through the same procedure, only at Step (b), apply the functor $\text{Hom}(\ ,G)$ instead of $\otimes G$, and obtain what are called the *cohomology groups of (X,A) with coefficient group G*. Much of the resulting theory parallels that of Chapter V. However, the geometric interpretation of cycles (or cocycles), etc. is somewhat different, and perhaps a bit more obscure. More importantly, it is possible to introduce additional operations into cohomology theory, most notably, what are called *cup products* and *Steenrod squares*. These new operations are additional invariants of homotopy type, and enable us to distinguish between spaces that we could not tell apart otherwise. Cup products are explained in the next chapter.

§2. Definition of Cohomology Groups—Proofs of the Basic Properties

For any pair (X,A) and any abelian group G, define

$$C^q(X,A;G) = \mathrm{Hom}(C_q(X,A),G),$$

and

$$\delta_q : C^q(X,A;G) \to C^{q+1}(X,A;G)$$

by $\delta_q = \mathrm{Hom}(\partial_{q+1}, 1_G)$. Then

$$C^*(X,A;G) = \{C^q(X,A;G), \delta_q\}$$

is a *cochain complex*, in accordance with the following definition:

Definition 2.1. A *cochain complex* K consists of a sequence of abelian groups $\{K^q\}$ and homomorphisms $\delta_q : K^q \to K^{q+1}$ defined for all q and subject to the condition that $\delta_{q+1} \delta_q = 0$ for all q. The homomorphism δ_q is called a *coboundary operator*.

An important example of a cochain complex is the following: Let $C = \{C_q, \partial_q\}$ be a chain complex; define $K^q = \mathrm{Hom}(C_q, G)$ and $\delta_q : K^q \to K^{q+1}$ by $\delta_q = \mathrm{Hom}(\partial_{q+1}, 1)$, where 1 denotes the identity homomorphism $G \to G$. Then $K = \{K^q, \delta_q\}$ is a cochain complex; we will denote this cochain complex by

$$K = \mathrm{Hom}(C,G).$$

On the other hand, if K is a cochain complex, then an analogous definition leads to a chain complex $\mathrm{Hom}(K,G)$.

Obviously, the theory of chain complexes and the theory of cochain complexes are isomorphic; to get from one to the other, change the sign of all the indices. The distinction between the two is made partly for tradition, and partly for convenience in the applications we have in mind. Corresponding to the notions of *chain map* and *chain homotopy* we have *cochain maps* and *cochain homotopies*: Let K and L be cochain complexes. A *cochain map* $f : K \to L$ is a sequence of isomorphisms

$$f^q : K^q \to L^q$$

which commute with the coboundary operators. If $f, g : K \to L$ are cochain maps, then a *cochain homotopy* D between f and g is a sequence of homomorphisms $D^q : K^q \to L^{q-1}$ such that

$$f^q - g^q = \delta^{q-1} D^q + D^{q+1} \delta^q.$$

We leave it to the reader to define the following two concepts:

(a) Suppose C and C' are chain complexes and $f:C \to C'$ is a chain map. Then a cochain map

$$\text{Hom}(f,1):\text{Hom}(C',G) \to \text{Hom}(C,G)$$

is defined.

(b) Assume C and C' are chain complexes, $f,g:C \to C'$ are chain maps, and $D:C \to C'$ is a chain homotopy between f and g. Then a cochain homotopy $\text{Hom}(D,1)$ is defined between the cochain maps $\text{Hom}(f,1)$ and $\text{Hom}(g,1)$.

If $K = \{K^q\}$ is a cochain complex with coboundary operator $\delta^q:K^q \to K^{q+1}$, then the following notation and terminology is standard:

$Z^q(K) = $ kernel δ^q, the *q-dimensional cocycles*,
$B^q(K) = $ image δ^{q-1}, the *q-dimensional coboundaries*,
$H^q(K) = Z^q(K)/B^q(K)$, the *q-dimensional cohomology group*.

Thus for any pair (X,A) and abelian group G, we have the cochain complex

$$C^*(X,A;G) = \text{Hom}(C(X,A),G)$$

and the associated cohomology groups

$$H^q(X,A;G) = H^q(C^*(X,A;G)).$$

Let $f:(X,A) \to (Y,B)$ be a continuous map of pairs; then we have the induced chain map,

$$f_{\#}:C(X,A) \to C(Y,B),$$

which gives rise to a cochain map

$$f^{\#} = \text{Hom}(f_{\#},1):C^*(Y,B;G) \to C^*(X,A;G)$$

and hence to an induced homomorphism on cohomology groups

$$f^*:H^q(Y,B;G) \to H^q(X,A;G)$$

for all q. Note that the induced homomorphism in cohomology goes the opposite way from that in homology; we are dealing with a *contravariant* functor.

If two maps $f_0,f_1:(X,A) \to (Y,B)$ are homotopic, then any homotopy $f:(X \times I, A \times I) \to (Y,B)$ gives rise to a chain homotopy $D:C(X,A) \to C(Y,B)$ between the chain maps $f_{0\#}$ and $f_{1\#}$. Hence $\text{Hom}(D,1)$ is a cochain homotopy between $f_0^{\#} = \text{Hom}(f_{0\#},1)$ and $f_1^{\#} = \text{Hom}(f_{1\#},1)$; it follows that the induced homomorphisms

$$f_0^*, f_1^*:H^q(Y,B;G) \to H^q(X,A;G)$$

are the same.

Next, we will discuss exact sequences. Let

$$E:0 \to C' \xrightarrow{i} C \xrightarrow{j} C'' \to 0$$

be a short exact sequence of chain complexes and chain maps. If we apply the functor $\text{Hom}(\ ,G)$, we do not obtain a short exact sequence of cochain complexes, in general. All we can be certain of is that the following sequence is exact:

$$\text{Hom}(C',G) \xleftarrow{\text{Hom}(i,1)} \text{Hom}(C,G) \xleftarrow{\text{Hom}(j,1)} \text{Hom}(C'',G) \leftarrow 0.$$

In general, $\text{Hom}(i,1)$ will *not* be an epimorphism. However, if the sequence E is *split exact*, then the sequence

$$0 \leftarrow \text{Hom}(C',G) \leftarrow \text{Hom}(C,G) \leftarrow \text{Hom}(C'',G) \leftarrow 0$$

will also be split exact, and we will get a corresponding long exact sequence of cohomology groups.

We can apply these considerations to the short exact sequence of chain complexes

$$0 \to C(A) \xrightarrow{i_\#} C(X) \xrightarrow{j_\#} C(X,A) \to 0$$

for any pair (X,A). This is a split exact sequence of chain complexes, hence we obtain a corresponding split exact sequence of cochain complexes

$$0 \leftarrow C^*(A;G) \xleftarrow{i^*} C^*(X;G) \xleftarrow{j^*} C^*(X,A;G) \leftarrow 0$$

for any abelian group G. It follows that there is a long exact sequence of cohomology groups:

$$\cdots \xleftarrow{\delta^*} H^q(A;G) \xleftarrow{i^*} H^q(X;G) \xleftarrow{j^*} H^q(X,A;G) \xleftarrow{\delta^*} H^{q-1}(A;G)$$

with all the usual properties.

For some purposes it is convenient to define *reduced* cohomology groups $\tilde{H}^0(X;G)$ in dimension 0. For this purpose, one uses the augmented chain complex $\tilde{C}(X)$ that is defined in §V.3. We define the *augmented cochain complex*

$$\tilde{C}^*(X;G) = \text{Hom}(\tilde{C}(X),G)$$

and the *reduced* cohomology groups

$$\tilde{H}^q(X;G) = H^q(\tilde{C}^*(X;G)).$$

One readily proves that for any nonempty space X and abelian group G,

$$\tilde{H}^q(X;G) = H^q(X;G) \quad \text{for} \quad q \neq 0$$

while for $q = 0$ we have a split short exact sequence,

$$0 \to G \xrightarrow{\varepsilon_*} H^0(X;G) \to \tilde{H}^0(X;G) \to 0.$$

We leave it for the reader to check that if P is a space consisting of a single point, then

$$\tilde{H}^q(P;G) = 0 \quad \text{for all } q.$$

From this it follows that

$$\varepsilon^*: G \to H^0(P;G)$$

is an isomorphism.

We will discuss the excision property and the Mayer–Vietoris sequence in cohomology later in this chapter.

§3. Coefficient Homomorphisms and the Bockstein Operator in Cohomology

Let $h: G_1 \to G_2$ be a homomorphism of abelian groups. Then for any chain complex C, we get an obvious cochain map

$$\text{Hom}(1,h): \text{Hom}(C,G_1) \to \text{Hom}(C,G_2)$$

and an induced map on cohomology groups. In particular, for any pair (X,A), we have the cochain map

$$\text{Hom}(1,h): C^*(X,A;G_1) \to C^*(X,A;G_2)$$

and the induced homomorphism

$$h_\#: H^q(X,A;G_1) \to H^q(X,A;G_2)$$

on cohomology groups. The reader should state and prove naturality properties of the coefficient homomorphism $h_\#$ analogous to Properties (a) and (b) of §V.5. In addition, he should prove that if G is a left module over some ring R, then $H^q(X,A;G)$ inherits a natural left R-module structure; in that case, the homomorphisms f^* and δ_* are homomorphisms of left R-modules.

Next, let

$$0 \to G' \overset{h}{\to} G \overset{k}{\to} G'' \to 0$$

be a short exact sequence of abelian groups. From this, we get the following sequence of cochain complexes:

$$0 \to C^*(X,A;G') \xrightarrow{\text{Hom}(1,h)} C^*(X,A;G) \xrightarrow{\text{Hom}(1,k)} C^*(X,A;G'') \to 0.$$

Since $C(X,A)$ is a chain complex of free abelian groups, it follows easily that this sequence of cochain complexes is exact. By the usual procedure, we get the following long exact sequence of cohomology groups:

$$\cdots \overset{\beta}{\to} H^q(X,A;G') \xrightarrow{h_\#} H^q(X,A;G) \xrightarrow{k_\#} H^q(X,A;G'') \overset{\beta}{\to} H^{q+1}(X,A;G').$$

Here β is the Bockstein operator in cohomology. It has naturality properties similar to that of the Bockstein operator in homology.

§4. The Universal Coefficient Theorem for Cohomology Groups

The object of this theorem is to express $H^q(X,A;G)$ in terms of integral homology groups of (X,A); it is analogous to Corollary V.6.3.

Let $K = \{K_n, \partial_n\}$ be an arbitrary chain complex, G an abelian group, $x \in H_n(K)$, and $u \in H^n(\text{Hom}(K,G))$. The *inner product* $\langle u,x \rangle$ of u and x is the element of G obtained according to the following simple prescription: Choose a representative cocycle $u' \in \text{Hom}(K_n, G)$ for u, and a representative cycle $x' \in K_n$ for x. Then

$$\langle u,x \rangle = u'(x') \in G.$$

It is easy to verify that this definition is independent of the choice of the representatives u' and x', and that the inner product is additive in each variable separately, i.e.,

$$\langle u_1 + u_2, x \rangle = \langle u_1, x \rangle + \langle u_2, x \rangle,$$
$$\langle u, x_1 + x_2 \rangle = \langle u, x_1 \rangle + \langle u, x_2 \rangle.$$

This inner product is one of the basic ideas of cohomology theory.

Using this inner product, we define a homomorphism

$$\alpha : H^n(\text{Hom}(K,G)) \to \text{Hom}(H_n(K), G)$$

by the following rule: for any $u \in H^n(\text{Hom}(K,G))$ and $x \in H_n(K)$,

$$(\alpha u)(x) = \langle u,x \rangle.$$

The homomorphism α has the following three naturality properties (cf. §V.6):

(a) If $f : K \to K'$ is a chain map, then the following diagram is commutative:

$$
\begin{array}{ccc}
H^q(\text{Hom}(K,G)) & \xrightarrow{\ \alpha\ } & \text{Hom}(H_q(K),G) \\
\Big\uparrow{\scriptstyle \text{Hom}(f,1)^*} & & \Big\uparrow{\scriptstyle \text{Hom}(f_*,1)} \\
H^q(\text{Hom}(K',G)) & \xrightarrow{\ \alpha'\ } & \text{Hom}(H_q(K'),G).
\end{array}
$$

(b) Let $E : 0 \to K' \to K \to K'' \to 0$ be a split exact sequence of chain complexes. Then the following sequence of cochain complexes is also exact,

$$0 \leftarrow \text{Hom}(K',G) \leftarrow \text{Hom}(K,G) \leftarrow \text{Hom}(K'',G) \leftarrow 0,$$

and the following diagram is commutative:

$$
\begin{array}{ccc}
H^q(\text{Hom}(K',G)) & \xrightarrow{\ \alpha'\ } & \text{Hom}(H_q(K'),G) \\
\Big\downarrow{\scriptstyle \delta^*} & & \Big\downarrow{\scriptstyle \text{Hom}(\partial_E,1)} \\
H^{q+1}(\text{Hom}(K'',G)) & \xrightarrow{\ \alpha''\ } & \text{Hom}(H_{q+1}(K''),G).
\end{array}
$$

(c) If $h:G_1 \to G_2$ is a homomorphism of coefficient groups, then the following diagram is commutative:

$$
\begin{array}{ccc}
H^q(\mathrm{Hom}(K,G_1)) & \xrightarrow{\ \alpha\ } & \mathrm{Hom}(H_q(K),G_1) \\
\Big\downarrow{h_{\#}} & & \Big\downarrow{\mathrm{Hom}(1,h)} \\
H^q(\mathrm{Hom}(K,G_2)) & \xrightarrow{\ \alpha\ } & \mathrm{Hom}(H_q(K'),G_2).
\end{array}
$$

Of course, we will mainly be interested in the homomorphism in case $K = C(X,A)$:

$$\alpha : H^q(X,A;G) \to \mathrm{Hom}(H_q(X,A);G).$$

We leave it to the reader to reformulate the naturality Properties (a), (b), and (c) above in an appropriate way for the cohomology of spaces.

In order to further investigate the properties of the homomorphism α, it is best to use homological algebra; in particular, it is necessary to make use of the functor $\mathrm{Ext}(A,B)$. To be concise, $\mathrm{Ext}(A,B)$ bears the same relation to $\mathrm{Hom}(A,B)$ that $\mathrm{Tor}(A,B)$ does to $A \otimes B$ (these are both examples of *first derived* functors). Although $\mathrm{Tor}(A,B)$ is symmetric in the two variables, there can be no question of $\mathrm{Ext}(A,B)$ being symmetrical, since it is contravariant in the first variable and covariant in the second variable.

In order to make use of the functor Ext, it is convenient to have available certain basic properties of divisible abelian groups.

Definition 4.1. An abelian group A is *divisible* if given any $a \in A$ and any nonzero integer n, there exists an element $x \in A$ such that $nx = a$.

EXAMPLE 4.1. The additive group of rational numbers is divisible. It is easily proved that any quotient group of a divisible group is divisible, and any direct sum of divisible groups is divisible. Thus we could construct many more examples.

In a certain sense, divisible groups have properties which are *dual* to those of free abelian groups. For example, any subgroup of a free abelian group is also free abelian, while any quotient group of a divisible group is divisible. Any free group F is *projective* (in the category of abelian groups), in the sense that given any epimorphism $h:A \to B$ and any homomorphism $g:F \to B$, there exists a homomorphism $f:F \to A$ such that the following diagram is commutative:

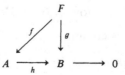

(the proof is easy).

Dually, an abelian group G is called *injective* if given any monomorphism $h:B \to A$ and any homomorphism $g:B \to G$, there exists a homomorphism

$f : A \rightarrow G$ such that the following diagram is commutative:

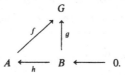

Note that this diagram is obtained from the previous one by reversing all the arrows.

Theorem 4.1. *An abelian group is injective if and only if it is divisible.*

The proof that an injective group is divisible is easy, and is left to the reader.

Assume that G is divisible; we will prove that it is injective. Let A, B, h, and g be as in the diagram above. We may as well assume that B is a subgroup of A, and h is the inclusion map. Consider all pairs (G_i, h_i) where G_i is a subgroup of A which contains B, and $h_i : G_i \rightarrow G$ is a homomorphism such that $h_i | B = g$. This family of pairs is nonvacuous, because (B, g) obviously satisfies the required conditions. Define $(G_i, h_i) < (G_j, h_j)$ if $G_i \subset G_j$ and $h_j | G_i = h_i$. Apply Zorn's lemma to this family with this ordering to conclude there exists a maximal pair (G_m, h_m). We assert $G_m = A$; for if $G_m \neq A$, let $a \in A - G_m$; using the fact that G is divisible, it is easily shown that h_m can be extended to the subgroup generated by G_m and a. But this contradicts maximality of G_m.

It is well known that every abelian group is isomorphic to a quotient of a free abelian group. The following is the dual property:

Proposition 4.2. *Any group is isomorphic to a subgroup of a divisible group.*

PROOF. There are various ways to prove this. One way is to express the given group G as the quotient group of a free group F:

$$G \approx F/R.$$

Obviously F can be considered as a subgroup of a divisible group D; for if $\{b_i\}$ is a basis for F, then we may take D as a rational vector space on the same basis. Then G is isomorphic to a subgroup of the divisible group D/R.

Q.E.D.

We will now list the basic properties of $\text{Ext}(A, B)$. For any abelian groups A and B, $\text{Ext}(A, B)$ is also an abelian group. If $f : A' \rightarrow A$ and $g : B \rightarrow B'$ are homomorphisms, then

$$\text{Ext}(f, g) : \text{Ext}(A, B) \rightarrow \text{Ext}(A', B')$$

is a homomorphism with the usual functorial properties.

There are two ways to define or construct $\text{Ext}(A,B)$:

(a) By means of a *free or projective resolution* of A. Choose a short exact sequence $0 \to F_1 \xrightarrow{d} F_0 \xrightarrow{\varepsilon} A \to 0$ with F_0 (and hence F_1) free abelian. Then the following sequence is exact:

$$0 \leftarrow \text{Ext}(A,B) \leftarrow \text{Hom}(F_1,B) \xleftarrow{\text{Hom}(d,1)} \text{Hom}(F_0,B) \xleftarrow{\text{Hom}(\varepsilon,1)} \text{Hom}(A,B) \leftarrow 0.$$

In other words, $\text{Ext}(A,B)$ is the cokernel of the homomorphism $\text{Hom}(d,1)$.

(b) By means of an *injective resolution* of B. Choose a short exact sequence $0 \to B \xrightarrow{\varepsilon} D_0 \xrightarrow{d} D_1 \to 0$ with D_0 (and hence D_1) divisible. (By the proposition above, such a sequence always exists.) Then the following sequence is exact:

$$0 \to \text{Hom}(A,B) \xrightarrow{\text{Hom}(1,\varepsilon)} \text{Hom}(A,D_0) \xrightarrow{\text{Hom}(1,d)} \text{Hom}(A,D_1) \to \text{Ext}(A,B) \to 0.$$

Thus $\text{Ext}(A,B)$ is the cokernel of the homomorphism $\text{Hom}(1,d)$.

Naturally, one must prove that the group $\text{Ext}(A,B)$ is independent of the projective resolution in (a), and of the injective resolution in (b). Also, it must be proved that the two definitions give rise to the same group. For information on these matters, the reader is referred to books on homological algebra (see the bibliography for Chapter V).

The definition of the induced homomorphism $\text{Ext}(f,g)$ is left to the reader.

From these definitions, the following two statements are obvious consequences:

(1) If A is a free abelian group, then $\text{Ext}(A,B) = 0$ for any group B.

(2) If B is a divisible group, then $\text{Ext}(A,B) = 0$ for any group A.

Using the definition (a) above, one readily shows that:

(3) $\text{Ext}(Z_n,B) \approx B/nB$,

$$\text{Hom}(Z_n,B) \approx \{x \in B \mid nx = 0\}.$$

By means of (1) and (3), the structure of $\text{Ext}(A,B)$ can be determined in case A is a finitely generated abelian group.

We conclude this summary of the principal properties of the functor Ext by mentioning the following two exact sequences. Let

$$0 \to A \xrightarrow{h} B \xrightarrow{k} C \to 0$$

be a short exact sequence of abelian groups, and let G be an arbitrary abelian group. Then the following two sequences are exact:

$$0 \to \text{Hom}(C,G) \xrightarrow{\text{Hom}(k,1)} \text{Hom}(B,G) \xrightarrow{\text{Hom}(h,1)} \text{Hom}(A,G)$$
$$\to \text{Ext}(C,G) \xrightarrow{\text{Ext}(k,1)} \text{Ext}(B,G) \xrightarrow{\text{Ext}(h,1)} \text{Ext}(A,G) \to 0,$$

(4.1)

$$0 \to \text{Hom}(G,A) \xrightarrow{\text{Hom}(1,h)} \text{Hom}(G,B) \xrightarrow{\text{Hom}(1,k)} \text{Hom}(G,C)$$
$$\to \text{Ext}(G,A) \xrightarrow{\text{Ext}(1,h)} \text{Ext}(G,B) \xrightarrow{\text{Ext}(1,k)} \text{Ext}(G,C) \to 0.$$

(4.2)

In these exact sequences, the *connecting homomorphisms*, $\text{Hom}(A,G) \to \text{Ext}(C,G)$ and $\text{Hom}(G,C) \to \text{Ext}(G,A)$ have all the naturality properties that one might expect.

With these preliminaries out of the way, we can now state the main result in this area:

Theorem 4.3 (*Universal coefficient theorem for cohomology*). *Let K be a chain complex of free abelian groups, and let G be an arbitrary abelian group. Then there exists a split exact sequence*

$$0 \to \mathrm{Ext}(H_{n-1}(K),G) \xrightarrow{\beta} H^n(\mathrm{Hom}(K,G)) \xrightarrow{\alpha} \mathrm{Hom}(H_n(K),G) \to 0.$$

The homomorphism β is natural, with respect to coefficient homomorphisms and chain maps. The splitting is natural with respect to coefficient homomorphisms but not with respect to chain maps.

PROOF. The proof we present is dual to that given in §V.6. For the reader who has some feeling for this duality, it is a purely mechanical exercise to transpose the previous proof to the present one.

First we need a lemma, which is the dual of Lemma V.6.1.

Lemma 4.4. *If G is a divisible group, then the homomorphism*

$$\alpha : H^n(\mathrm{Hom}(K,G)) \to \mathrm{Hom}(H_n(K),G)$$

is an isomorphism for any chain complex K.

The proof of this lemma is a nice exercise, involving the various definitions and the fact that divisible groups are injective.

Now we will prove the theorem. Let

$$0 \to G \xrightarrow{\varepsilon} D_0 \xrightarrow{d} D_1 \to 0$$

be a short exact sequence with D_0 and D_1 divisible (see Property (b) above). Consider the corresponding long exact sequence in cohomology, and the following commutative diagram:

$$
\begin{array}{ccccccc}
\cdots \xrightarrow{\beta_0} & H^n(\mathrm{Hom}(K,G)) & \xrightarrow{\varepsilon_*} & H^n(\mathrm{Hom}(K,D_0)) & \xrightarrow{d_*} & H^n(\mathrm{Hom}(K,D_1)) & \xrightarrow{\beta_0} \\
 & \downarrow{\scriptstyle \alpha} & & \downarrow{\scriptstyle \alpha_0} & & \downarrow{\scriptstyle \alpha_1} & \\
0 \longrightarrow & \mathrm{Hom}(H_n(K),G) & \xrightarrow[\mathrm{Hom}(1,\varepsilon)]{} & \mathrm{Hom}(H_n(K),D_0) & \xrightarrow[\mathrm{Hom}(1,d)]{} & \mathrm{Hom}(H_n(K),D_1). &
\end{array}
$$

The bottom lime is exact by the standard properties of the functor Hom, and the diagram is commutative by the naturality properties of α. Also, α_0 and α_1 are isomorphisms, since D_0 and D_1 are divisible groups. From this diagram one deduces that α is an epimorphism, and kernel α = kernel ε_*.

Next, one considers the following similar diagram:

$$
\begin{array}{ccccccc}
\cdots \xrightarrow{\varepsilon_*} & H^{n-1}(\mathrm{Hom}(K,D_0)) & \xrightarrow{d_*} & H^{n-1}(\mathrm{Hom}(K,D_1)) & \xrightarrow{\beta_0} & H^n(\mathrm{Hom}(K,G)) & \xrightarrow{\varepsilon_*} \cdots \\
 & \downarrow{\scriptstyle \alpha_0} & & \downarrow{\scriptstyle \alpha_1} & & \boxed{1} & \\
\cdots \longrightarrow & \mathrm{Hom}(H_{n-1}(K),D_0) & \xrightarrow[\mathrm{Hom}(1,d)]{} & \mathrm{Hom}(H_{n-1}(K),D_1) & \longrightarrow & \mathrm{Ext}(H_{n-1}(K),G) & \longrightarrow 0.
\end{array}
$$

Once again the bottom line is exact, and the diagram is commutative; as before, α_0 and α_1 are isomorphisms. One now proves that there is a unique homomorphism

$$\beta : \mathrm{Ext}(H_{n-1}(K),G) \to H^n(\mathrm{Hom}(K,G))$$

which makes the square labelled 1 commutative. Then one proves that β is a monomorphism, and image $\beta = \mathrm{image}\ \beta_0$. Since image $\beta_0 = \mathrm{kernel}\ \varepsilon_\#$, it follows that image $\beta = \mathrm{kernel}\ \alpha$.

It remains to prove that the short exact sequence of the theorem splits. This can be done by the method used in the proof of Theorem V.6.2, modified to cover the case at hand. The details are left to the reader.

Corollary 4.5. *For any pair (X,A) and any abelian group G there exists a split short exact sequence:*

$$0 \to \mathrm{Ext}(H_{n-1}(X,A), G) \xrightarrow{\beta} H^n(X,A;G) \xrightarrow{\alpha} \mathrm{Hom}(H_n(X,A), G) \to 0.$$

The homomorphisms α and β are natural with respect to homomorphisms induced by continuous maps of pairs and coefficient homomorphisms. The splitting can be chosen to be natural with respect to coefficient homomorphisms, but not with respect to homomorphisms induced by continuous maps.

EXERCISES

4.1. Let (X,A) be a pair such that $H_n(X,A)$ is a finitely generated abelian group for all n. Prove that $H^n(X,A;\mathbf{Z})$ is also finitely generated for all n, and that

$$\mathrm{rank}(H^n(X,A;\mathbf{Z})) = \mathrm{rank}(H_n(X,A)),$$
$$\mathrm{Torsion}(H^n(X,A;\mathbf{Z})) \approx \mathrm{Torsion}(H_{n-1}(X,A)).$$

4.2. Prove that $\alpha : H^n(X,A;G) \to \mathrm{Hom}(H_n(X,A), G)$ is an isomorphism for $n = 0, 1$ (for any pair (X,A) and any group G).

4.3. For any pair (X,A), prove that $H^1(X,A;\mathbf{Z})$ is a torsion-free abelian group.

4.4. Let X be a finite regular graph. Express the structure of the cohomology groups $H^n(X,G)$ in terms of the Euler characteristic and number of components of X.

4.5. Describe the structure of the cohomology groups $H^q(S^n;G)$ and $H^q(E^n,S^{n-1};G)$ for all q, n, and G.

4.6. Let X be an n-dimensional pseudomanifold as defined in §IV.8. Determine the structure of $H^n(X;G)$.

4.7. Let X be a compact connected 2-dimensional manifold. Determine the structure of $H^n(X;G)$ for all n and G (use the classification theorem for such manifolds to express your final result).

4.8. Let $K = \{K^n\}$ be a finite dimensional CW-complex on the space X. Prove that there is an isomorphism $H^n(X;G) \approx H^n(\mathrm{Hom}(C(K),G))$ for all n and G (here

$C(K) = \{H_q(K^q, K^{q-1})\}$ is a chain complex described in §V.7. Prove also that this isomorphism has the following naturality property: Let L be a CW-complex on Y and $f : X \to Y$ a continuous map which is *cellular*, i.e., $f(K^n) \subset L^n$ for all n. Then there is an induced chain map $f_\# : C(K) \to C(L)$, and the following diagram is commutative:

$$H^n(X; G) \approx H^n(\mathrm{Hom}(C(K), G))$$

$$\Big\uparrow f^* \qquad\qquad \Big\uparrow \mathrm{Hom}(f_\#, 1)^*$$

$$H^n(Y; G) \approx H^n(\mathrm{Hom}(C(L), G)).$$

4.9. Consider continuous maps $f : P^2 \to S^2$, where P^2 denotes the real projective plane. By considering the induced homomorphism $f^* : H^2(S^2; \mathbf{Z}) \to H^2(P^2; \mathbf{Z})$, show that there are at least two homotopy classes of such maps (cf. the example in §V.7. Use the results of Exercise 4.8).

4.10. Show that the homomorphism $f^* : H^n(Y, B; G) \to H^n(X, A; G)$ induced by a continuous map $f : (X, A) \to (Y, B)$ is *not* determined by knowledge of the homomorphisms on homology

$$f_* : H_q(X, A) \to H_q(Y, B)$$

for all q.

4.11. Prove that the splitting of the short exact sequence of Corollary 4.5 can *not* be chosen to be natural with respect to homomorphisms induced by continuous maps.

§5. Geometric Interpretation of Cochains, Cocycles, etc.

In homology theory it is not difficult to have some geometric intuition about chains, cycles, bounding cycles, etc. This geometric intuition is often of assistance in leading one to the correct solution of problems. Unfortunately, these things are more complicated for cohomology theory.

In order to understand the situation better, let us first reconsider homology theory. Let $K = \{K^n\}$ be a CW-complex on the space X, and let $u \in C_n(K, G)$; then u has a unique expression of the form

$$u = \sum_i g_i e_i^n,$$

where $g_i \in G$ and the e_i^n are oriented n-cells of K. It is natural to associate with the chain u the subset

$$|u| = \bigcup_i \overline{e}_i^n,$$

where the union is over all cells e_i^n such that the corresponding coefficient $g_i \neq 0$. If $u = 0$, we define $|u| = \varnothing$. The set $|u|$ is called the *support of u*. It has the following properties:

(a) $|u|$ is a compact subset of X.
(b) $|u| = \varnothing$ if and only if $u = 0$.
(c) $|u \pm v| \subset |u| \cup |v|$.
(d) $|d_n u| \subset |u|$.

Of course the chain u is not determined by the set $|u| \subset X$ (except in the case where $G = Z_2$), but the structure of the set $|u|$ is a vital piece of information about u. One thinks of u as determined by $|u|$ and the coefficients g_i which are assigned to various oriented subsets of $|u|$.

There is also a natural way to define the support of a singular chain u in an arbitrary topological space X. Let $u \in C_n(X,G)$; if $u = 0$, we define $|u| = \varnothing$. If $u \neq 0$, then u has a unique expression as a finite linear combination of nondegenerate singular n-cubes with nonzero coefficients,

$$u = \sum_i g_i T_i \qquad g_i \in G$$

and it is natural to define

$$|u| = \bigcup_i T_i(I^n).$$

It is clear that Properties (a)–(d) above continue to hold. However, it is also clear that in this situation $|u|$ does not give as much information about u as it did in the previous situation. The reason is that two quite different nondegenerate n-cubes may have the same image set, i.e., we may have n-cubes

$$T_1, T_2 : I^n \to X$$

such that $T_1(I^n) = T_2(I^n)$, yet $T_1 \neq T_2$.

We will now try to define the support of a cochain so that Properties (a)–(d) will hold. First of all, it is convenient to formulate the definition of a cochain in a slightly different, but equivalent, way. This alternate definition is based on the following principle: *Let F be a free abelian group with basis $B \subset F$, and let G be an arbitrary abelian group. Then there is a natural 1-1 correspondence between homomorphisms $u:F \to G$ and arbitrary functions $f:B \to G$.* This correspondence is established by assigning to each such homomorphism u the function $f = u|B$, the restriction of u to B, and to each such function f its unique linear extension u.

Let us apply this principle to the n-cochains of a CW-complex K on the space X. Let $u \in C^n(K,G) = \mathrm{Hom}(C_n(K),G)$. The chain group $C_n(K)$ has as natural basis the set of n-cells $\{e_i^n\}$, where a definite orientation has been chosen for each such cell. Thus we can think of u as a function which assigns to each such oriented n-cell e_i^n an element $u(e_i^n) \in G$. In view of the previous definition for support of a chain, it seems natural to define $|u|$ to be the union of all closed n-cells $\bar{e}_i^n$, such that $u(\bar{e}_i^n) \neq 0$. However, experience has shown that this definition definitely does not work! The main trouble is that the analogue of Condition (d) above does not hold.

We will indicate a way to correct this deficiency for the case of cochains in a *regular* CW-complex. Recall that given a regular CW-complex, for each cell e_i^n there exists a characteristic map

$$f_i : (E^n, S^{n-1}) \to (\bar{e}_i^n, \dot{e}_i^n)$$

which is a homeomorphism. Of course, if $n > 0$, there will exist for each cell e_i^n infinitely many such maps which are homeomorphisms, and there is no reason to prefer one over another. We will assume that for each e_i^n one such characteristic map has been chosen, and call it the *preferred* characteristic map. By means of this preferred characteristic map, geometric concepts which are valid for E^n can be carried over to $\bar{e}_i^n$. In particular, we wish to carry over the following two concepts from E^n to $\bar{e}_i^n$:

(1) The center of the cell E^n is the origin, $(0, 0, \ldots, 0)$. By definition, the *center* of e_i^n is the image of the center of E^n under the preferred characteristic map.

(2) If A is any subset of S^{n-1}, the *cone over* A, denoted by $\Gamma(A)$, is the following subset of E^n:

$$\Gamma(A) = \{t \cdot a \,|\, a \in A \text{ and } 0 \le t \le 1\},$$

i.e., $\Gamma(A)$ is the union of all straight line segments joining the origin to points $a \in A$. Analogously, if A is any subset of $\dot{e}_i^n$, then $\Gamma(A)$ is a subset of $\bar{e}_i^n$, defined using the preferred characteristic map for the cell e_i^n. Note that if A is a closed set, then so is $\Gamma(A)$. More generally, if A is a subset of the $(n-1)$-skeleton K^{n-1} then we define $\Gamma(A)$ to be the union of A and the sets $\Gamma(A \cap \dot{e}_i^n)$ for all n-cells e_i^n. $\Gamma(A)$ is a subset of K^n, and if A is closed, so is $\Gamma(A)$ because of the weak topology. We can iterate this procedure, defining

$$\Gamma^2(A) = \Gamma(\Gamma(A)),$$
$$\Gamma^n(A) = \Gamma(\Gamma^{n-1}(A)),$$
$$\Gamma^\infty(A) = \bigcup_{n=1}^\infty \Gamma^n(A).$$

We will mainly be interested in this operation for the case of a finite-dimensional CW-complex. Then $\Gamma^\infty(A)$ is attained after a finite number of iterations.

Now let $u \in C^n(K, G)$; consider u as a function defined on oriented n-cells e_i^n with values in G. Define A to the set of all center points of all cells e_i^n such that $u(e_i^n) \ne 0$. Then A is a closed, discrete subset of X; however, it is not compact, in general. We define

$$|u| = \Gamma^\infty(A).$$

If K is finite dimensional, it is clear that $|u|$ is a closed subset of X. One can also verify the analogue of Conditions (b), (c), and (d) above:

$$|u| = \varnothing \quad \text{if and only if} \quad u = 0,$$
$$|u \pm v| \subset |u| \cup |v|,$$
$$|\delta(u)| \subset |u|.$$

Although rather complicated, this seems to be the proper definition. The fact that $|u|$ is noncompact in general is not a defect in our definition, it is an inherent property of the cohomology theory we are using. It is possible to define a cohomology theory based on cochains with "compact supports," but we will not do this for the present.

Note that if K is a CW-complex of dimension N, and $u \in C^k(K,G)$, then $|u|$ is a set of dimension $\leq N - k$. Thus as k increases, the dimension of $|u|$ decreases. This is just the opposite of what happens with chains.

There is also a definition of support of singular cochains in a general space which we will now consider, although it is less satisfactory than that we have just given.

If $u \in C^n(X,G) = \text{Hom}(C_n(X),G)$, then u is a homomorphism of $C_n(X) = Q_n(X)/D_n(X)$ into G. Hence we can regard u as a function which is defined on singular n-cubes with values in G, and vanishes on all degenerate singular n-cubes. Rather than defining $|u|$, it will be more convenient to define the complementary set: A point x does *not* belong to $|u|$ if and only if there is an open neighborhood U of x such that $u(T) = 0$ for all singular n-cubes $T : I^n \to U$. From this definition it is clear that the complementary set is open, hence $|u|$ is closed. We also have the following properties:

$$u = 0 \quad \text{implies} \quad |u| = \varnothing,$$
$$|u \pm v| \subset |u| \cup |v|,$$
$$|\delta u| \subset |u|.$$

Unfortunately, we can have nonzero cochains u such that $|u| = \varnothing$. This defect can be remedied by factoring out all such cochains (i.e., passing to a quotient group). By using Theorem II.6.3 it can be proved that this process does not change the resulting cohomology theory. However, we will have no need to pursue this matter further, (cf. Massey, [1], Lemma 8.16, p. 260).

§6. Proof of the Excision Property; The Mayer–Vietoris Sequence

Let (X,A) be a pair and let W be a subset of A. We then have the following split exact sequence of chain complexes (cf. §V.6):

$$0 \to C(X - W, A - W) \xrightarrow{i_*} C(X,A) \to \frac{C(X,A)}{C(X - W, A - W)} \to 0.$$

Note that these are all chain complexes of free abelian groups. By passing to the long exact homology sequence, we see that $i_*: H_q(X - W, A - W) \rightarrow H_q(X,A)$ is an isomorphism for all q if and only if $H_q(C(X,A)/C(X - W, A - W)) = 0$ for all q.

We may also apply the functor $\text{Hom}(\ ,G)$ to the above split exact sequence of chain complexes, obtaining the following exact sequence:

$$0 \leftarrow C^*(X - W, A - W; G) \xleftarrow{i_\#} C^*(X,A;G)$$

$$\leftarrow \text{Hom}\left(\frac{C(X,A)}{C(X - W, A - W)}, G\right) \leftarrow 0.$$

Passing to cohomology, we see that $i^*: H^q(X,A;G) \rightarrow H^q(X - W, A - W; G)$ is an isomorphism for all q if and only if $H^q(\text{Hom}(C(X,A)/C(X - W, A - W), G)) = 0$ for all q. Making use of Theorem 4.3, one concludes that the excision property for integral homology implies a corresponding property for cohomology: *If W is a subset of A such that $\overline{W} \subset$ interior A, then $i^*: H^q(X,A;G) \rightarrow H^q(X - W, A - W; G)$ is an isomorphism for all q.*

Let $\mathcal{U}$ be an open covering of X, or more generally, a family of sets whose interiors cover X. It is known that the inclusion

$$\sigma: C(X,A,\mathcal{U}) \rightarrow C(X,A)$$

induces an isomorphism on homology (Theorem II.6.3). By the same type of argument as that just given, it can be shown that the induced homomorphism on cochain complexes

$$\text{Hom}(\sigma,1): C^*(X,A;G) \rightarrow C^*(X,A,\mathcal{U};G)$$

also induces an isomorphism on passage to cohomology. This fact can be used to prove the existence of the Mayer–Vietoris sequence for cohomology as follows. Let A and B be subsets of X such that

$$X = (\text{interior } A) \cup (\text{interior } B).$$

Then we may take $\mathcal{U} = \{A,B\}$, and $\sigma: C(X,\mathcal{U}) \rightarrow C(X)$ will have the properties described above. In §III.5 we introduced the commutative diagram of chain complexes

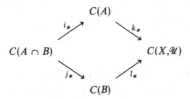

and the following short exact sequence

$$0 \rightarrow C(A \cap B) \xrightarrow{\Phi} C(A) \oplus C(B) \rightarrow C(X,\mathcal{U}) \rightarrow 0$$

in order to prove the Mayer–Vietoris sequence for homology theory. Recall that Φ and Ψ are defined by

$$\Phi(x) = (i_\# x, j_\# x), \qquad \Psi(u,v) = k_\#(u) - l_\#(v).$$

Also, $C(X,\mathcal{U})$ is a chain complex of free abelian groups, hence the short exact sequence splits. Therefore we may apply the functor $\mathrm{Hom}(\ ,G)$ to obtain the following short exact sequence of cochain complexes:

$$0 \leftarrow C^*(A \cap B; G) \leftarrow C^*(A; G) \oplus C^*(B; G) \leftarrow C^*(X,\mathcal{U}; G) \leftarrow 0.$$

It is readily verified that homomorphisms $\mathrm{Hom}(\Phi,1)$ and $\mathrm{Hom}(\Psi,1)$ have the following expression in terms of $i^\#, j^\#, k^\#,$ and $l^\#$:

$$\mathrm{Hom}(\Psi,1)(x) = (k^\#(x), -l^\#(x)),$$

$$\mathrm{Hom}(\Phi,1)(u,v) = i^\# u + j^\# v.$$

Therefore we may pass to the corresponding long exact sequence of cohomology groups, and make use of the isomorphism $H^q(X,\mathcal{U}; G) \approx H^q(X; G)$ to obtain the *Mayer–Vietoris sequence in cohomology*:

$$\cdots \xleftarrow{\psi} H^{q+1}(X;G) \xleftarrow{\Delta} H^q(A \cap B; G) \xleftarrow{\varphi} H^q(A;G) \oplus H^q(B;G) \xleftarrow{\psi} H^q(X;G) \xleftarrow{\Delta} \cdots .$$

Here

$$\psi(x) = (k^*(x), -l^*(x)),$$

$$\varphi(u,v) = i^*(u) + j^*(v).$$

It should be remarked that there are other ways of deriving the Mayer–Vietoris sequence for cohomology.

EXERCISES

6.1. Let $K = \{K_q, \partial_q\}$ be a chain complex such that each K_q is a vector space over a commutative field F, and each ∂_q is linear over F. Define the cochain complex $\mathrm{Hom}_F(K,V)$, where V is a vector space over F, and the natural homomorphism

$$\alpha : H^q(\mathrm{Hom}_F(K,V)) \rightarrow \mathrm{Hom}_F(H^q(K),V).$$

Prove that α is an isomorphism.

6.2. Let $\{X_1, X_2\}$ be an excisive couple in the space X, as defined in §IV.6. Prove that the inclusion map $i : (X_1, X_1 \cap X_2) \rightarrow (X_1 \cup X_2, X_2)$ induces an isomorphism

$$i^* : H^q(X_1 \cup X_2, X_2; G) \rightarrow H^q(X_1, X_1 \cap X_2; G)$$

for all q and all groups G.

We will conclude this chapter by pointing out one basic property of homology theory which does not have an obvious analog for cohomology. The property we have in mind was stated earlier as Proposition III.6.1. This proposition says, in essence, that for any pair (X,A), the homology group $H_n(X,A)$ is the direct limit of the groups $H_n(C,D)$, where (C,D) ranges over all compact pairs contained in (X,A). It is tempting to conjecture that the cohomology group $H^n(X,A;G)$ is the *inverse* limit of the groups $H^n(C,D;G)$. However, counterexamples can be given to show that this is false. A special case of this question comes up in §3 of the Appendix.

Bibliography for Chapter VII

[1] W. S. Massey, *Homology and Cohomology Theory: An Approach Based on Alexander–Spanier Cochains*, Marcel Dekker, Inc., New York, 1978, Chapter 8, §8.

CHAPTER VIII

Products in Homology and Cohomology

§1. Introduction

The most important product is undoubtedly the so-called *cup product*: It assigns to any elements $u \in H^p(X;G_1)$ and $v \in H^q(X;G_2)$ an element $u \cup v \in H^{p+q}(X;G_1 \otimes G_2)$. This product is bilinear (or distributive), and is natural with respect to homomorphisms induced by continuous maps. It is an additional element of structure on the cohomology groups that often allows one to distinguish between spaces of different homotopy types, even though they have isomorphic homology and cohomology groups. This additional structure also imposes restrictions on the possible homomorphisms which can be induced by continuous maps.

Another product we shall consider is called the *cap product*. It assigns to elements $u \in H^p(X;G_1)$ and $v \in H_q(X;G_2)$ an element $u \cap v \in H_{q-p}(X; G_1 \otimes G_2)$. It is also bilinear and natural. While the cap product is not as important as the cup product, it is needed for the statement and proof of the Poincaré duality theorem in the next chapter.

We will also consider two other products: A *cross product* which is closely related to the cup product, and a *slant product*, which has strong connections with the cap product. The main reasons for considering these two additional products is for the light they throw on the cup and cap product.

In order to make effective use of cup products, it is necessary to have ways of computing them for various spaces. Unfortunately, this is a rather difficult topic; any systematic discussion of it would be rather lengthy. In Chapter X we will use the Poincaré duality theorem to determine cup products in projective spaces; then we can use these products to prove some interesting theorems (Borsuk–Ulam theorem, nontriviality of the Hopf maps, etc.). In the present chapter we will mainly be concerned with a systematic discussion of the basic properties of these various products.

Because this chapter is rather long and does not have many examples, it may be best to skim throught it on a first reading. Then the reader can return to it later to study more carefully the various details as they are needed.

§2. The Inner Product

In §VII.4 we defined the so-called inner product, and used it to define a natural homomorphism $\alpha : H^n(\text{Hom}(K,G)) \to \text{Hom}(H_n(K),G)$. The various naturality properties of the homomorphism α could also be interpreted as naturality properties of the inner product.

It will be convenient to generalize the definition of the inner product slightly for later use in this chapter. Let G_1 and G_2 be arbitrary abelian groups, and let K be a chain complex. Then for any elements $u \in H^q(\text{Hom}(K,G_1))$ and $v \in H_q(K \otimes G_2)$, the inner product $\langle u,x \rangle \in G_1 \otimes G_2$ is defined as follows. Choose a representative cocycle $u' \in \text{Hom}(K_q,G_1)$ for u, and a representative cycle

$$x' = \sum_{i=1}^{k} x_i \otimes g_i, \qquad x_i \in K_q, \, g_i \in G_2$$

for x. Then

$$\langle u,x \rangle = \sum_{i=1}^{k} u'(x_i) \otimes g_i \in G_1 \otimes G_2.$$

This more general version of the inner product has essentially the same properties as the original version.

§3. An Overall View of the Various Products

To define products, one needs to make use of the natural chain homotopy equivalences of Chapter VI,

$$\zeta : C(X) \otimes C(Y) \to C(X \times Y)$$
$$\xi : C(X \times Y) \to C(X) \otimes C(Y),$$

especially the later. We will continue to use the above notation for these chain maps, as in Chapter VI.

First, we introduce the cross product. Recall that if $f : G \to G'$ and $g : H \to H'$ are homomorphisms of abelian groups, then $f \otimes g : G \otimes H \to G' \otimes H'$ denotes the tensor product of the two homomorphisms. Using this notation, if $u \in C^p(X,G_1) = \text{Hom}(C_p(X),G_1)$ and $v \in C^q(Y,G_2) = \text{Hom}(C_q(Y),G_2)$, then

$u \otimes v \in \mathrm{Hom}(C_p(X) \otimes C_q(Y), G_1 \otimes G_2)$. We may consider $u \otimes v$ as an element of $\mathrm{Hom}((C(X) \otimes C(Y))_{p+q}, G_1 \otimes G_2)$ if we understand that $u \otimes v$ is the zero homomorphism on $C_i(X) \otimes C_j(Y)$, except when $i = p$ and $j = q$. Let

$$\xi^\# = \mathrm{Hom}(\xi, 1): \mathrm{Hom}(C(X) \otimes C(Y), G_1 \otimes G_2) \to \mathrm{Hom}(C(X \times Y), G_1 \otimes G_2)$$
$$= C^*(X \times Y; G_1 \otimes G_2).$$

Then we define $u \times v \in C^{p+q}(X \times Y; G_1 \otimes G_2)$ by

$$u \times v = \xi^\#(u \otimes v).$$

It is readily verified that

$$\delta(u \times v) = (\delta u) \times v + (-1)^p u \times \delta v.$$

From this coboundary formula, the following facts follow:

  (1) If u and v are cocycles, then so is $u \times v$.

  (2) If u_1 and u_2 are cocycles which are cohomologous, then $u_1 \times v$ and $u_2 \times v$ are cohomologous for any cocycle v.

  (3) Similarly, if v_1 and v_2 are cohomologous cocycles, then $u \times v_1$ and $u \times v_2$ are cohomologous for any cocycle u.

From these three statements it is clear that we can pass to cohomology classes, and thus define a *cross product* which assigns to any cohomology class $x \in H^p(X; G_1)$ and $y \in H^q(Y; G_2)$ a cohomology class $x \times y \in H^{p+q}(X \times Y; G_1 \otimes G_2)$. The two most important properties of this cross product are the following:

(1) *Bilinearity.* $(x_1 + x_2) \times y = x_1 \times y + x_2 \times y$ and $x \times (y_1 + y_2) = x \times y_1 + x \times y_2$.

(2) *Naturality.* If $f: X' \to X$ and $g: Y' \to Y$ are continuous maps, $x \in H^p(X; G_1)$ and $y \in H^q(Y; G_2)$, then

$$(f^*x) \times (g^*y) = (f \times g)^*(x \times y).$$

Later on we will generalize the definition of the cross product to relative cohomology groups, and prove various additional properties.

Next, we will define the *cup product* in terms of the cross product. For any space X, let d_X or d for short, denote the diagonal map $X \to X \times X$ defined by $d(x) = (x, x)$. If $u \in H^p(X, G_1)$ and $v \in H^q(X, G_2)$, define $u \cup v \in H^{p+q}(X, G_1 \otimes G_2)$ by

$$u \cup v = d^*(u \times v).$$

We see immediately that the cup product has the following two basic properties:

(1) *Bilinearity.* $(u_1 + u_2) \cup v = u_1 \cup v + u_2 \cup v$ and $u \cup (v_1 + v_2) = u \cup v_1 + u \cup v_2$.

(2) *Naturality.* If $f: X' \to X$ is a continuous map, $u \in H^p(X, G_1)$ and $v \in H^q(X, G_2)$, then

$$f^*(u \cup v) = (f^*u) \cup (f^*v).$$

We have just defined the cup product in terms of the cross product, using the diagonal map d. Conversely, it is possible to derive the cross product from the cup product. To clarify this point, let us assume that the cup product is given, which is bilinear and natural as just described. Define a new cross product, $u \mathbin{\#} v$ by the formula

$$u \mathbin{\#} v = (p_1^* u) \cup (p_2^* v)$$

for any $u \in H^p(X, G_1)$ and $v \in H^q(Y, G_2)$. Here $p_1 \colon X \times Y \to X$ and $p_2 \colon X \times Y \to Y$ are the projections. Then it follows easily that this new cross product is also bilinear and natural, in the same sense as the original cross product. If we use this new cross product to define a new cup product by the formula

$$u \cup' v = d^*(u \mathbin{\#} v)$$

for any $u \in H^p(X, G_1)$ and $v \in H^q(X, G_2)$, then we find that $u \cup' v = u \cup v$, i.e., the new cup product is the same as the old. This may be proved by the following computation:

$$\begin{aligned}
u \cup' v &= d^*(u \mathbin{\#} v) = d^*((p_1^* u) \cup (p_2^* v)) \\
&= (d^* p_1^* u) \cup (d^* p_2^* v) \\
&= (p_1 d)^* u \cup (p_2 d)^* v = u \cup v.
\end{aligned}$$

Similarly, we find that

$$\begin{aligned}
u \mathbin{\#} v &= (p_1^* u) \cup (p_2^* v) \\
&= d_{X \times Y}^*((p_1^* u) \times (p_2^* v)) \\
&= d_{X \times Y}^*(p_1 \times p_2)^*(u \times v) \\
&= [(p_1 \times p_2) d_{X \times Y}]^*(u \times v) = u \times v
\end{aligned}$$

for any $u \in H^p(X, G_1)$ and $v \in H^q(Y, G_2)$.

We can reformulate what we have just proved as follows: the formulas

$$u \cup v = d^*(u \times v),$$
$$u \times v = (p_1^* u) \cup (p_2^* v)$$

establish a 1-1 correspondence between cross products and cup products (which are required to be bilinear and natural).

From this point of view, the theory of cup products and the theory of cross products are logically equivalent. However, cup products are more useful, while cross products have a more direct and simpler definition. Later on we will consider other properties of cross and cup products, such as associativity, commutativity, and existence of a unit. We will also extend the definitions to relative cohomology groups, and consider their behavior under the coboundary operator of the exact cohomology sequence of a pair (X, A). Naturally, the exposition of the properties of cup products will parallel that of cross products.

Remark on Terminology. Cross products are sometimes called *exterior cohomology products* and cup products are then called *interior cohomology products*.

Next, we will discuss slant products, and cap products, which are derived from slant products by means of the diagonal map.

First we define a homomorphism

$$\text{Hom}(C^p(Y),G_1) \otimes [C(X) \otimes C(Y)]_q \otimes G_2 \to C_{q-p}(X) \otimes G_1 \otimes G_2,$$

denoted by $\varphi \otimes u \to \varphi\backslash\backslash u$, as follows:

$$\varphi\backslash\backslash a \otimes b \otimes g = (-1)^{|\varphi||a|}a \otimes \varphi(b) \otimes g$$

for any $\varphi \in \text{Hom}(C^p(Y),G_1)$, $a \in C(X)$, $b \in C(Y)$ and $g \in G_2$. Here the notation $|\varphi|$ means the degree of φ, $|a|$ means the degree of a, etc., and we make the convention that $\varphi(b) = 0$ unless $b \in C^p(Y)$. We can verify the formula

$$\partial(\varphi\backslash\backslash a \otimes b \otimes g) = (\delta\varphi)\backslash\backslash a \otimes b \otimes g + (-1)^{|\varphi|}\varphi\backslash\backslash\partial(a \otimes b \otimes g)$$

provided we follow the convention that

$$(\delta\varphi)(b) = (-1)^{|\varphi|}\varphi(\partial b).$$

We next define a homomorphism

$$\text{Hom}(C^p(Y),G_1) \otimes C_q(X \times Y, G_2) \to C_{q-p}(X, G_1 \otimes G_2),$$

denoted by $u \otimes v \to u\backslash v$, by using the Eilenberg–Zilber chain map ξ:

$$u\backslash v = u\backslash\backslash\xi(v).$$

Once again we have the formula

$$\partial(u\backslash v) = (\delta u)\backslash v + (-1)^{|u|}u\backslash\partial(v).$$

Hence we can pass to homology classes and get a homomorphism

$$H^p(Y,G_1) \otimes H_q(X \times Y, G_2) \to H_{q-p}(X, G_1 \otimes G_2),$$

denoted by $u \otimes v \to u\backslash v$, which is called the *slant product*. In addition to the obvious bilinearity of the slant product, it satisfies the following naturality condition: Let $f: X \to X'$ and $g: Y \to Y'$ be continuous maps. Then for any $u \in H^p(Y',G_1)$ and $v \in H_q(X \times Y, G_2)$ we have

$$f_*((g^*u)\backslash v) = u\backslash(f \times g)_*v.$$

This naturality relation can be indicated by the following diagram:

$$
\begin{array}{ccc}
H^p(Y) \otimes H_q(X \times Y) & \longrightarrow & H_{q-p}(X) \\
\big\uparrow{\scriptstyle g^*} & \big\downarrow{\scriptstyle (f \times g)_*} & \big\downarrow{\scriptstyle f_*} \\
H^p(Y') \otimes H_q(X' \times Y') & \longrightarrow & H_{q-p}(X')
\end{array}
$$

although this is not a commutative diagram in the conventional sense.

Remark: One can reformulate the slant product so as to obtain commutative diagrams in the usual sense. Recall that there is a natural adjoint associativity isomorphism

$$\text{Hom}(B \otimes A, C) \approx \text{Hom}(A, \text{Hom}(B,C))$$

for any abelian groups A, B, and C. Thus we can consider the slant product as a homomorphism

$$H_q(X \times Y) \rightarrow \text{Hom}(H^p(Y), H_{q-p}(X)).$$

Then the naturality condition gives rise to the following diagram, which is commutative in the usual sense:

$$\begin{array}{ccc}
H_q(X \times Y) & \longrightarrow & \text{Hom}(H^p(Y), H_{q-p}(X)) \\
\downarrow{\scriptstyle (f \times g)_*} & & \downarrow{\scriptstyle \text{Hom}(f^*, g_*)} \\
H_q(X' \times Y') & \longrightarrow & \text{Hom}(H^p(Y'), H_{q-p}(X')).
\end{array}$$

However, most people find this formulation of the slant product rather awkward to work with.

We can now define the cap product. It is a homomorphism

$$H^p(X, G_1) \otimes H_q(X, G_2) \rightarrow H_{q-p}(X, G_1 \otimes G_2),$$

denoted by $u \otimes v \rightarrow u \cap v$, and defined by

$$u \cap v = u \backslash d_*(v)$$

where $d: X \rightarrow X \times X$ is the diagonal map. It is bilinear, and natural in the following sense. Let $f: X \rightarrow X'$ be a continuous map. Then for any $u \in H^p(X')$, $v \in H_q(X)$ we have

$$f_*((f^*u) \cap v) = u \cap f_*v.$$

The corresponding diagram is the following:

$$\begin{array}{ccc}
H^p(X) \otimes H_q(X) & \stackrel{\cap}{\longrightarrow} & H_{q-p}(X) \\
\uparrow{\scriptstyle f^*} \quad \downarrow{\scriptstyle f_*} & & \downarrow{\scriptstyle f_*} \\
H^p(X') \otimes H_q(X') & \stackrel{\cap}{\longrightarrow} & H_{q-p}(X').
\end{array}$$

Once again, this could be made into a conventional commutative diagram by using the Hom functor rather than $\otimes$.

We have just shown how to derive the cap product from the slant product. Conversely, the slant product can be derived from the cap product, as follows. For any $u \in H^p(Y, G_1)$ and $v \in H_q(X \times Y, G_2)$, define

$$u \backslash v = p_{1*}((p_2^* u) \cap v),$$

where p_1 and p_2 are the projections of $X \times Y$ on the first and second factors respectively. By the same methods used in the discussion of cross and cup products, one can prove that our formulas establish a 1-1 correspondence between slant and cap products (which are required to be natural and to be bilinear). Thus the theories of these two different kinds of products should be logically equivalent. Actually, cap products will be needed in our discussion of the Poincaré duality theorem for manifolds; however, the definition of slant products is a bit simpler.

Remark. We have based our discussion of the cup and cap product on the use of the Eilenberg–Zilber natural chain homotopy equivalence

$$\xi: C(X \times Y) \to C(X) \otimes C(Y)$$

together with the diagonal map $d: X \to X \otimes X$. An alternative procedure would be to use natural diagonal maps $\Delta^X: C(X) \to C(X) \otimes C(X)$ as described in Exercise VI.5.2. For the connection between ξ and Δ, see Exercise VI.5.3. The choice of which method to use is largely a matter of taste. However, there is some advantage to having both cross and cup products, and the relationship between them.

§4. Extension of the Definition of the Various Products to Relative Homology and Cohomology Groups

The main difficulty in extending cross and slant products to relative cohomology and homology groups is the problem of extending the Eilenberg–Zilber chain homotopy equivalence ξ to relative groups; this problem was already encountered in the discussion in §VI.6 of the homology groups of product spaces. The main result of that discussion may be summarized as follows: Let (X,A) and (Y,B) be pairs. Then the chain map ξ induces a chain homotopy equivalence

$$\frac{C(X)}{C(A)} \otimes \frac{C(Y)}{C(B)} \xleftarrow{\xi} \frac{C(X \times Y)}{C(X \times B) + C(A \times Y)}.$$

If we assume that $\{X \times B, A \times Y\}$ is an excisive couple in $X \times Y$, then the homomorphism

$$k: \frac{C(X \times Y)}{C(X \times B) + C(A \times Y)} \to \frac{C(X \times Y)}{C(X \times B \cup A \times Y)}$$

induces isomorphisms on homology and cohomology with any coefficients.

In view of this, when we want to define cross or slant products in the homology and/or cohomology of pairs (X,A) and (Y,B), we will always assume that $\{X \times B, A \times Y\}$ is an excisive couple in $X \times Y$. With this added assumption, our previous definitions generalize very easily. The details are as follows.

Cross Product. If $u \in C^p(X,A;G_1) = \mathrm{Hom}(C_p(X)/C_p(A),G_1)$ and $v \in C^q(Y,B;G_2) = \mathrm{Hom}(C_q(Y)/C_q(B),G_2)$, then

$$u \otimes v \in \mathrm{Hom}\left(\left(\frac{C(X)}{C(A)} \otimes \frac{C(Y)}{C(B)}\right)_{p+q}, G_1 \otimes G_2\right)$$

$$= \mathrm{Hom}\left(\left(\frac{C(X) \otimes C(Y)}{C(X) \otimes C(B) + C(A) \otimes C(Y)}\right)_{p+q}, G_1 \otimes G_2\right)$$

and

$$\xi^{\#}(u \otimes v) \in \mathrm{Hom}\left(\left(\frac{C(X \times Y)}{C(X \times B) + C(A \times Y)}\right)_{p+q}, G_1 \otimes G_2\right).$$

Passing to cohomology groups, and applying the isomorphism $(k*)^{-1}$, we obtain the cross product in cohomology, which is a homomorphism

$$H^p(X,A;G_1) \otimes H^q(Y,B;G_2) \xrightarrow{\times} H^{p+q}(X \times Y; A \times Y \cup X \times B; G_1 \otimes G_Y).$$

The *naturality condition* now reads as follows: Let $f:(X,A) \to (X',A')$ and $g:(Y,B) \to (Y',B')$ be continuous maps of pairs. Then the following diagram is commutative:

$$
\begin{array}{ccc}
H^p(X',A') \otimes H^q(Y',B') & \xrightarrow{\quad \times \quad} & H^{p+q}(X' \times Y', A' \times Y' \cup X' \times B') \\
\downarrow{\scriptstyle f^* \otimes g^*} & & \downarrow{\scriptstyle (f \times g)^*} \\
H^p(X,A) \otimes H^q(Y,B) & \xrightarrow{\quad \times \quad} & H^{p+q}(X \times Y, A \times Y \cup X \times B).
\end{array}
$$

In symbols,

$$(f^*u) \times (g^*v) = (f \times g)^*(u \times v)$$

for any $u \in H^p(X',A';G_1)$ and $v \in H^q(Y',B';G_2)$. It is assumed, of course, that $\{A \times Y, X \times B\}$ and $\{A' \times Y', X' \times B'\}$ are excisive couples.

Slant product. First, one defines the homomorphism

$$\mathrm{Hom}\left(\frac{C^p(Y)}{C^p(B)}, G_1\right) \otimes \left[\frac{C(X)}{C(A)} \otimes \frac{C(Y)}{C(B)}\right]_q \otimes G_2 \to \frac{C_{q-p}(X)}{C_{q-p}(A)} \otimes G_1 \otimes G_2,$$

denoted by $\varphi \otimes u \to \varphi\backslash\backslash u$, by the formula

$$\varphi\backslash\backslash a \otimes b \otimes g = (-1)^{|\varphi||a|} a \otimes \varphi(b) \otimes g$$

exactly as in §3. Then one defines a homomorphism

$$\mathrm{Hom}\left(\frac{C^p(Y)}{C^p(B)}, G_1\right) \otimes \frac{C_q(X \times Y)}{C_q(A \times Y) + C_q(X \times B)} \otimes G_2 \to C_{q-p}(X,A; G_1 \otimes G_2),$$

denoted by $\varphi \otimes u \to \varphi\backslash u$, by the formula

$$\varphi\backslash u = \varphi\backslash\backslash\xi(u).$$

Passing to homology and cohomology, and using the isomorphism $(k_*)^{-1}$, we obtain the *slant product*, a homomorphism

$$H^p(Y,B;G_1) \otimes H_q(X \times Y; A \times Y \cup X \times B; G_2) \to H_{q-p}(X,A; G_1 \otimes G_2)$$

which is denoted by $u \otimes v \to u \backslash v$. The *naturality* condition is expressed by the following diagram:

$$\begin{array}{ccc}
H^p(Y',B') \otimes H_q(X' \times Y', A' \times Y' \cup X' \times B') & \longrightarrow & H_{q-p}(X',A') \\
\downarrow{g^*} & \uparrow{(f \times g)_*} & \uparrow{f_*} \\
H^p(Y,B) \otimes H_q(X \times Y, A \times Y \cup X \times B) & \longrightarrow & H_{q-p}(X,A).
\end{array}$$

Here $f:(X,A) \to (X',A')$ and $g:(Y,B) \to (Y',B')$ are continuous maps of pairs, and it is assumed that $\{A \times Y, X \times B\}$ and $\{A' \times Y', X' \times B'\}$ are excisive couples.

We will now take up the problem of defining the cup and cap product for relative cohomology and homology groups. Here the situation is slightly different. For the cup product, the object is to define a homomorphism

$$H^p(X,A;G_1) \otimes H^q(X,B;G_2) \overset{\cup}{\to} H^{p+q}(X, A \cup B; G_1 \otimes G_2)$$

under a reasonable set of assumptions; and for the cap product, one wishes to define a homomorphism

$$H^p(X,A;G_1) \otimes H_q(X, A \cup B; G_2) \to H_{q-p}(X, B; G_1 \otimes G_2)$$

under minimal hypotheses. The cup product will be defined from the cross product, and the cap product will be defined from the slant product by use of the diagonal map $d: X \to X \times X$.

Cup Products. Let us consider a triad $(X;A,B)$ consisting of a topological space X and arbitrary subspaces A and B. We have the following two chain maps, induced by obvious inclusions:

$$k: \frac{C(X \times X)}{C(A \times X) + C(X \times B)} \to \frac{C(X \times X)}{C(A \times X \cup X \times B)},$$

$$l: \frac{C(X)}{C(A) + C(B)} \to \frac{C(X)}{C(A \cup B)}.$$

If we attempt to define the cup product using the cross product and the diagonal map, we are led to the following commutative diagram:

$$\begin{array}{ccc}
H^p(X,A) \otimes H^q(X,B) & & \\
\downarrow{\times} & & \\
H^{p+q}\left(\mathrm{Hom}\left(\dfrac{C(X \times X)}{C(A \times X) + C(X \times B)}, G_1 \otimes G_2\right)\right) & \overset{d_1^*}{\longrightarrow} & H^{p+q}\left(\mathrm{Hom}\left(\dfrac{C(X)}{C(A) + C(B)}, G_1 \otimes G_2\right)\right) \\
\uparrow{k^*} & & \uparrow{l^*} \\
H^{p+q}(X \times X, X \times B \cup A \times X; G_1 \otimes G_2) & \overset{d_2^*}{\longrightarrow} & H^{p+q}(X, A \cup B; G_1 \otimes G_2).
\end{array}$$

Here d_1^* and d_2^* are induced by the diagonal map $d: X \to X \times X$. From this diagram it is clear that to define cup products, we may either assume that $\{A \times X, X \times B\}$ is an excisive couple, in which case k^* will be an isomorphism, or we may assume that $\{A,B\}$ is an excisive couple in X, in which case l^* will be an isomorphism. It is preferable and customary to make the latter assumption for a couple of reasons. First of all in the important special case $A = B$, $\{A,B\}$ is always an excisive couple, while $\{A \times X, X \times B\}$ need not be excisive (as far as is known). Secondly, for some of our later results about cup products, we will need to assume that $\{A,B\}$ is an excisive couple in X for other reasons. Thus we may as well assume it is excisive at the beginning. Therefore *in order to define cup products*

$$H^p(X,A) \otimes H^q(X,B) \overset{\cup}{\to} H^{p+q}(X, A \cup B)$$

we will always assume that $\{A,B\}$ is an excisive couple in X. This has the following slight disadvantage: In order to have the relation

$$u \cup v = d^*(u \times v)$$

hold true, it is necessary to assume that *both* $\{A,B\}$ and $\{X \times B, A \times X\}$ are excisive couples.

Cap Product. The discussion is analogous to that just given for the cup product. Let A and B be arbitrary subsets of X; then we have the following commutative diagram:

$$H^p(X,A;G_1) \otimes H_q\left(\frac{C(X \times X)}{C(X \times B \cup A \times X)} \otimes G_2\right) \overset{1 \otimes d_{2*}}{\longleftarrow} H^p(X,A;G_1) \otimes H_q\left(\frac{C(X)}{C(A \cup B)} \otimes G_2\right)$$

$$\uparrow{\scriptstyle 1 \otimes k_*} \qquad\qquad\qquad\qquad\qquad\qquad \uparrow{\scriptstyle 1 \otimes l_*}$$

$$H^p(X,A;G_1) \otimes H_q\left(\frac{C(X \times X)}{C(X \times B) + C(A \times X)} \otimes G_2\right) \overset{1 \otimes d_{1*}}{\longleftarrow} H^p(X,A;G_1) \otimes H_q\left(\frac{C(X)}{C(A) + C(B)} \otimes G_2\right)$$

$$\searrow{\scriptstyle \text{slant product}} \qquad\qquad\qquad \swarrow{\scriptstyle \text{cap product}}$$

$$H_{q-p}(X,B; G_1 \otimes G_2)$$

This diagram is entirely analogous to the preceding one, and the symbols for the various maps have the same meaning. In order to *define the cap product, we will assume that $\{A,B\}$ is an excisive couple in X.* Then the cap product is a homomorphism

$$H^p(X,A;G_1) \otimes H_q(X, A \cup B;G_2) \overset{\cap}{\to} H_{q-p}(X, B; G_1 \otimes G_2)$$

which is the composition of $(1 \otimes l_*)^{-1}$, $1 \otimes d_{1*}$, and the slant product in the above diagram. If in addition we assume that $\{X \times B, A \times X\}$ is an excisive couple in $X \times X$, then the following relation holds between the slant and cap products:

$$u \cap v = u\backslash(d_{2*}v).$$

§5. Associativity, Commutativity, and Existence of a Unit for the Various Products

In order to discuss these questions, it is necessary to discuss the associativity, commutativity, and existence of a unit for the Eilenberg–Zilber chain homotopy equivalence $\xi:C(X \times Y) \to C(X) \otimes C(Y)$. In order to discuss the associativity of ξ, consider the following diagram:

$$
\begin{array}{ccc}
C(X \times Y \times Z) & \xrightarrow{\;\xi^{X \times Y, Z}\;} & C(X \times Y) \otimes C(Z) \\
\big\downarrow{\scriptstyle \xi^{X, Y \times Z}} & & \big\downarrow{\scriptstyle \xi^{X, Y} \otimes 1} \\
C(X) \otimes C(Y \times Z) & \xrightarrow[\;1 \otimes \xi^{Y, Z}\;]{} & C(X) \otimes C(Y) \otimes C(Z).
\end{array}
$$

Recall that the proof of the existence of the chain map ξ required the choice of a certain chain $e_n \in [C(I^n) \otimes C(I^n)]_n$ for each positive integer n. It is too much to expect that the above diagram would be commutative for an arbitrarily constructed chain map ξ. However, using the method of acyclic models, it is easy to prove that the two different chain maps in this diagram from $C(X \times Y \times Z)$ to $C(X) \otimes C(Y) \otimes C(Z)$ are chain homotopic, (in fact, by a *natural* chain homotopy). Hence on passage to homology we *do* obtain a commutative diagram.

EXERCISES

5.1. Prove that the natural chain map $\eta:C(X \times Y) \to C(X) \otimes C(Y)$ (explicitly defined in §VI.5) is associative, i.e., if it is substituted for ξ in the diagram above, one obtains a commutative diagram.

5.2. Prove that the natural chain map $\zeta:C(X) \otimes C(Y) \to C(X \times Y)$ defined in §VI.3 is associative (in the sense discussed above).

In order to discuss the commutativity of ξ, consider the following diagram:

$$
\begin{array}{ccc}
C(X \times Y) & \xrightarrow{\;\xi^{X, Y}\;} & C(X) \otimes C(Y) \\
\big\downarrow{\scriptstyle t_*} & & \big\downarrow{\scriptstyle T} \\
C(Y \times X) & \xrightarrow{\;\xi^{Y, X}\;} & C(Y) \otimes C(X).
\end{array}
$$

In this diagram, $t:X \times Y \to Y \times X$ is defined by $t(x,y) = (y,x)$, and T is defined by

$$
T(a \otimes b) = (-1)^{pq} b \otimes a
$$

for any $a \in C_p(X)$ and $b \in C_q(Y)$. It is readily checked that T is a chain map. Therefore $T\xi^{X,Y}$ and $\xi^{Y,X}t_{\#}$ are both natural chain maps $C(X \times Y) \to C(Y) \otimes C(X)$, and by the method of acyclic models they can be proven chain homotopic (by a natural chain homotopy). It is interesting to note that there is one rather important difference between the question of the associativity and the question of the commutativity of ξ: As we saw in Exercise 5.1, it is possible to choose ξ so that it will be associative. However, it is known that it is *not* possible to choose a natural chain map ξ which is commutative. This follows from the fact that the Steenrod squaring operations exist and are nonzero (see Exercise VI.5.4 and the reference to Spanier's book given there). This is one of the mysterious "facts of life" in algebraic topology.

Next we will discuss the property of the Eilenberg–Zilber map ξ that guarantees the existence of units for cross and cup products. For this purpose, let us regard the additive group of integers $\mathbf{Z}$ as a chain complex which is "concentrated in degree 0," i.e., as a chain complex C such that $C_0 = \mathbf{Z}$, and $C_q = 0$ for $q \neq 0$. Then the augmentation $\varepsilon : C_0(X) \to \mathbf{Z}$ can be looked on as a chain map $\varepsilon : C(X) \to \mathbf{Z}$. With these conventions, consider the following two diagrams:

$$
\begin{array}{ccc}
C(X \times Y) & \xrightarrow{\ P_{1*}\ } & C(X) \\
\downarrow{\scriptstyle \xi^{X,Y}} & & \| \\
C(X) \otimes C(Y) & \xrightarrow[1 \otimes \varepsilon]{} & C(X) \otimes \mathbf{Z}
\end{array}
\qquad
\begin{array}{ccc}
C(X \times Y) & \xrightarrow{\ P_{2*}\ } & C(Y) \\
\downarrow{\scriptstyle \xi^{X,Y}} & & \| \\
C(X) \otimes C(Y) & \xrightarrow[\varepsilon \otimes 1]{} & \mathbf{Z} \otimes C(Y).
\end{array}
$$

Once again, by the use of acyclic models it can be proved that these two diagrams are homotopy commutative, (In these diagrams, $p_1 : X \times Y \to X$ and $p_2 : X \times Y \to Y$ denote projections on the first and second factors respectively.)

EXERCISE

5.3 Verify that if we substitute the explicit map η defined in §VI.5 for ξ in the above diagrams, they become commutative.

With these preliminaries out of the way, we can state our various associative laws, commutative laws, etc. The verifications of these properties will be left to the reader for the most part. First we will list the various associative laws.

Associative law for cross products. Let $u \in H^p(X,A;G_1)$, $v \in H^q(Y,B;G_2)$, and $w \in H^r(Z,C;G_3)$. Then

$$u \times (v \times w) = (u \times v) \times w$$

provided enough couples are assumed excisive to insure that all x-products are defined.

Associative law for cup products. Let $u \in H^p(X,A;G_1)$, $v \in H^q(X,B;G_2)$ and $w \in H^r(X,C;G_3)$. Then

$$u \cup (v \cup w) = (u \cup v) \cup w$$

provided enough couples in X are assumed excisive for everything to be well defined.

Associative law for slant products. Let $u \in H^p(Y,B;G_1)$, $v \in H^q(Z,C;G_2)$ and $w \in H_r((X,A) \times (Y,B) \times (Z,C);G_3)$, where we set

$$(X,A) \times (Y,B) = (X \times Y, X \times B \cup A \times Y)$$

etc., for the sake of brevity. Then

$$(u \times v)\backslash w = u\backslash(v\backslash w),$$

provided enough couples in the various product spaces are assumed excisive so that everything is well defined.

Associative law for cap products. Assume that $u \in H^p(X,A;G_1)$, $v \in H^q(X,B;G_2)$, and $w \in H_r(X,A \cup B \cup C;G_3)$. Then

$$(u \cup v) \cap w = u \cap (v \cap w)$$

in $H_{r-p-q}(X,C;G_1 \otimes G_2 \otimes G_3)$, provided $\{A,B\}$, $\{B,A \cup C\}$, $\{A \cup B,C\}$, and $\{A,C\}$ are excisive couples in X.

The fact that one has to make so many awkward assumptions about excisive couples in order to state an associative law must be considered a defect of singular homology and cohomology theory. Fortunately, in practice one does not usually have trouble about this, because it will be clear from the context in many cases, that all the couples involved are automatically excisive. This will be true if all the subspaces are open sets, or if all are subcomplexes of CW-complexes, for example.

Next, we will take up the commutative laws.

Commutative law for cross products. Let $u \in H^p(X,A;G_1)$ and $v \in H^q(Y,B;G_2)$. Then

$$t^*(u \times v) = (-1)^{pq} v \times u,$$

where $t:(Y,B) \times (X,A) \to (X,A) \times (Y,B)$ is defined by $t(y,x) = (x,y)$. Of course one must assume that $\{X \times B, A \times Y\}$ is an excisive couple.

Commutative law for cup products. Let $u \in H^p(X,A;G_1)$ and $v \in H^q(X,B;G_2)$. Then

$$u \cup v = (-1)^{pq} v \cup u$$

provided $\{A,B\}$ is an excisive couple.

There is no commutative law for slant or cap products; they do not lend themselves to any such law. This is *not* to say that the homotopy commutativity of the Eilenberg–Zilber chain homotopy equivalence ζ does not affect these products, however.

Existence of Units. For any space X, the augmentation $\varepsilon: C_0(X) \to \mathbf{Z}$ may be considered to be a 0-cochain, which is a cocycle. We will denote its cohomology class by $1 \in H^0(X;\mathbf{Z})$, or 1_X to be more explicit. For cross products, we have the following equations:

$$u \times 1_Y = p_1^*(u), \qquad u \in H^p(X,A;G),$$
$$1_X \times v = p_2^*(v), \qquad v \in H^q(Y,B;G).$$

In these equations, p_1 and p_2 denote the projections on the first and second factors of the product space, as usual.

For cup products, the equations are even simpler: for any $u \in H^p(X,A;G)$,

$$1_X \cup u = u \cup 1_X = u.$$

For slant products, we have

$$1_Y \backslash v = p_{1*}(v)$$

for any $v \in H_q(X \times Y; A \times Y; G)$, while for cap products,

$$1_X \cap v = v$$

for any $v \in H_q(X,B;G)$.

Note that 1_X acts as both a left and right unit for cross and cup products, while for slant and cap products, we have only a left unit. Also note that there is *no* unit in $H^0(X,A)$ if A is nonempty.

§6. Digression: The Exact Sequence of a Triple or a Triad

In order to describe in a most concise way the behavior of the various products with respect to the boundary operator $\partial_* : H_q(X,A) \to H_{q-1}(A)$ or the coboundary operator $\delta^* : H^p(A) \to H^{p+1}(X,A)$, it is convenient to make use of the exact homology (or cohomology) sequence of a triad.

First of all, let (X,A,B) be a *triple*, i.e., X is a topological space, and $X \supset A \supset B$. Then we have the following split exact sequence of chain complexes:

$$0 \to C(A,B) \xrightarrow{i_*} C(X,B) \xrightarrow{j_*} C(X,A) \to 0.$$

Since this sequence is split exact, if we apply the functor $\otimes G$ or $\mathrm{Hom}(\ ,G)$, we again obtain a short exact sequence of chain or cochain complexes. We may then pass to the corresponding long exact homology and cohomology sequences:

$$\cdots \xrightarrow{\partial_*} H_q(A,B;G) \xrightarrow{i_*} H_q(X,B;G) \xrightarrow{j_*} H_q(X,A;G) \xrightarrow{\partial_*} H_{q-1}(A,B;G) \to \cdots,$$

$$\cdots \xleftarrow{\delta^*} H^q(A,B;G) \xleftarrow{i^*} H^q(X,B;G) \xleftarrow{j^*} H^q(X,A;G) \xleftarrow{\delta^*} H^{q-1}(A,B;G) \leftarrow \cdots.$$

Note: The exact homology or cohomology sequence of a triple can also be derived directly from the basic concepts of singular homology theory, without going back to chain complexes; cf. Eilenberg and Steenrod, [2], Chapter I, §10.

Next, let $(X;A,B)$ be a *triad*, i.e., A and B are arbitrary subsets of X (no inclusion relations are assumed between A and B). Assume that $\{A,B\}$ is an excisive couple in X; it follows that the inclusion maps

$$k_1 : (A, A \cap B) \to (A \cup B, B)$$
$$k_2 : (B, A \cap B) \to (A \cup B, A)$$

induce isomorphisms on homology and cohomology groups with any co-efficients. If we substitute the (co-) homology groups of $(A, A \cap B)$ for those of $(A \cup B, B)$ in the exact (co-) homology sequence of the triple $(X, A \cup B, B)$, (using the isomorphism induced by k_1), we obtain one of the (co-) homology sequences of the triad $(X; A, B)$. To obtain the other (co-) homology sequence of this triad, use the isomorphism induced by k_2 to substitute the (co-) homology groups of $(B, A \cap B)$ for those of $(A \cup B, A)$ in the exact (co-) homology sequence of the triple $(X, A \cup B, A)$. The resulting homology sequences are as follows:

$$\cdots \overset{\Delta_*}{\to} H_n(A, A \cap B) \to H_n(X, B) \to H_n(X, A \cup B) \overset{\Delta_*}{\to} H_{n-1}(A, A \cap B) \cdots,$$

$$\cdots \overset{\Delta_*}{\to} H_n(B, A \cap B) \to H_n(X, A) \to H_n(X, A \cup B) \overset{\Delta_*}{\to} H_{n-1}(B, A \cap B) \cdots.$$

The coefficient group has been omitted from the notation. The homomorphisms Δ_* will be referred to as *the boundary operators of the triad* $(X; A, B)$; they are defined so as to make the following two diagrams commutative:

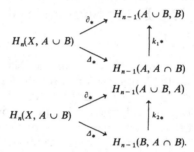

Analogously, we will denote the coboundary operators of the exact co-homology sequences of this triad as follows:

$$\Delta^*: H^{n-1}(A, A \cap B) \to H^n(X, A \cup B),$$
$$\Delta^*: H^{n-1}(B, A \cap B) \to H^n(X, A \cup B).$$

We have introduced the exact homology and cohomology sequences of a triad for a very specific purpose in connection with the various products. However these exact sequences, and the exact sequence of a triple, are of interest in their own right.

There is one other exact sequence which it is convenient to introduce now, known as the *relative Mayer–Vietoris sequence*. It will be needed in Chapter IX. Let $(X; A, B)$ be a triad, and assume that $\{A, B\}$ is an excisive couple in X. We will use the following notation for inclusion maps:

$$i: (X, A \cap B) \to (X, A),$$
$$j: (X, A \cap B) \to (X, B),$$
$$k: (X, A) \to (X, A \cup B),$$
$$l: (X, B) \to (X, A \cup B).$$

Consider the following sequence of chain complexes and chain maps:

$$0 \to C(X, A \cap B) \overset{\Phi}{\to} C(X,A) \oplus C(X,B) \overset{\Psi}{\to} \frac{C(X)}{C(A) + C(B)} \to 0.$$

Here the chain maps Φ and ψ are defined as follows:

$$\Phi(x) = (i_\# x, j_\# x),$$
$$\Psi(u,v) = k_\#(u) - l_\#(v).$$

It is not difficult to prove that this sequence is exact; in fact, it is even split exact, because all the chain complexes consist of free abelian groups. If we pass to the corresponding long exact homology sequence, and substitute $C(X, A \cup B)$ for $C(X)/[C(A) + C(B)]$, we obtain the relative Mayer–Vietoris sequence of the triad $(X; A, B)$:

$$\cdots \to H_n(X, A \cap B) \overset{\varphi}{\to} H_n(X,A) \oplus H_n(X,B) \overset{\psi}{\to} H_n(X, A \cup B)$$
$$\overset{\Delta}{\to} H_{n-1}(X, A \cap B) \overset{\varphi}{\to} \cdots .$$

Of course there is a dual exact sequence of cohomology groups.

§7. Behavior of Products with Respect to the Boundary and Coboundary Operator of a Pair

We will content ourselves with stating the main properties involved, leaving the proofs to the reader.

(a) *Cross Products.* Assume that (X,A) and (Y,B) are pairs such that $\{A \times Y, X \times B\}$ is an excisive couple. Then the following two diagrams are commutative:

$$
\begin{array}{ccc}
H^p(A) \otimes H^q(Y,B) & \overset{\times}{\longrightarrow} & H^{p+q}(A \times Y, A \times B) \\
\downarrow{\scriptstyle \delta^* \otimes 1} & & \downarrow{\scriptstyle \Delta_*} \\
H^{p+1}(X,A) \otimes H^q(Y,B) & \overset{\times}{\longrightarrow} & H^{p+q+1}(X \times Y, X \times B \cup A \times Y)
\end{array}
$$

$$
\begin{array}{ccc}
H^p(X,A) \otimes H^q(B) & \overset{\times}{\longrightarrow} & H^{p+q}(X \times B, A \times B) \\
\downarrow{\scriptstyle (-1)^p \otimes \delta^*} & & \downarrow{\scriptstyle \Delta^*} \\
H^p(X,A) \otimes H^{q+1}(Y,B) & \overset{\times}{\longrightarrow} & H^{p+q+1}(X \times Y, X \times B \cup A \times Y).
\end{array}
$$

These two relations may also be expressed by equations, as follows: For any $u \in H^p(A)$ and $v \in H^q(Y,B)$,

$$(\delta^*u) \times v = \Delta^*(u \times v).$$

For any $u \in H^p(X,A)$ and $v \in H^q(B)$,

$$(-1)^p(u \times \delta^*v) = \Delta^*(u \times v).$$

Obviously, the second relation can be derived from the first by use of the commutative law.

  (b) *Cup Products.* Assume that $\{A,B\}$ is an excisive couple in X. Then we have the following two diagrams to describe the relations involved (they are not commutative diagrams in the usual sense; cf. the discussion of naturality of the slant and cap products).

$$
\begin{array}{ccc}
H^p(A) \otimes H^q(A, A \cap B) & \overset{\cup}{\longrightarrow} & H^{p+q}(A, A \cap B) \\
\downarrow{\scriptstyle \delta^*} \qquad \uparrow{\scriptstyle k^*} & & \downarrow{\scriptstyle \Delta^*} \\
H^{p+1}(X,A) \otimes H^q(X,B) & \overset{\cup}{\longrightarrow} & H^{p+q+1}(X, A \cup B)
\end{array}
$$

$$
\begin{array}{ccc}
H^p(B, A \cap B) \otimes H^q(B) & \overset{\cup}{\longrightarrow} & H^{p+q}(B, A \cap B) \\
\uparrow{\scriptstyle l^*} \quad \downarrow{\scriptstyle \delta^*} \quad {\scriptstyle (-1)^p} & & \downarrow{\scriptstyle \Delta^*} \\
H^p(X,A) \otimes H^{q+1}(X,B) & \overset{\cup}{\longrightarrow} & H^{p+q+1}(X, A \cup B).
\end{array}
$$

These relations may also be stated in equations, as follows. If $u \in H^p(A)$ and $v \in H^q(X,B)$, then

$$(\delta^*u) \cup v = \Delta^*(u \cup k^*v).$$

For the second relation, if $u \in H^p(X,A)$ and $v \in H^q(B)$, then

$$(-1)^p u \cup \delta^*v = \Delta^*((l^*u) \cup v).$$

EXERCISE

7.1. Under the above assumptions, prove that we have the following commutative diagram:

$$
\begin{array}{ccccccc}
H^p(X,A) & \overset{j^*}{\longrightarrow} & H^p(X) & \overset{i^*}{\longrightarrow} & H^p(A) & \overset{\delta^*}{\longrightarrow} & H^{p+1}(X,A) \\
\downarrow{\scriptstyle \cup v} & & \downarrow{\scriptstyle \cup v} & & \downarrow{\scriptstyle \cup k^*v} & & \downarrow{\scriptstyle \cup v} \\
H^{p+q}(X, A \cup B) & \longrightarrow & H^{p+q}(X,B) & \longrightarrow & H^{p+q}(A, A \cap B) & \overset{\Delta^*}{\longrightarrow} & H^{p+q+1}(X, A \cup B).
\end{array}
$$

Here $v \in H^q(X,B)$.

  (c) *Slant products.* Assume, as in (a), that (X,A) and (Y,B) are pairs such that $\{X \times B, A \times Y\}$ is an excisive couple. Then the relations are expressed

by the following two diagrams of which the first is a commutative diagram
in the usual sense:

$$H^p(Y,B) \otimes H_q(X \times Y, A \times Y \cup X \times B) \xrightarrow{\text{slant}} H_{q-p}(X,A)$$

$$\downarrow {\scriptstyle (-1)^p \otimes \varDelta_*} \qquad\qquad\qquad\qquad\qquad \downarrow {\scriptstyle \partial_*}$$

$$H^p(Y,B) \otimes H_{q-1}(A \times Y, A \times B) \xrightarrow{\text{slant}} H_{q-p-1}(A)$$

$$H^p(Y,B) \otimes H_q(X \times Y, A \times Y \cup X \times B)$$

$$\uparrow {\scriptstyle \delta^*} \qquad\qquad \downarrow {\scriptstyle \varDelta_*} \qquad {\scriptstyle (-1)^p} \qquad \searrow {\scriptstyle \text{slant}} \qquad H_{q-p}(X,A).$$

$$H^{p-1}(B) \otimes H_{q-1}(X \times B, A \times B) \qquad \nearrow {\scriptstyle \text{slant}}$$

The second diagram expresses the fact that the homomorphisms δ^* and $\varDelta_*$
are adjoint in a certain sense. These relations may be expressed in equations
as follows: Let $u \in H^p(Y,B)$ and $v \in H_q(X \times Y, A \times Y \cup X \times B)$. Then

$$\partial_*(u \backslash v) = (-1)^p u \backslash \varDelta_* v.$$

For the second relation, let $u \in H^{p-1}(B)$ and $v \in H_q(X \times Y, A \times Y \cup X \times B)$.
Then

$$(\delta^* u) \backslash v + (-1)^{p-1} u \backslash \varDelta_* v = 0$$

(d) *Cap Products.* Assume that $\{A,B\}$ is an excisive couple in X. Then the
following diagram is commutative, up to the sign $(-1)^p$:

$$H^p(X,A) \otimes H_q(X, A \cup B) \xrightarrow{\cap} H_{q-p}(X,B)$$

$$\downarrow {\scriptstyle k^* \otimes \varDelta_*} \qquad\qquad\qquad\qquad \downarrow {\scriptstyle \partial_*}$$

$$H^p(B, A \cap B) \otimes H_{q-1}(B, A \cap) \xrightarrow{\cap} H_{q-p-1}(B).$$

This relation may be expressed by the following equation:

$$(-1)^p(k^* u) \cap (\varDelta_* v) = \partial_*(u \cap v)$$

for any $u \in H^p(X,A)$ and $v \in H_q(X, A \cup B)$. A second relation is indicated by
the following diagram:

$$H^p(X,A) \otimes H^q(X, A \cup B) \xrightarrow{\qquad} H_{q-p}(X,B)$$

$$\uparrow {\scriptstyle \delta^*} \qquad\qquad \downarrow {\scriptstyle \varDelta_*} \qquad {\scriptstyle (-1)^p} \qquad \uparrow {\scriptstyle k_*}$$

$$H^{p-1}(A) \otimes H_{q-1}(A, A \cap B) \xrightarrow{\qquad} H_{q-p}(A, A \cap B).$$

Equivalently,

$$(\delta^* u) \cap v + (-1)^{p-1} k_*(u \cap \varDelta_* v) = 0$$

for any $u \in H^{p-1}(A)$ and any $v \in H_q(X, A \cup B)$.

7.2. Prove that the following diagram is commutative:

$$
\begin{array}{ccccccc}
H_q(B,A \cap B) & \longrightarrow & H_q(X,A) & \longrightarrow & H_q(X, A \cup B) & \xrightarrow{\;\Delta_*\;} & H_{q-1}(B, A \cap B) \\
\downarrow{\scriptstyle (k^*u)\cap} & & \downarrow{\scriptstyle u\cap} & & \downarrow{\scriptstyle u\cap} \quad (-1)^p & & \downarrow{\scriptstyle (k^*u)\cap} \\
H_{q-p}(B) & \longrightarrow & H_{q-p}(X) & \longrightarrow & H_{q-p}(X,B) & \xrightarrow{\;\partial_*\;} & H_{q-p-1}(B).
\end{array}
$$

Here $u \in H^p(X,A)$.

7.3. Prove that corresponding homomorphisms in the following two exact sequences are "adjoints" of each other, with respect to the indicated cap product:

$$
\begin{array}{ccccccc}
H^p(A) & \longleftarrow & H^p(X) & \longleftarrow & H^p(X,A) & \xleftarrow{\;\delta^*\;} & H^{p-1}(A) \\
\otimes & & \otimes & & \otimes & & \otimes \\
H_q(A, A \cap B) & \longrightarrow & H_q(X,B) & \longrightarrow & H_q(X, A \cup B) & \longrightarrow & H_{q-1}(A, A \cap B) \\
\downarrow{\scriptstyle \cap} & & \downarrow{\scriptstyle \cap} & & \downarrow{\scriptstyle \cap} & & \downarrow{\scriptstyle \cap} \\
H_{q-p}(A, A \cap B) & \longrightarrow & H_{q-p}(X,B) & = & H_{q-p}(X,B) & \longleftarrow & H_{q-p}(A, A \cap B).
\end{array}
$$

§8. Relations Involving the Inner Product

These relations involve the inner product, which was defined in §2, and the cross, slant, cup, and cap products.

(a) Assume that (X,A) and (Y,B) are pairs such that $\{A \times Y, X \times B\}$ is an excisive couple in $X \times Y$. In Chapter VI we defined the homomorphism

$$\alpha: H_p(X,A) \otimes H_q(Y,B) \to H_{p+q}((X,A) \times (Y,B)).$$

Let $a \in H_p(X,A)$, $b \in H_q(Y,B)$, $u \in H^p(X,A;G_1)$, and $v \in H^q(Y,B;G_2)$. Then

$$(-1)^{pq}\langle u \times v, \alpha(a \otimes b)\rangle = \langle u,a\rangle \otimes \langle v,b\rangle.$$

The proof of this relation is easy.

(b) Assume, as in (a) that $\{A \times Y, X \times B\}$ is an excisive couple in $X \times Y$. Let $u \in H^p(X,A;G_1)$, $v \in H^q(Y,B;G_2)$, and $w \in H_{p+q}(X \times Y, A \times Y \cup X \times B;G_3)$. Then

$$\langle u \times v, w\rangle = \langle u, v\backslash w\rangle.$$

(c) Assume that $\{A,B\}$ is an excisive couple in X. Let $u \in H^p(X,A;G_1)$, $v \in H^q(X,B;G_2)$, and $w \in H_{p+q}(X, A \cup B;G_3)$. Then

$$\langle u \cup v, w\rangle = \langle u, v \cap w\rangle.$$

A noteworthy special case of this relation occurs when $A = \varnothing$, $p = 0$, $G_1 = \mathbf{Z}$, and $u = 1 \in H^0(X;\mathbf{Z})$. Then $1 \cup v = v$, and it is easily verified that $\langle 1, v \cap w \rangle = \varepsilon_*(v \cap w)$. Thus under these hypotheses, we obtain the relation

$$\langle v,w \rangle = \varepsilon_*(v \cap w)$$

which expresses the inner product in terms of the cap product and the augmentation.

The proof of Relations (b) and (c) are easy. In the case where $G_1 = G_2 = G_3 = F$, where F is a field, and all the homology and cohomology groups involved are finite-dimensional vector spaces over F, Relation (b) shows that cross products are determined by slant products, and vice-versa. Similarly, Relation (c) shows that under these hypotheses, cup products are determined by cap products, and vice-versa (cf. Exercise VII.6.1).

§9. Cup and Cap Products in a Product Space

Let $u \in H^p(X,A)$, $v \in H^q(X,B)$, $w \in H^r(Y,C)$, and $x \in H^s(Y,D)$ (the coefficient groups are omitted from the notation). Then

$$(u \times w) \cup (v \times x) = (-1)^{qr}(u \cup v) \times (w \cup x) \qquad (9.1)$$

provided we assume enough couples are excisive so that everything is well defined. In particular, this would be the case if A, B, C, and D were all empty.

Probably the easiest way to prove Equation (9.1) is to make use of the relation between cup and cross products explained in §3. If everything is expressed in terms of cup products, this relation becomes almost obvious. Therefore the details are left to the reader.

To state an analogous relation for cap products, we must use the homomorphism

$$\alpha : H_p(X,A) \otimes H_q(Y,B) \to H_{p+q}(X \times Y, A \times Y \cup X \times B)$$

defined in §§VI.4 and IV.6. This can be extended in an obvious way to a homomorphism

$$\alpha : H_p(X,A;G_1) \otimes H_q(Y,B;G_2) \to H_{p+q}(X \times Y, A \times Y \cup X \times B, G_1 \otimes G_2)$$

with arbitrary coefficients G_1 and G_2. Assume that $u \in H^p(X,A_1)$, $v \in H^q(Y,B_1)$, $a \in H_r(X, A_1 \cup A_2)$ and $b \in H_s(Y,B_1 \cup B_2)$. Then

$$(u \times v) \cap \alpha(a \otimes b) = (-1)^{qr}\alpha((u \cap a) \otimes (v \cap b)) \qquad (9.2)$$

provided enough couples are assumed excisive. A detailed proof of this relation is written out in Dold [1], pp. 240–241.

This completes our survey of the main properties of the four products.

§10. Remarks on the Coefficients for the Various Products—The Cohomology Ring

In all four products, we started out with homology or cohomology classes u and v with coefficient groups G_1 and G_2 respectively, and the product always had coefficient group $G_1 \otimes G_2$. Sometimes it is convenient to assume given a homomorphism $h: G_1 \otimes G_2 \to G$ and to systematically apply the coefficient homomorphism $h_\#$ to the resulting product. For example, if R is a ring; and $h: R \otimes R \to R$ is the homomorphism induced by the multiplication, then we get a cup product which assigns to elements $u \in H^p(X;R)$ and $v \in H^q(X;R)$ an element $u \cup v \in H^{p+q}(X;R)$. With this multiplication, the direct sum

$$H^*(X;R) = \sum_n H^n(X;R)$$

becomes a kind of ring which is called a *graded ring*, because the underlying additive group is the direct sum of a sequence of subgroups, indexed by the integers. In fact, $H^*(X;R)$ is the prototype of a graded ring. If R has a unit $1 \in R$, then $H^*(X;R)$ has a unit $1_X \in H^0(X;R)$; it is represented by the cocycle

$$C_0(X) \xrightarrow{\varepsilon} \mathbf{Z} \xrightarrow{e} R,$$

where ε is the augmentation and e is the unique ring homomorphism defined by $e(1) = 1$. If the ring R is commutative, then $H^*(X;R)$ is commutative in the graded sense (sometimes called skew-commutative or anticommutative):

$$u \cup v = (-1)^{pq} v \cup u$$

for any $u \in H^p(X;R)$ and $v \in H^q(X;R)$. In this case, $H^*(X;R)$ is a *graded algebra* over the commutative ring R.

We mention two more examples like this, leaving the reader to fill in the details of the definitions, etc. For both examples, let R be a ring with unit, M a left R-module, and $h: R \otimes M \to M$ the homomorphism defining the module structure.

EXAMPLE 10.1. The cap product assigns to any elements $u \in H^p(X;R)$ and $v \in H_q(X;M)$ an element $u \cap v \in H_{q-p}(X;M)$. Using this cap product, the direct sum

$$H_*(X;M) = \sum_n H_n(X;M)$$

becomes a *graded left module* over the graded ring $H^*(X;R)$.

EXAMPLE 10.2. Let (X,A) be an arbitrary pair. The cup product assigns to elements $u \in H^p(X;R)$ and $v \in H^q(X,A;M)$ an element $u \cup v \in H^{p+q}(X,A;M)$. This makes

$$H^*(X,A;M) = \sum_n H^n(X,A;M)$$

into a *graded left module* over $H^*(X;R)$. Moreover, each of the homomorphisms of the exact sequence of the pair (X,A),

$$j^*:H^*(X,A;M) \to H^*(X;M),$$
$$i^*:H^*(X;M) \to H^*(A;M),$$

and

$$\delta^*:H^*(A;M) \to H^*(X,A;M)$$

are homomorphisms of graded left $H^*(X;R)$-modules (the definition of the module structure on $H^*(X;M)$ and $H^*(A;M)$ is left to the reader). In this example, the homomorphisms j^* and i^* have degree 0, while δ^* has degree $+1$.

§11. The Cohomology of Product Spaces (The Künneth Theorem for Cohomology)

By combining the Künneth Theorem of §VI.6 with the universal coefficient theorem for cohomology of §VII.4, one can express the cohomology groups of a product space, $H^n(X \times Y;G)$ in terms of the homology groups of the factors, $H_p(X)$ and $H_q(Y)$, (in principle, at least). What we are now interested in is the expression of $H^n(X \times Y;G)$ in terms of the *cohomology* groups of the factors, $H^p(X)$ and $H^q(Y)$. The point is that we can use such an expression together with the relations given in §9 to obtain information about cup and cap products in $X \times Y$ terms of these products in the factors, X and Y.

The cross product defines a homomorphism

$$H^p(X,\mathbf{Z}) \otimes H^q(Y,\mathbf{Z}) \to H^{p+q}(X \times Y;\mathbf{Z}).$$

This definition can be extended in an obvious way to a homomorphism

$$\sum_{p+q=n} H^p(X;\mathbf{Z}) \otimes H^q(Y;\mathbf{Z}) \to H^n(X \times Y;\mathbf{Z}).$$

One would then hope to prove that this homomorphism is a monomorphism, and that the cokernel is isomorphic to something of the form

$$\sum_{p,q} \mathrm{Tor}(H^p(X;\mathbf{Z}),H^q(Y;\mathbf{Z}))$$

just as in the case of homology. Unfortunately, simple examples show that this is too much to hope for: If X and Y are discrete spaces having infinitely many points, no such theorem holds. However, if X or Y is a *finite* discrete space, then there is no problem.

This is the key to the situation: one must impose some sort of finiteness condition on at least one of the factors.

Before we can state and prove such a theorem, we need some algebraic preliminaries. First of all, recall that if F is a free abelian group of *finite* rank then $\mathrm{Hom}(F,\mathbf{Z})$ is also a free abelian group (of the same rank). It may be proved that if F is a free abelian of infinite rank, then $\mathrm{Hom}(F,\mathbf{Z})$ is *not* free. However, we will have no need for this result. It follows that if $K = \{K_q, \partial_q\}$ is a chain complex such that K_q is free abelian of finite rank for each q, then $\mathrm{Hom}(K,\mathbf{Z})$ is a cochain complex of free abelian groups.

Secondly, recall that we introduced earlier the natural homomorphism $\mathrm{Hom}(A,A') \otimes \mathrm{Hom}(B,B') \to \mathrm{Hom}(A \otimes B, A' \otimes B')$ which assigns to homomorphisms $f : A \to A'$ and $g : B \to B'$ the tensor product of the two homomorphisms, $f \otimes g : A \otimes B \to A' \otimes B'$. In general, the abelian groups $\mathrm{Hom}(A,A') \otimes \mathrm{Hom}(B,B')$ and $\mathrm{Hom}(A \otimes B, A' \otimes B')$ are *not* isomorphic. However, in the special case where A is free abelian of finite rank and $A' = \mathbf{Z}$, it is readily verified that the natural homomorphism is an isomorphism of $\mathrm{Hom}(A,\mathbf{Z}) \otimes \mathrm{Hom}(B,B')$ onto $\mathrm{Hom}(A \otimes B, B')$. We can now extend this result to chain complexes. Suppose that $K = \{K_q, \partial_q\}$ is a positive chain complex such that each K_q is free abelian of finite rank, that $C = \{C_q, \partial_q\}$ is another positive chain complex, and G is an abelian group. Then the natural chain map

$$\mathrm{Hom}(K,\mathbf{Z}) \otimes \mathrm{Hom}(C,G) \to \mathrm{Hom}(K \otimes C, G)$$

is an isomorphism of chain complexes.

Finally, we need the following lemma of a rather technical nature:

Lemma 11.1. *Let (X,A) be a pair such that $H_q(X,A)$ is finitely generated for all q. Then there exists a chain complex $K = \{K_q, \partial_q\}$ such that each K_q is a free abelian group of finite rank, and a chain homotopy equivalence $f : K \to C(X,A)$.*

PROOF. For each q, choose an epimorphism e_q of a finitely generated free abelian group F_q onto $H_q(X,A)$; denote the kernel by R_{q+1}, and let $d_{q+1} : R_{q+1} \to F_q$ denote the inclusion homomorphism. Then

$$0 \leftarrow H_q(X,A) \overset{e_q}{\leftarrow} F_q \overset{d_{q+1}}{\longleftarrow} R_{q+1} \leftarrow 0$$

is a short exact sequence, and both F_q and R_{q+1} are free abelian of finite rank. Define $K_q = F_q \oplus R_q$ for all q, and $\partial_q : K_q \to K_{q-1}$ by

$$\partial_q | F_q = 0,$$
$$\partial_q | R_q = d_q.$$

Then $K = \{K_q, \partial_q\}$ is a chain complex such that each K_q is free abelian of finite rank. It is an easy exercise to prove that there exist homomorphisms

$$\varphi_q : F_q \to Z_q(X,A)$$
$$\psi_{q+1} : R_{q+1} \to B_q(X,A)$$

for all q such that the following diagram is commutative:

$$\begin{array}{ccccccccc} 0 & \longleftarrow & H_q(X,A) & \xleftarrow{\;e_q\;} & F_q & \xleftarrow{\;d_{q+1}\;} & R_{q+1} & \longleftarrow & 0 \\ & & \| & & \downarrow{\scriptstyle\varphi_q} & & \downarrow{\scriptstyle\psi_{q+1}} & & \\ 0 & \longleftarrow & H_q(X,A) & \longleftarrow & Z_q(X,A) & \longleftarrow & B_q(X,A) & \longleftarrow & 0. \end{array}$$

Next, we may choose a homomorphism $\theta_{q+1}\colon R_{q+1} \to C_{q+1}(X,A)$ such that the following diagram is commutative:

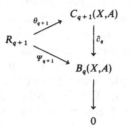

Now define $f_q\colon K_q \to C_q(X,A)$ by

$$f_q|F_q = \varphi_q,$$
$$f_q|R_q = \theta_q.$$

It is readily checked that $f = \{f_q\}$ is a chain map, and that the induced homomorphism

$$f_*\colon H_q(K) \to H_q(X,A)$$

is an isomorphism for all q. Therefore f is a chain homotopy equivalence, by Theorem V.2.3.   Q.E.D.

Now that we have these technical details behind us, we can state the desired theorem:

Theorem 11.2. *Let (X,A) and (Y,B) be pairs such that the following two conditions hold: $H_q(X,A)$ is finitely generated for all q, and $\{X \times B, A \times Y\}$ is an excisive couple in $X \times Y$. Then the cross product defines a homomorphism $\alpha\colon \sum_{p+q=n} H^p(X,A;\mathbf{Z}) \otimes H^q(Y,B;G) \to H^n(X \times Y; A \times Y \cup X \times B; G)$ which is a monomorphism onto a direct summand and the cokernel is naturally isomorphic to $\sum_{p+q=n+1} \mathrm{Tor}(H^p(X,A;\mathbf{Z}), H^q(Y,B;G))$.*

We will indicate the main steps in the proof, leaving the verification of details to the reader.

By Lemma 11.1, there exists a chain complex K of finitely generated free abelian groups and a chain homotopy equivalence $f\colon K \to C(X,A)$. It follows that $\mathrm{Hom}(K,\mathbf{Z})$ is a cochain complex of free abelian groups, and

$\mathrm{Hom}(f,1)\colon\mathrm{Hom}(C(X,A),\mathbf{Z}) \to \mathrm{Hom}(K,\mathbf{Z})$ is also a chain homotopy equivalence. Now consider the following commutative diagram of cochain complexes and cochain maps:

$$\mathrm{Hom}(K \otimes C(Y,B),G) \xleftarrow{\quad \mathrm{Hom}(f \otimes 1,1) \quad} \mathrm{Hom}(C(X,A) \otimes C(Y,B),G)$$
$$\uparrow{\scriptstyle a} \qquad\qquad\qquad\qquad\qquad\qquad\qquad\qquad \uparrow$$
$$\mathrm{Hom}(K,\mathbf{Z}) \otimes \mathrm{Hom}(C(Y,B),G) \xleftarrow{\quad \mathrm{Hom}(f,1) \otimes 1 \quad} \mathrm{Hom}(C(X,A),\mathbf{Z}) \otimes \mathrm{Hom}(C(Y,B),G).$$

In this diagram, the symbol 1 refers to an appropriate identity map. By the discussion preceding Lemma 11.1, the arrow labelled a denotes an isomorphism. Since f is a chain homotopy equivalence, it follows that the horizontal arrows denote cochain homotopy equivalences. Hence on passage to cohomology, all four arrows in this diagram would induce isomorphisms. To complete the proof, one applies the Künneth theorem to the tensor product $\mathrm{Hom}(K,\mathbf{Z}) \otimes \mathrm{Hom}(C(Y,B),G)$. This is legitimate, since $\mathrm{Hom}(K,\mathbf{Z})$ is a cochain complex of free abelian groups. The remaining details may be left to the reader.   Q.E.D.

Corollary 11.3. *Let X and Y be topological spaces such that $H_q(X)$ is finitely generated for all q and such that at least one of the two spaces has all cohomology groups torsion-free. Then*

$$\alpha\colon \sum_{p+q=n} H^p(X;\mathbf{Z}) \otimes H^q(Y;\mathbf{Z}) \to H^n(X \times Y;\mathbf{Z})$$

is an isomorphism for all n. In this case the cohomology ring $H^*(X \times Y;\mathbf{Z})$ is completely determined by $H^*(X;\mathbf{Z})$ and $H^*(Y;\mathbf{Z})$.

The last sentence of this corollary follows from the relations for cup products in a product space given in §9. It also inspires the following definition.

Definition 11.1. Let $A^* = \sum_i A^i$ and $B^* = \sum_j B^j$ be graded rings. The *tensor product* $A^* \otimes B^*$ is the graded ring defined as follows:

$$(A^* \otimes B^*)^n = \sum_{i+j=n} A^i \otimes B^j \quad \text{(direct sum)}.$$

The multiplication is defined as follows:

$$(u_1 \otimes v_1) \cdot (u_2 \otimes v_2) = (-1)^{p_2 q_1}(u_1 u_2) \otimes (v_1 v_2),$$

where $u_i \in A^{p_i}$ and $v_j \in B^{q_j}$ for $i,j = 1,2$. With this structure $A^* \otimes B^*$ is also a graded ring.

Using this definition, the corollary above can be restated as follows: *Let X and Y be topological spaces such that $H_q(X)$ is finitely generated for all q, and at least one of the two spaces has all cohomology groups torsion-free. Then* *the cohomology ring $H^*(X \times Y; \mathbf{Z})$ is naturally isomorphic to the tensor product of the cohomology rings of the factors*:

$$\alpha: H^*(X; \mathbf{Z}) \otimes H^*(Y; \mathbf{Z}) \approx H^*(X \times Y; \mathbf{Z}).$$

EXAMPLE 11.1. The cohomology ring of an n-sphere, $H^*(S^n; \mathbf{Z})$ is easily determined. We know that $H^0(S^n; \mathbf{Z})$ is an infinite cyclic group generated by the unit, $1 \in H^0(S^n; \mathbf{Z})$, and $H^n(S^n; \mathbf{Z})$ is also infinite cyclic with generator u; all other cohomology groups are 0. The cup products are completely determined by the equations

$$u \cup 1 = 1 \cup u = u.$$

We can now use the above rules to determine the cohomology ring $H^*(S^m \times S^n; \mathbf{Z})$. Let $u \in H^m(S^m; \mathbf{Z})$ and $v \in H^n(S^n; \mathbf{Z})$ denote generators of these infinite cyclic groups. Then $H^*(S^m \times S^n; \mathbf{Z})$ is the direct sum of four infinite cyclic groups, with generators 1×1 (the unit), $u \times 1$, $1 \times v$, and $u \times v$. There is one nontrivial product:

$$(u \times 1) \cup (1 \times v) = u \times v.$$

EXERCISES

11.1. Let A be a retract of X with retraction $r: X \to A$ and inclusion map $i: A \to X$. Consider the induced homomorphisms

$$r^*: H^*(A; \mathbf{Z}) \to H^*(X; \mathbf{Z})$$
$$i^*: H^*(X; \mathbf{Z}) \to H^*(A; \mathbf{Z}).$$

Prove that kernel i^* is an ideal in the graded ring $H^*(X; \mathbf{Z})$, and image r^* is a subring.

11.2. Let X and Y be spaces with chosen basepoints, $x_0 \in X$ and $y_0 \in Y$. Define

$$X \vee Y = (X \times \{y_0\}) \cup (\{x_0\} \times Y).$$

It is sometimes called the 1-*point union* of X and Y. Assuming that X and Y are arcwise connected, express the structure of the cohomology ring $H^*(X \vee Y; \mathbf{Z})$ in terms of $H^*(X; \mathbf{Z})$ and $H^*(Y; \mathbf{Z})$. (Assume that also x_0 and y_0 have "nice" neighborhoods in X and Y respectively, as described in Problem III.5.2.)

11.3. Let m and n be positive integers, $X = S^m \times S^n$, and $Y = S^m \vee S^n \vee S^{m+n}$. Prove that $H_q(X; G) \approx H_q(Y; G)$ and $H^q(X; G) \approx H^q(Y; G)$ for any abelian group G and integer q; then prove that X and Y are *not* of the same homotopy type.

We conclude this lengthy chapter with an analogue of Corollary 11.3 for the case where we use cohomology with *coefficients in a commutative field F*. The result is easy to state, and of rather wide generality.

Theorem 11.4. *Let (X,A) and (Y,B) be pairs such that $\{X \times B, A \times Y\}$ is an excisive couple, and $H_q(X,A; F)$ is a finite-dimensional vector space over F for* *all q. Then the x-product defines a natural isomorphism*

$$\alpha: \sum_{p+q=n} H^p(X,A;F) \underset{F}{\otimes} H^q(Y,B;F) \to H^n(X \times Y, A \times Y \cup X \times B; F).$$

Corollary 11.5. *Let X be a space such that $H_q(X;F)$ has finite rank over F for all q. Then for any space Y, the cohomology algebra $H^*(X \times Y; F)$ is naturally isomorphic to the tensor product:*

$$\alpha: H^*(X;F) \underset{F}{\otimes} H^*(Y;F) \approx H^*(X \times Y; F).$$

The proof of this theorem and corollary is actually somewhat simpler than the proof of Theorem 11.2 and Corollary 11.3 because one has to deal with vector spaces over F rather than abelian groups. It is also necessary to use relations such as the following:

$$C(X) \otimes C(Y) \otimes F \approx C(X,F) \underset{F}{\otimes} C(Y,F)$$

$$\text{Hom}(C(X),F) \approx \text{Hom}_F(C(X,F),F).$$

Once again, the details are left to the reader.

Bibliography for Chapter VIII

[1] A. Dold, *Lectures on Algebraic Topology*, Springer-Verlag, New York, 1972.
[2] S. Eilenberg and N. Steenrod, *Foundations of Algebraic Topology*, Princeton University Press, Princeton, 1952.

CHAPTER IX

Duality Theorems for the Homology of Manifolds

§1. Introduction

An n-dimensional manifold is a Hausdorff space such that every point has an open neighborhood which is homeomorphic to Euclidean n-space, $\mathbf{R}^n$ (see Massey, [6], Chapter I). One of the main goals of this chapter will be to prove one of the oldest results of algebraic topology, the famous Poincaré duality theorem for compact, orientable manifolds. It is easy to state the Poincaré duality theorem but the proof is lengthy.

If a compact connected n-dimensional manifold M can be subdivided into cells so as to be a regular cell complex, then it is a pseudomanifold, and the results of §IV.8 are applicable. Thus if it is orientable, $H_n(M,\mathbf{Z})$ will be an infinite cyclic group. One of our first goals will be to prove that this result is still true even if the manifold is not a regular cell complex. To "orient" such a manifold means to choose a generator μ of the group $H_n(M,\mathbf{Z})$. The Poincaré duality theorem then asserts that *the homomorphism of $H^q(M^n,G)$ into $H_{n-q}(M^n,G)$, defined by $x \to x \cap \mu$ for any $x \in H^q(M^n,G)$, is an isomorphism for all integers q and all coefficient groups G*! This is a rather severe restriction on the homology and cohomology groups of a compact, orientable manifold. By using the relation $(x \cup y) \cap u = x \cap (y \cap \mu)$, we will be able to show that the Poincaré duality theorem has strong implications for cup products in a manifold.

We will also prove a duality theorem relating the homology and cohomology groups of a manifold with boundary. Finally, we will discuss the famous Alexander duality theorem. This relates the cohomology groups of a closed subset X of Euclidean n-space, $\mathbf{R}^n$, and the homology groups of the complementary set $\mathbf{R}^n - X$. It is a far-reaching generalization of the results proved in III.6 (i.e., the Jordan–Brouwer separation theorem, etc).

The method of proof we use for the Poincaré duality theorem was first described by J. Milnor in some mimeographed lecture notes in 1964; see also the appendix to [8]. The basic idea of Milnor's proof is very natural and may be explained as follows. It follows from the definition that any n-manifold is a union of certain open subsets, each of which is homeomorphic to $\mathbf{R}^n$. Thus it seems natural to try to prove the theorem first for $\mathbf{R}^n$, and then to use Mayer–Vietoris sequences to extend to the case of a finite union of open subsets, each of which is homeomorphic to $\mathbf{R}^n$. Finally we can extend to the case of an infinite union of such open sets by a direct limit argument. The only trouble with this idea is that the Poincaré duality theorem as formulated above applies *only* to compact manifolds. Thus it will be necessary to state and prove a more general version of the Poincaré duality theorem which is also applicable to noncompact manifolds. The reader must not let the technical complications involved in stating and proving this more general version obscure the basic ideas involved.

§2. Orientability and the Existence of Orientations for Manifolds

Let M be an arbitrary n-dimensional manifold; we emphasize that M need not be compact or connected; in fact we do not even need to assume that M is paracompact! For any point $x \in M$, consider the local homology groups $H_i(M, M - \{x\})$ (cf. the exercises in §III.2). Using the fact that x has a neighborhood homeomorphic to $\mathbf{R}^n$ and the excision property, we see that

$$H_i(M, M - \{x\}) \approx H_i(\mathbf{R}^n, \mathbf{R}^n - \{x\}).$$

Hence if we use integer coefficients, $H_i(M, M - \{x\})$ is infinite cyclic for $i = n$, and zero for $i \neq n$. A choice of a generator for the infinite cyclic group $H_n(M, M - \{x\}; \mathbf{Z})$ will be referred to as a *local orientation* of M at x.

Definition 2.1. An *orientation* of an n-dimensional manifold M is a function μ which assigns to each point $x \in M$ a local orientation $\mu_x \in H_n(M, M - \{x\}; \mathbf{Z})$ subject to the following continuity condition: Given any point $x \in M$, there exists a neighborhood N of x and an element $\mu_N \in H_n(M, M - N)$ such that $i_*(\mu_N) = \mu_y$ for any $y \in N$, where $i_* : H_n(M, M - N) \to H_n(M, M - \{y\})$ denotes the homomorphism induced by inclusion.

In order to better understand this continuity condition, recall that any point $x \in M$ has an open neighborhood U which is homeomorphic to $\mathbf{R}^n$. By the excision property, for any $y \in U$,

$$H_n(U, U - \{y\}) \approx H_n(M, M - \{y\}).$$

However, if x and y are any two points of $\mathbf{R}^n$, there is a canonical isomorphism $H_n(\mathbf{R}^n, \mathbf{R}^n - \{x\}) \approx H_n(\mathbf{R}^n, \mathbf{R}^n - \{y\})$ defined by choosing a closed ball $E^n \subset \mathbf{R}^n$ large enough so that x and y are both in the interior of E^n, and noting that in the following diagram,

$$H_n(\mathbf{R}^n, \mathbf{R}^n - \{x\}) \xleftarrow{\;i_*\;} H_n(\mathbf{R}^n, \mathbf{R}^n - E^n)$$
$$\downarrow{\scriptstyle j^*}$$
$$H_n(\mathbf{R}^n, \mathbf{R}^n - \{y\})$$

both i_* and j_* are isomorphisms. Moreover, the isomorphism between $H_n(\mathbf{R}^n, \mathbf{R}^n - \{x\})$ and $H_n(\mathbf{R}^n, \mathbf{R}^n - \{y\})$ that we thus obtain is independent of the choice of the ball E^n.

Terminology. The manifold M is said to be *orientable* if it admits at least one orientation; otherwise, it is called *nonorientable*. A pair consisting of a manifold M and an orientation is called an *oriented manifold*.

EXAMPLE 2.1. (a) Euclidean n-space, $\mathbf{R}^n$, is orientable (use the fact mentioned above that there exists a canonical isomorphism $H_n(\mathbf{R}^n, \mathbf{R}^n - \{x\}) \approx H_n(\mathbf{R}^n, \mathbf{R}^n - \{y\})$ for any two point $x, y \in \mathbf{R}^n$). (b) Similarly, the n-sphere, S^n, is orientable according to our definition. (c) If M is an n-manifold, μ is an orientation for M, and N is an open subset of M, then μ restricted to N is an orientation of the n-manifold N. (d) Let M be an n-dimensional manifold with orientation μ and N and n-dimensional manifold with orientation v. Let $\mu \times v$ denote the function which assigns to each point $(x,y) \in M \times N$ the homology class

$$\mu_x \times v_y \in H_{m+n}(M \times N, M \times N - \{(x,y)\}).$$

Using the Künneth theorem, it is seen that $\mu_x \times v_y$ is a generator of the homology group in question. It is also easy to verify that the required continuity condition holds, and thus $\mu \times v$ is an orientation for $M \times N$. Thus the product of two orientable manifolds is orientable.

In dealing with questions such as these, we will need to frequently consider for any subset A of the manifold M, the homology groups $H_i(M, M - A)$. If $B \subset A$, it will be convenient to denote the corresponding homomorphism $H_i(M, M - A) \to H_i(M, M - B)$ by the symbol ρ_B; for any homology class $u \in H_i(M, M - A)$, $\rho_B(u)$ can be thought of as the "restriction" of u to a homology group associated with B.

Let M be an n-dimensional manifold with orientation μ; it would be advantageous if there were a global homology class $\mu_M \in H_n(M, \mathbf{Z})$ such that for any $x \in M$,

$$\mu_x = \rho_x(\mu_M).$$

Unfortunately, this can not be true if M is noncompact, as the reader can easily verify by using Proposition III.6.1. The closest possible approximation to such a result is the following theorem. It will play a crucial role in the statement and proof of the Poincaré duality theorem:

Theorem 2.1. *Let M be an n-manifold with orientation μ. Then for each compact set $K \subset M$ there exists a unique homology class $\mu_K \in H_n(M, M - K)$ such that*

$$\rho_x(\mu_K) = \mu_x$$

for each $x \in K$.

Note that if M is a *compact* manifold, this theorem assures us of the existence of a unique global homology class $\mu_M \in H_n(M, \mathbf{Z})$ such that for any point $x \in M$,

$$\mu_x = \rho_x(\mu_M).$$

PROOF. The uniqueness of μ_K is a direct consequence of a more general lemma below (Lemma 2.2). Therefore we will concentrate on the existence proof. Obviously, if the compact set K is contained in a sufficiently small neighborhood of some point, the continuity condition in the definition of μ assures us of the existence of μ_K. Next, suppose that $K = K_1 \cup K_2$, where K_1 and K_2 are compact subsets of M, and both μ_{K_1} and μ_{K_2} are assumed to exist. Then $\{M - K_1, M - K_2\}$ is an excisive couple, and hence we have a relative Mayer–Vietoris sequence (cf. §VIII.6):

$$H_{n+1}(M, M - K_1 \cap K_2) \overset{\Delta}{\to} H_n(M, M - K)$$
$$\overset{\varphi}{\to} H_n(M, M - K_1) \oplus H_n(M, M - K_2)$$
$$\overset{\psi}{\to} H_n(M, M - K_1 \cap K_2).$$

Recall that the homomorphisms φ and ψ are defined by

$$\varphi(u) = (\rho_{K_1}(u), \rho_{K_2}(u))$$
$$\psi(v_1, v_2) = \rho_{K_1 \cap K_2}(v_1) - \rho_{K_1 \cap K_2}(v_2)$$

for any $u \in H_n(M, M - K)$, $v_1 \in H_n(M, M - K_1)$, and $v_2 \in H_n(M, M - K_2)$. By the uniqueness of $\mu_{K_1 \cap K_2}$, we see that

$$\rho_{K_1 \cap K_2}(\mu_{K_1}) = \rho_{K_1 \cap K_2}(\mu_{K_2})$$
$$= \mu_{K_1 \cap K_2},$$

and hence

$$\psi(\mu_{K_1}, \mu_{K_2}) = 0.$$

It follows from Lemma 2.2 below that $H_{n+1}(M, M - (K_1 \cap K_2)) = 0$; hence by exactness there is a unique homology class $\mu_K \in H_n(M, M - K)$ such that

$$\varphi(\mu_K) = (\mu_{K_1}, \mu_{K_2}).$$

It is readily verified that this homology class μ_K satisfies the desired condition $\rho_x(\mu_K) = \mu_x$ for any $x \in K$.

Next, assume that $K = K_1 \cup K_2 \cup \cdots \cup K_r$, where each K_i is a compact subset of M, and μ_{K_i} exists. By an obvious induction on r, using what we have just proved, we can conclude that μ_K exists. But any compact subset K of M can obviously be expressed as a finite union of subsets K_i, each of which is sufficiently small so that the corresponding homology class μ_{K_i} exists. Hence μ_K exists, as was to be proved.   Q.E.D.

It remains to state and prove Lemma 2.2.

Lemma 2.2. *Let M be an n-dimensional manifold and G an abelian group.*
(a) *For any compact set $K \subset M$ and all $i > n$,*

$$H_i(M, M - K; G) = 0.$$

(b) *If $u \in H_n(M, M - K; G)$ and $\rho_x(u) = 0$ for all $x \in K$, then $u = 0$.*

PROOF. The method of proof is to start with the case $M = R^n$ and then to progress to successively more complicated cases, ending with the general case.

Case 1: $M = \mathbf{R}^n$ and K is a compact, convex subset of $\mathbf{R}^n$. To prove this case, choose a large ball $E^n \subset \mathbf{R}^n$ such that K is contained in the interior of E^n. For any $x \in K$, consider the following commutative diagram:

$$H_i(M, M - K) \xrightarrow{\rho_x} H_i(M, M - \{x\})$$

$$\diagdown_1 \qquad \diagup_2$$

$$H_i(E^n, S^{n-1})$$

Then it is readily proved that Arrows 1 and 2 are isomorphisms. Hence ρ_x is an isomorphism for all i which suffices to prove the lemma in this case.

Case 2: $K = K_1 \cup K_2$, where K, K_1, and K_2 are compact subsets of M and it is assumed that the lemma is true for K_1, K_2, and $K_1 \cap K_2$. In order to prove this case, we will again use the relative Mayer–Vietoris sequence of the triad $(M; M - K_1, M - K_2)$. The proof of this case is based on the following portion of this Mayer–Vietoris sequence:

$$H_{i+1}(M, M - K_1 \cap K_2) \xrightarrow{\Delta} H_i(M, M - K)$$
$$\xrightarrow{\varphi} H_i(M, M - K_1) \oplus H_i(M, M - K_2).$$

The proof of Parts (a) and (b) of the lemma for this case is quite easy, and may be left to the reader.

Case 3: $M = \mathbf{R}^n$ and $K = K_1 \cup K_2 \cup \cdots \cup K_r$, where each K_i is compact and convex. This case is proved by induction on r, using cases 1 and 2 (the fact that the intersection of convex sets is convex is used).

Case 4: $M = \mathbf{R}^n$, and K is an arbitrary compact subset. We assert that for any $u \in H_i(\mathbf{R}^n, \mathbf{R}^n - K)$, there exists an open set N containing K and an element $u' \in H_i(\mathbf{R}^n, \mathbf{R}^n - N)$ such that $k_*(u') = u$, where

$$k:(\mathbf{R}^n, \mathbf{R}^n - N) \to (\mathbf{R}^n, \mathbf{R}^n - K)$$

is the inclusion map. To prove this assertion, recall that there exists a compact pair $(X,A) \subset (\mathbf{R}^n, \mathbf{R}^n - K)$, and a homology class $v \in H_i(X,A)$ such that the inclusion homomorphism $H_i(X,A) \to H_i(\mathbf{R}^n, \mathbf{R}^n - K)$ maps v onto u (see Proposition III.6.1). Now we may choose N to be any open neighborhood of K which is disjoint from A, and the assertion will certainly be true.

Given the open neighborhood N of K, we may find a finite collection $\{B_1, B_2, \ldots, B_r\}$ of closed balls such that $B_j \subset N$ for $1 \leq j \leq r$, and the union of the B_j's covers K. We may also assume that $K \cap B_j \neq \varnothing$ for $1 \leq j \leq r$. Now consider the following commutative diagram:

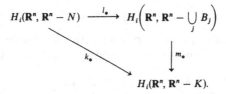

We will use this diagram to prove the lemma for this case. The proof of Part (a) for this case is very easy: If $i > n$, then $H_i(\mathbf{R}^n, \mathbf{R}^n - \bigcup_j B_j) = 0$ by Case 3, and hence the given element $k_*(u') = u \in H_i(\mathbf{R}^n, \mathbf{R}^n - K)$ must be zero also. The proof of Part (b) is only slightly more difficult. Assume $u \in H_n(\mathbf{R}^n, \mathbf{R}^n - K)$, $\rho_x(u) = 0$ for all $x \in K$, and that N and $u' \in H_n(\mathbf{R}^n, \mathbf{R}^n - N)$ have been chosen so that $u = k_*(u')$. Let $u'' = l_*(u') \in H_n(\mathbf{R}^n, \mathbf{R}^n - \bigcup_j B_j)$ in the above diagram. We assert that $\rho_y(u'') = 0$ for each $y \in B_1 \cup B_2 \cup \cdots \cup B_r$. To see this, assume that $y \in B_i$; choose a point $x \in B_i \cap K$. Consider the following commutative diagram:

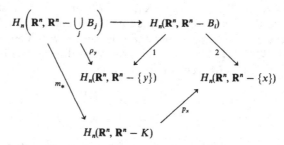

All homomorphisms in this diagram are induced by inclusion maps, and the homomorphisms denoted by Arrows 1 and 2 are isomorphisms, (by Case 1). Since $m_*(u'') = u$, and $\rho_x(u) = 0$, it readily follows that $\rho_y(u'') = 0$ as desired. Therefore we can conclude by Case 3 that $u'' = 0$, and hence $u = m_*(u'')$ is also zero.

Case 5: M is arbitrary, but the compact set K is assumed to be "small" enough so that there exists an open set U which is homeomorphic to $\mathbf{R}^n$

and $U \supset K$. In this case $H_i(M, M - K) \approx H_i(U, U - K)$ by the excision property; hence we can apply Case 4 to reach the desired conclusion.

Case 6: The general case. In this case, K is a finite union of compact subsets,

$$K = K_1 \cup K_2 \cup \cdots \cup K_r,$$

where each K_i is small enough so that Case 5 applies. Hence we can make an induction on r, using Case 2, to compute the proof of the lemma.   Q.E.D.

In order to study the homology of arbitrary manifolds (i.e., orientable or nonorientable) it is desirable to go through similar considerations with Z_2 coefficients. Let M be an arbitrary n-dimensional manifold, and $x \in M$. The local homology group $H_n(M, M - \{x\}; Z_2)$ is cyclic of order 2, hence it has a unique generator $\mu_x \in H_n(M, M - \{x\}; Z_2)$ (no choice is involved). It is readily seen that the function μ which assigns to each $x \in M$ the element μ_x satisfies the continuity condition occurring in the definition of an orientation: Each point $x \in M$ has a neighborhood N for which there exists an element $\mu_N \in H_n(M, M - N; Z_2)$ such that $\rho_y(\mu_N) = \mu_y$ for all $y \in N$. It is convenient to refer to μ as the "mod 2 orientation of M."

Theorem 2.3. *Let M be an arbitrary n-dimensional manifold (i.e., M need not be orientable). Then for each compact set $K \subset M$ there exists a unique homology class $\mu_K \in H_n(M, M - K; Z_2)$ such that*

$$\rho_x(\mu_K) = \mu_x$$

for any $x \in K$, where μ_x denotes the unique nonzero element of the local homology group $H_n(M, M - \{x\}; Z_2)$.

The proof may be patterned on that of Theorem 2.1; the details are left to the reader.

EXERCISES

In these exercises, it is assumed that the reader is familiar with the theory of covering spaces; see Massey [6], Chapter V.

2.1. Let $(\tilde{X}, p)$ be a covering space of X, where X and $\tilde{X}$ are both locally arcwise connected Hausdorff spaces. Prove that X is an n-dimensional manifold if and only if $\tilde{X}$ is an n-dimensional manifold.

2.2. Let $(\tilde{M}, p)$ be a covering space of M, where $\tilde{M}$ and M are both connected n-manifolds. Assume that M is orientable. Prove that $\tilde{M}$ is orientable, and that every covering transformation (i.e., automorphism) of $(\tilde{M}, p)$ is orientation preserving (the definition of *orientation preserving* is the obvious one).

2.3. Let $(\tilde{M}, p)$ be a regular covering space of M. Assume $\tilde{M}$ is a connected, orientable n-manifold, and that every covering transformation of $(\tilde{M}, p)$ is orientation preserving. Prove that M is orientable.

2.4. Let M^n be a compact connected orientable n-manifold and let $T:M^n \to M^n$ be a homeomorphism. How can one determine whether or not T is orientation preserving, in terms of knowledge about the induced homomorphism $T_*:H_n(M^n) \to H_n(M^n)$?

2.5. For which integers n is real projective n-space, RP^n, orientable, and for which n is it nonorientable? (For the definition of RP^n, see §IV.3 or Massey [6], p. 216; see also Statement (h) in §III.2.)

§3. Cohomology with Compact Supports

In order to state and prove the Poincaré duality theorem for noncompact manifolds, it is necessary to use a new kind of cohomology theory, called *cohomology with compact supports*. On compact spaces, this new cohomology theory reduces to the usual kind of cohomology.

Recall that $C^*(X,A;G)$ is a subcomplex of $C^*(X;G)$; it is (by definition) the kernel of the cochain map

$$i^\# : C^*(X;G) \to C^*(A;G).$$

Definition 3.1. A cochain $u \in C^q(X,G)$ has compact support if and only if there exists a compact set $K \subset X$ such that $u \in C^q(X, X - K; G)$.

Note that the set of cochains $u \in C^q(X;G)$ which have compact support is a subgroup of $C^q(X;G)$, which we will denote by $C_c^q(X;G)$. Also, if u has compact support, so does its coboundary, $\delta(u)$, hence we obtain the cochain complex

$$C_c^*(X;G) = \{C_c^q(X,G),\delta\}.$$

We denote the q-dimensional cohomology group of this complex by $H_c^q(X;G)$; it is called the *q-dimensional cohomology group of X with compact supports*. Obviously, if X is compact, $C_c^*(X) = C^*(X)$, and $H_c^q(X) = H^q(X)$. If X is noncompact, $H_c^q(X,G)$ is obviously a topological invariant of the space X; however, it is definitely *not* a homotopy type invariant of X. We will have examples to illustrate this point later. It is only an invariant of what is called the *proper* homotopy type of X; see Massey [7], p. 38.

One could now systematically develop the various properties of cohomology with compact supports. The reader who is interested in seeing this done is referred to the 1948/49 Cartan seminar notes [2], Exposé V, §6, Exposé VIII, §4 and 5, and Exposé IX, §4; see also various books on sheaf theory. We will not do this, because the singular cohomology theory with compact supports does not have such nice properties; the Čech–Alexander–Spanier cohomology with compact supports is a much more elegant theory;

cf. Massey [7]. We will confine ourselves to elaborating those properties of cohomology with compact supports that are actually needed in this chapter.

There is an alternative definition of cohomology with compact supports, based on the notion of *direct limit*; the reader who is not already familiar with direct limits can quickly learn all that is needed from the Appendix to [7]. We will now proceed to explain this alternative definition.

First of all, note that the compact subsets of any topological space X are partially ordered by inclusion; even more, they are *directed* by the inclusion relation, because the union of any two compact subsets is compact.

Next, observe that the cochain group $C_c^q(X)$ may be looked on as the union of the subgroups $C^q(X, X - K)$, where K ranges over all compact subsets of X. In other words,

$$C_c^q(X; G) = \text{dir lim } C^q(X, X - K; G),$$

where the direct limit is taken over the above mentioned directed set, consisting of all compact subsets $K \subset X$. Now the operation of taking homology groups of a cochain complex commutes with the passage to the direct limit; therefore

$$H_c^q(X; G) = \text{dir lim } H^q(X, X - K; G),$$

where again the direct limit is taken over all compact subsets $K \subset X$. This is the definition that we will actually use for $H_c^q(X, G)$.

EXERCISES

3.1. Determine the structure of the groups $H_c^i(\mathbf{R}^n; G)$ for all i. (**Caution:** Even though $\mathbf{R}^n$ is contractible, these cohomology groups are not all trivial. Note also the structure of $H_c^0(\mathbf{R}^n)$.)

3.2. Let X be an arcwise connected Hausdorff space which is noncompact. What is the structure of $H_c^0(X; G)$ for any coefficient group G?

3.3. A continuous map $f: X \to Y$ is said to be *proper* if the inverse image under f of any compact subset of Y is compact. Let $f: X \to Y$ be a proper continuous map, and let $f^\#: C^p(Y, G) \to C^p(X, G)$ denote the induced homomorphism on cochains. Prove that $f^\#(C_c^p(Y)) \subset C_c^p(X)$, and hence f induces a homomorphism of $H_c^p(Y)$ into $H_c^p(X)$.

§4. Statement and Proof of the Poincaré Duality Theorem

Let M be an n-dimensional manifold with orientation μ; we stress that we do not need to assume that M is compact, connected, or even paracompact. Moreover, we do not need to make any hypotheses of triangulability or differentiability.

Because of the choice of orientation μ, there is singled out a unique homology class $\mu_K \in H_n(M, M - K; \mathbf{Z})$ for each compact subset K (see Theorem 2.1). Hence the cap product with μ_K defines a homomorphism

$$H^q(M, M - K; G) \to H_{n-q}(M; G)$$

by the formula

$$x \to x \cap \mu_K$$

for any $x \in H^q(M, M - K; G)$. Here the coefficient group G is arbitrary, and the cap product is defined using the natural isomorphism $G \otimes \mathbf{Z} \approx G$. Because of the naturality of the cap product, the homomorphisms thus defined for different compact sets are compatible in the following sense: if K and L are compact, and $K \subset L$, then the following diagram is commutative:

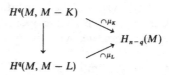

(here the homomorphism denoted by the vertical arrow is induced by inclusion). Now it is a basic property of direct limits that any such compatible family of homomorphisms induces a homomorphism of the direct limit; thus we have a well defined homomorphism

$$P : H_c^q(M; G) \to H_{n-q}(M; G)$$

(the letter P stands for Poincaré).

Theorem 4.1 (Poincaré duality). *Let M be an oriented n-dimensional manifold and G an arbitrary abelian group. Then the homomorphism*

$$P : H_c^q(M; G) \to H_{n-q}(M; G)$$

is an isomorphism for all q.

We will give the proof of this theorem now, postponing the discussion of examples, special cases, and applications to later. As in the proof of Lemma 2.2, there are several cases, starting with $M = \mathbf{R}^n$, and ending with the general case.

Case 1: $M = \mathbf{R}^n$. Let B_k denote the closed ball in $\mathbf{R}^n$ with center at the origin and radius k. Clearly, the sequence of closed balls

$$B_1, B_2, B_3, \ldots.$$

is cofinal in the directed set of all compact subsets of $\mathbf{R}^n$. It follows that

$$H_c^q(\mathbf{R}^n; G) = \text{dir lim } H^q(\mathbf{R}^n, \mathbf{R}^n - B_k; G).$$

Note also that the homomorphism

$$H^q(\mathbf{R}^n, \mathbf{R}^n - B_k) \rightarrow H^q(\mathbf{R}^n, \mathbf{R}^n - B_{k+1})$$

is an isomorphism for all k and q; hence it follows that

$$H_c^q(\mathbf{R}^n; G) = \begin{cases} G & \text{for } q = n, \\ 0 & \text{for } q \neq n. \end{cases}$$

In view of the known structure of $H_{n-q}(\mathbf{R}^n; G)$, we see that it is indeed true that the groups $H_{n-q}(\mathbf{R}^n; G)$ and $H_c^q(\mathbf{R}^n; G)$ are isomorphic for all G and q. It only remains to prove that

$$P: H_c^n(\mathbf{R}^n; G) \rightarrow H_0(\mathbf{R}^n; G)$$

is an isomorphism; in view of the definition of P, it suffices to prove that for any closed n-dimensional ball $B \subset \mathbf{R}^n$, the homomorphism

$$H^n(\mathbf{R}^n, \mathbf{R}^n - B; G) \rightarrow H_0(\mathbf{R}^n; G)$$

defined by $x \rightarrow x \cap \mu_B$ is an isomorphism. Now μ_B is a generator of the infinite cyclic group $H_n(\mathbf{R}^n, \mathbf{R}^n - B; \mathbf{Z})$. We will complete the proof by using the following relation:

$$\varepsilon_*(x \cap \mu_B) = \langle x, \mu_B \rangle$$

(see §VIII.8). Since $\mathbf{R}^n$ is arcwise connected, the homomorphism

$$\varepsilon_*: H_0(\mathbf{R}^n; G) \rightarrow G$$

is an isomorphism. Moreover, by the universal coefficient theorem for cohomology (see §VII.4), the homomorphism

$$\alpha: H^n(\mathbf{R}^n, \mathbf{R}^n - B; G) \rightarrow \text{Hom}(H_n(\mathbf{R}^n, \mathbf{R}^n - B); G)$$

is also an isomorphism. Using the definition of α in terms of the scalar product, the desired conclusion follows.

Case 2: Assume $M = U \cup V$, where U and V are open subsets, of M, and that Poincaré duality holds for U, V, and $U \cap V$ (it is assumed, of course, that the orientation for U is the restriction of μ to U, and similarly for V and $U \cap V$). In this situation, we can construct a Mayer–Vietoris exact sequence for cohomology with compact supports:

$$\cdots \rightarrow H_c^{q-1}(M) \rightarrow H_c^q(U \cap V) \rightarrow H_c^q(U) \oplus H_c^q(V) \rightarrow H_c^q(M) \rightarrow \cdots.$$

To construct this sequence, let $K \subset U$ and $L \subset V$ be compact sets; we then have the following relative Mayer–Vietoris sequence, which is exact:

$$\xrightarrow{\Delta} H^q(M, M - K \cap L) \xrightarrow{\varphi} H^q(M, M - K) \oplus H^q(M, M - L) \xrightarrow{\psi} H^q(M, M - K \cup L)$$

(we have used this Mayer–Vietoris sequence a couple of times previously in this chapter). Now by the excision property, we have the following isomorphisms:

$$H^q(M, M - K \cap L) \approx H^q(U \cap V, U \cap V - K \cap L),$$
$$H^q(M, M - K) \approx H^q(U, U - K),$$

and

$$H^q(M, M - L) \approx H^q(V, V - L).$$

Next, note that as K ranges over all compact subsets of U and L ranges over all compact subsets of V, $K \cap L$ ranges over all compact subsets of $U \cap V$ and $K \cup L$ ranges over all compact subsets of M. Hence as we pass to the direct limit over all such ordered pairs (K, L), the direct limit of the relative Mayer–Vietoris sequences gives the desired result.

We now have the following diagram:

$$\cdots \longrightarrow H_c^q(U \cap V) \longrightarrow H_c^q(U) \oplus H_c^q(V) \longrightarrow H_c^q(M) \longrightarrow \cdots$$
$$\downarrow \qquad\qquad\qquad \downarrow \qquad\qquad\qquad \downarrow$$
$$\cdots \longrightarrow H_{n-q}(U \cap V) \longrightarrow H_{n-q}(U) \oplus H_{n-q}(V) \longrightarrow H_{n-q}(M) \longrightarrow \cdots .$$

The top line of this diagram is the Mayer–Vietoris sequence we have just constructed, and the bottom line is the usual Mayer–Vietoris sequence in homology. The vertical arrows are the Poincaré duality homomorphisms for $U \cap V$, U, V, and M respectively. We assert that *every square of this diagram is commutative*. As a general rule, it is fairly easy to check whether or not a diagram such as this is commutative. But this seems to be an exception to the general rule! The proof of commutativity is lengthy, to say the least. The complete details are given in the appendix to this chapter (see Lemma 8.2).

In any event, once we have proved commutativity for this diagram, the proof that M satisfies Poincaré duality in this case is an obvious consequence of the five-lemma.

Case 3: M is the union of a nested family of open subsets $\{U_\lambda\}$ and it is assumed that the Poincaré duality theorem holds for each of the U_λ. In order to prove this case, it is necessary to make use of a natural homomorphism

$$\tau : H_c^q(U; G) \to H_c^q(X; G)$$

which is defined as follows for any open subset U of the Hausdorff space X. If K is any compact subset of U, then the excision property guarantees us an isomorphism

$$H^q(U, U - K) \approx H^q(X, X - K).$$

Passing to the direct limit over all compact sets $K \subset U$, we obtain the desired homomorphism (it is not an isomorphism in general, because not every compact subset of X is contained in U). Two of the most important properties of τ are the following:

(a) If $U = X$, then τ is the identity homomorphism.

(b) If $U \subset V \subset X$, and U and V are open subsets of X, then the following diagram is commutative:

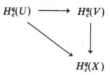

In addition, if U is an open subset of the oriented n-manifold M, then the following diagram is commutative:

$$
\begin{array}{ccc}
H_c^q(U) & \xrightarrow{\ \tau\ } & H_c^q(M) \\
\downarrow{\scriptstyle P} & & \downarrow{\scriptstyle P} \\
H_{n-q}(U) & \xrightarrow{\ i_*\ } & H_{n-q}(M)
\end{array}
$$

Here $i: U \to M$ denotes the inclusion homomorphism, and as usual, the orientation of U is assumed to be the restriction of the orientation of M. The proof of the commutativity of this diagram is an easy consequence of the definition of P and the naturality of the cap product.

With these preliminaries taken care of, we can now easily prove Case 3. Because the open subsets U_λ are nested, we can form the direct limits,

$$\text{dir lim } H_c^q(U_\lambda),$$

and

$$\text{dir lim } H_{n-q}(U_\lambda).$$

In the first case, it is understood that the homomorphisms in the direct system of groups $\{H_c^q(U_\lambda)\}$ are the τ's corresponding to any inclusion, while in the second case, they are the i_*'s corresponding to any inclusion. Next, observe that the homomorphisms

$$\tau_\lambda: H_c^q(U_\lambda) \to H_c^q(M)$$
$$i_{\lambda*}: H_{n-q}(U_\lambda) \to H_{n-q}(M)$$

(which are defined for all λ) constitute a compatible collection of homomorphisms, and hence define homomorphisms of the direct limit groups:

$$\text{dir lim } H_c^q(U_\lambda) \to H_c^q(M)$$
$$\text{dir lim } H_{n-q}(U_\lambda) \to H_{n-q}(M).$$

We assert that these homomorphisms are both isomorphisms; this is a consequence of the fact that any compact subset of M is contained in some U_λ. Finally, the Poincaré duality homomorphism $P: H_c^q(U_\lambda) \to H_{n-q}(U_\lambda)$ is assumed to be an isomorphism for each λ; it follows by passage to the direct limit that $P: H_c^q(M) \to H_{n-q}(M)$ is also an isomorphism.

Case 4: M is an open subset of $\mathbf{R}^n$. If M is convex, then it is homeomorphic to $\mathbf{R}^n$, and Case 1 applies. If M is not convex, then we make use of the fact that the topology of $\mathbf{R}^n$ has a countable basis consisting of open n-dimensional balls. Hence M is a countable union of open balls:

$$M = \bigcup_{i=1}^{\infty} B_i.$$

Let

$$M_k = \bigcup_{i=1}^{k} B_i.$$

The theorem must be true for each M_k, by an obvious induction on k (use Case 2). Then we can apply Case 3 to conclude that the theorem is true for

$$M = \bigcup_{k=1}^{\infty} M_k.$$

Case 5: The general case. Let M be an arbitrary oriented n-manifold. Consider the family of all open subsets U of M such that Poincaré duality holds for U. This family is obviously nonempty. In view of Case 3, we can apply Zorn's lemma to this family to conclude that there exists a maximal open set V belonging to it. If $V \neq M$, then there is an open subset $B \subset M$ such that B is homeomorphic to $\mathbf{R}^n$, and B is not contained to V. We could then apply Cases 2 and 4 to conclude that Poincaré duality also holds for $V \cup B$, contradicting the maximality of V. Thus $V = M$, and we are through.   Q.E.D.

Next we will take up the mod 2 version of the Poincaré duality theorem. While this version is weaker in that it only applies to homology and cohomology groups with $\mathbf{Z}_2$ coefficients, it has the advantage that it applies to all manifolds, whether orientable or not.

We will use the hypotheses and notation of Theorem 2.3: M is an arbitrary n-dimensional manifold; for each point $x \in M$, μ_x denotes the unique non-zero element of the local homology group $H_n(M, M - \{x\}; \mathbf{Z}_2)$, and for each compact subset K, μ_K denotes the unique element of $H_n(M, M - K; \mathbf{Z}_2)$ such that $\rho_x(\mu_K) = \mu_x$ for all $x \in K$. Let G be a vector space over $\mathbf{Z}_2$. Define a homomorphism

$$H^q(M, M - K; G) \to H_{n-q}(M, G)$$

by the formula

$$x \to x \cap \mu_K$$

for any $x \in H^q(M, M - K; G)$ (use the natural isomorphism $G \otimes \mathbf{Z}_2 = G$ to define this cap product). The homomorphisms thus defined for all compact sets $K \subset M$ are compatible, and hence define a homomorphism of the direct limit group,

$$P_2: H_c^q(M;G) \rightarrow H_{n-q}(M;G)$$

which we will refer to as the mod 2 Poincaré duality homomorphism.

Theorem 4.2. *For any n-dimensional manifold M and any $\mathbf{Z}_2$-vector space G, the mod 2 Poincaré duality homomorphism P_2 is an isomorphism of $H_c^q(M;G)$ onto $H_{n-q}(M;G)$.*

The proof is almost word for word the same as that of Theorem 4.1; the necessary modifications are rather obvious.

EXERCISES

4.1. Let K be a compact, *connected* subset of M, and $u \in H_n(M, M - K; \mathbf{Z})$. Prove that if $\rho_x(u) = 0$ for some $x \in K$, then $\rho_x(u) = 0$ for all $x \in K$. Deduce that $H_n(M, M - K; \mathbf{Z})$ is either infinite cyclic or 0.

4.2. Let K be a compact, *connected* subset of M, $u \in H_n(M, M - K; \mathbf{Z})$ and let $x \in K$ be such that $\rho_x(u)$ is k times a generator of $H_n(M, M - \{x\})$. Prove that for any $y \in K$, $\rho_y(u)$ is also k times a generator of $H_n(M, M - \{y\})$.

4.3. Assume that M is connected, and that for each compact $K \subset M$, $H_n(M, M - K; \mathbf{Z}) \neq \{0\}$. Prove that M is orientable.

4.4. Let M be a compact, connected, nonorientable manifold. Prove that $H_n(M; \mathbf{Z}) = 0$.

4.5. Use the Poincaré duality theorem to prove that if M is a connected, noncompact orientable n-dimensional manifold, then $H_q(M, G) = 0$ for all $q \geq n$ and all coefficient groups G.

4.6. For any abelian group G, let $_2G = \{g \in G \mid 2g = 0\}$. Recall that there is a natural isomorphism $\alpha: H_n(M, M - \{x\}; \mathbf{Z}) \otimes G \rightarrow H_n(M, M - \{x\}; G)$ for any point x of the n-manifold M (see §V.6). Show that if $g \in {}_2G$, the element $g_x = \alpha(\mu_x \otimes g) \in H_n(M, M - \{x\}; G)$ is independent of the choice of the local orientation $\mu_x \in H_n(M, M - \{x\}; \mathbf{Z})$. Then prove that for each compact set $K \subset M$ and $g \in {}_2G$, there exists a unique homology class $g_K \in H_n(M, M - K; G)$ such that $\rho_x(g_K) = g_x$ for any $x \in K$.

4.7. Let M be an n-dimensional manifold. Assume that for each compact set $K \subset M$ there is chosen an element $h_K \in H_n(M, M - K; G)$ such that $\rho_x(h_K) = h_x$ for any $x \in K$, and that $2h_x \neq 0$ for all $x \in M$. Prove that the manifold M is orientable. (*Hint*: Show that there exists a fixed element $h \in G$ and unique local orientations μ_x for all $x \in M$ such that $h_x = \alpha(\mu_x \otimes h)$. Note that h can *not* be an element of $_2G$.)

4.8. Let M be a compact, connected n-manifold and G an abelian group. Prove that $H_n(M; G)$ is isomorphic to G if M is orientable, and is isomorphic to $_2G$ if M is nonorientable. (Use the results of the preceding exercises.)

4.9. Let M be a compact, connected n-manifold. Prove that $H_{n-1}(M;\mathbf{Z})$ is torsion-free if M is orientable, and that the torsion subgroup of $H_{n-1}(M;\mathbf{Z})$ is cyclic of order 2 if M is nonorientable. (Use Exercise 4.8 and the universal coefficient theorem. You may make use of the fact that all the integral homology groups of M are finitely generated; see Lemma 5.2 of the next section.)

§5. Applications of the Poincaré Duality Theorem to Compact Manifolds

Let M be a compact manifold with orientation μ; in this case, by Theorem 2.3, there exists a unique homology class $\mu_M \in H_n(M;\mathbf{Z})$ such that $\rho_x(\mu_M) = \mu_x$ for all $x \in M$; μ_M is often referred to as *the fundamental homology class of the oriented manifold M*. The Poincaré duality isomorphism

$$P : H^q(M;G) \to H_{n-q}(M;G)$$

is defined by

$$P(x) = x \cap \mu_M$$

for any $x \in H^q(M;G)$.

We can draw some immediate conclusions from this. For example, if M is assumed to be connected, then $H_n(M;G)$ is isomorphic to G. Similarly, $H_{n-1}(M;\mathbf{Z}) \approx H^1(M;\mathbf{Z}) \approx \mathrm{Hom}(H_1(M;\mathbf{Z}),\mathbf{Z})$ is a torsion-free group.

In case M is compact but not necessarily orientable, we can obtain similar results with $\mathbf{Z}_2$ coefficients. There is a unique mod 2 *fundamental class*, $\mu_M \in H_n(M;\mathbf{Z}_2)$ and the mod 2 Poincaré duality isomorphism

$$P_2 : H^q(M;\mathbf{Z}_2) \to H_{n-q}(M;\mathbf{Z}_2)$$

is defined by

$$P_2(x) = x \cap \mu_M.$$

From this isomorphism, we deduce that the rank of the vector space $H_n(M;\mathbf{Z}_2)$ (over $\mathbf{Z}_2$) is equal to the number of components of M.

We will now use the Poincaré duality theorem to deduce some restrictions on cup products in the cohomology of a manifold.

Theorem 5.1. *Let M be a compact oriented n-manifold and F a field. Then the bilinear form*

$$H^q(M;F) \otimes H^{n-q}(M;F) \to F$$

defined by

$$u \otimes v \to \langle u \cup v, \mu_M \rangle$$

for any $u \in H^q(M;F)$ and $v \in H^{n-q}(M;F)$ is nonsingular.

PROOF. The relation

$$\langle u \cup v, \mu_M \rangle = \langle u, v \cap \mu_M \rangle$$

can be interpreted as a commutativity relation, as indicated by the following diagram:

$$H^q(M;F) \otimes H^{n-q}(M,F)$$

$$\downarrow I \otimes P \qquad \overset{1}{\searrow} \quad F.$$

$$H^q(M;F) \otimes H_q(M,F) \quad \overset{2}{\nearrow}$$

In this diagram, arrow 1 denotes the bilinear form of the theorem, arrow 2 denotes the bilinear form defined by $x \otimes y \to \langle x,y \rangle$ for any $x \in H^q(M;F)$ and $y \in H_q(M,F)$, I denotes the identity map, and P the Poincaré duality isomorphism. The bilinear form denoted by arrow 2 is nonsingular, because of the isomorphism

$$H^q(M;F) \approx \operatorname{Hom}_F(H_q(M,F),F).$$

Since P is an isomorphism, it follows that the bilinear form denoted by arrow 1 is also nonsingular.   Q.E.D.

If the manifold M is nonorientable, this theorem will still be true provided we assume that F is a field of characteristic two, e.g., $F = \mathbf{Z}_2$.

It would be nice to have an analogue of Theorem 5.1 for the case of cohomology with integer coefficients, rather than coefficients in a field. Since the groups $H^q(M;\mathbf{Z})$ and $\operatorname{Hom}(H_q(M,\mathbf{Z}),\mathbf{Z})$ are *not* isomorphic in general, some modifications are necessary in order to obtain a valid theorem. One way to proceed is the following: For any space X, define $B_q(X)$, the *q-dimensional Betti group of X*, to be the quotient group of $H_q(X;\mathbf{Z})$ modulo its torsion subgroup. Similarly, define $B^q(X)$ to be the quotient group of $H^q(X;\mathbf{Z})$ modulo its torsion subgroup. If $H_{q-1}(X;\mathbf{Z})$ is a finitely generated abelian group, then

$$B^q(X) \approx \operatorname{Hom}(B_q(X),\mathbf{Z});$$

this is a direct consequence of the short exact sequence

$$0 \to \operatorname{Ext}(H_{q-1}(X),\mathbf{Z}) \to H^q(X;\mathbf{Z}) \overset{\alpha}{\to} \operatorname{Hom}(H_q(X),\mathbf{Z}) \to 0.$$

Lemma 5.2. *Let M be a compact manifold; then the integral homology group $H_q(M)$ is finitely generated for all q.*

If M could be given the structure of a CW-complex, then compactness would imply that this CW-complex was finite, and the theorem would follow. However it is not known at present whether or not all compact manifolds are CW-complexes. Fortunately, there is a way to avoid this difficulty. By results in Chapter IV, §8 of Dold, [4], a compact manifold is what is called an ENR (short for Euclidean neighborhood retract). Then proposition V.4.11 on p. 103 of Dold [4] asserts that the homology groups of an ENR are finitely generated.   Q.E.D.

Note: This follows from Poincaré duality in case M is orientable; cf. Spanier, [9] Corollary 11 at bottom of p. 298.

As a consequence of this lemma, we see that for any compact manifold M, we have a natural isomorphism

$$B^q(M) \approx \mathrm{Hom}(B_q(M),\mathbf{Z}).$$

This isomorphism is defined as follows. Let $H^q(M;\mathbf{Z}) \otimes H_q(M,\mathbf{Z}) \to \mathbf{Z}$ be a bilinear form defined by $x \otimes y \to \langle x,y \rangle$ for $x \in H^q(M;\mathbf{Z})$ and $y \in H_q(M;\mathbf{Z})$. It is obvious that if either x or y has finite order, then $\langle x,y \rangle = 0$. Hence there is an induced bilinear form on quotient groups:

$$B^q(M) \otimes B_q(M) \to \mathbf{Z}.$$

This bilinear form defines the desired isomorphism.

Now let us consider the bilinear form

$$H^q(M;\mathbf{Z}) \otimes H^{n-q}(M;\mathbf{Z}) \to \mathbf{Z}$$

defined by

$$u \otimes v \to \langle u \cup v, \mu_M \rangle$$

(this is similar to the bilinear form defined in Theorem 5.1). Once again, if u or v has finite order, then $\langle u \cup v, \mu_M \rangle = 0$. Hence there is an induced bilinear form

$$B^q(M) \otimes B^{n-q}(M) \to \mathbf{Z}.$$

Theorem 5.3. *Let M be a compact, connected, oriented n-manifold. Then the bilinear form*

$$B^q(M) \otimes B^{n-q}(M) \to \mathbf{Z}$$

defined above is nonsingular, and induces an isomorphism of $B^q(M)$ onto $\mathrm{Hom}(B^{n-q}(M),\mathbf{Z})$ for all q.

The proof is very similar to that of Theorem 5.1, and may be left to the reader.

For the present, we will given one application of these theorems. Further applications will be found in the next chapter.

Proposition 5.4. *Let M be a compact, orientable manifold of dimension $n = 4k + 2$, and let F be a field of characteristic $\neq 2$. Then $H^{2k+1}(M;F)$ is a vector space over F whose dimension is even.*

PROOF. By Theorem 5.1, the bilinear form

$$H^{2k+1}(M;F) \otimes H^{2k+1}(M;F) \to F$$

defined by

$$u \otimes v \to \langle u \cup v, \mu_M \rangle$$

is nonsingular. Moreover, by the commutative law for cup products,

$$u \cup v = -v \cup u$$

for any $u, v \in H^{2k+1}(M;F)$. It follows that the bilinear form is skew-symmetric; but it is a standard theorem of algebra that nonsingular skew-symmetric bilinear forms can only exist on vector spaces of even dimension (over a field of characteristic $\neq 2$). For a proof of this theorem, see Jacobson, [5], Section 6.2.

As an example of this proposition, consider compact orientable 2-manifolds.

EXERCISES

5.1. Let M be a compact, orientable n-manifold. Prove that the homology groups $H_q(M;\mathbf{Z})$ and $H_{n-q}(M;\mathbf{Z})$ have the same ranks. Also, show that the torsion subgroup of $H_q(M;\mathbf{Z})$ is isomorphic to the torsion subgroup of $H_{n-q-1}(M;\mathbf{Z})$.

5.2. Prove that the Euler characteristic of a compact n-manifold is 0 for n odd.

5.3. Prove that the Euler characteristic of a compact orientable manifold of dimension $4k+2$ is even.

5.4. Let M_1 and M_2 be compact, orientable n-manifolds, and let $f:M_1 \to M_2$ be a continuous map such that the induced homomorphism

$$f_*:H_n(M_1;\mathbf{Z}) \to H_n(M_2;\mathbf{Z})$$

is an isomorphism. Prove that for any coefficient group G the induced homomorphism

$$f_*:H_q(M_1;G) \to H_q(M_2;G)$$

is an epimorphism and the kernel of f_* is a direct summand of $H_q(M_1;G)$. Similarly, prove that

$$f^*:H^q(M_2;G) \to H^q(M_1;G)$$

is a monomorphism, and the image is a direct summand of $H^q(M_1;G)$.

5.5. Let M be a compact, connected, orientable n-manifold and $f:M \to M$ a continuous map such that $f_*:H_n(M,\mathbf{Z}) \to H_n(M;\mathbf{Z})$ is an isomorphism. Prove that the induced homomorphisms $f_*:H_q(M,G) \to H_q(M,G)$ and $f^*:H^q(M,G) \to H^q(M,G)$ are isomorphisms for all q and any group G. (*Hint*: Do the case $G = \mathbf{Z}$ first.)

5.6. Given any even integer n, show how to construct a compact connected, orientable manifold M of dimension $4k+2$ such that the rank of the vector space $H^{2k+1}(M;F)$ is n. (*Hint*: Consider first the case of 2-manifolds, i.e., $k=0$. For larger values of k, proceed by analogy with the case $k=0$, recalling the classification theorem for 2-manifolds.)

5.7. Let X be a Hausdorff space, and let K be a compact subset of X. Consider the cup product:

$$H^p(X;G_1) \otimes H^q(X, X - K;G_2) \overset{\cup}{\to} H^{p+q}(X, X - K;G_1 \otimes G_2).$$

Prove that passing to the direct limit over all compact subsets K of X, defines a homomorphism

$$H^p(X;G_1) \otimes H^q_c(X;G_2) \to H^{p+q}_c(X;G_1 \otimes G_2).$$

(This is called the *cup product homomorphism*.)

5.8. (a) Let M be an oriented n-manifold. For any compact set $K \subset M$, let $\mu_K \in H_n(M, M - K)$ denote the unique homology class such that $\rho_x(\mu_K) = \mu_x$ for any $x \in K$. Given $u \in H^n_c(M;G)$, choose a compact set $K \subset M$ such that there exists a representative $u' \in H^n(M, M - K; G)$ for u. Show that the element

$$\langle u', \mu_K \rangle \in G$$

is independent of the choice of the representative u' for u, and that this process defines a homomorphism $H^n_c(M;G) \to G$, sometimes called *integration over M*.

(b) Show that the following diagram is commutative

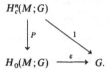

$$H^n_c(M;G)$$

Here arrow 1 denotes integration over M.

(c) Prove that for any elements $u \in H^p(M;G_1)$ and $v \in H^q_c(M;G_2)$, the following equation holds:

$$u \cap P(v) = P(u \cup v).$$

Here the cup product is that defined in Exercise 5.7, and P denotes the Poincaré duality isomorphism.

(d) Let F be a field. Define a bilinear form

$$\varphi: H^{n-p}(M;F) \otimes H^p_c(M;F) \to F$$

by setting $\varphi(u \otimes v) =$ the integral of $u \cup v$ over M. Prove that this bilinear form is nonsingular and that it defines an isomorphism

$$H^{n-p}(M;F) \approx \mathrm{Hom}_F(H^p_c(M;F),F).$$

§6. The Alexander Duality Theorem

Let A be a subset of a topological space X; by a *neighborhood N of A in X*, we mean a subset N of X which contains A in its interior. The neighborhoods of A (ordered by inclusion) constitute a directed set, since the intersection of any two neighborhoods of A is again a neighborhood of A. Consider the direct system of groups $\{H^q(N)\}$, where N ranges over all neighborhoods of A in X (the homomorphisms are those induced by the inclusion relations, of course). For each such N, the inclusion $A \subset N$ induces a homomorphism $H^q(N) \to H^q(A)$, and the collection of all such homomorphisms is obviously compatible. Hence there is induced a homomorphism

$$\mathrm{dir}\ \lim\ H^q(N) \to H^q(A).$$

The subspace A is said to be *tautly imbedded* in X (or simply *taut* in X) with respect to singular cohomology if this homomorphism is an isomorphism for all q and all coefficient groups. This concept was introduced by Spanier [9], p. 189. We need it for our discussion of the Alexander duality theorem.

EXAMPLE 6.1. Let A denote the subset of the plane $\mathbf{R}^2$ consisting of the union of the graph of the function $y = \sin(1/x)$ (for $x \neq 0$) and the y-axis. We assert that A is not taut in $\mathbf{R}^2$. In order to prove this, note that the open neighborhoods of A are cofinal in the family of all neighborhoods of A. Furthermore, the open, arcwise connected neighborhoods are cofinal in the family of all open neighborhoods. It follows that the direct limit, dir lim $H^0(N; \mathbf{Z})$, is infinite cyclic. On the other hand, $H^0(A; \mathbf{Z})$ is free abelian of rank 3 (there are three arc-components).

As another example, let P denote the subset of A consisting of one point, the origin. Then it is readily verified that P is not taut in A.

In some sense, these two examples are rather pathological. We will see shortly that any "nice" subset of a nice space is tautly imbedded. We will be mainly interested in the case where X is a manifold. Then it will turn out that the question of whether or not a subset A of X is taut or not depends only on A! Obviously, the question only depends on arbitrarily small neighborhoods of A in X, but we are asserting something stronger than this.

The situation may be explained in more detail as follows. This book has been concerned exclusively with singular homology and cohomology theory. However, there is also another type of cohomology theory, called Čech–Alexander–Spanier cohomology theory. For any pair (X,A), any integer q and any abelian group G, there is defined the q-dimensional Čech–Alexander–Spanier cohomology group, which we denote by $\bar{H}^q(X,A; G)$. Just as for the singular cohomology theory, a continuous map $f:(X,A) \to (Y,B)$ induces homomorphisms $f^*: \bar{H}^q(Y,B; G) \to \bar{H}^q(X,A; G)$ for all q. The basic properties of the Čech–Alexander–Spanier cohomology theory are exactly the same as those of the singular cohomology theory; the reader may find more details in Spanier, [9], Chapter 6, Sections 4 and 5, or Massey [7], Chapter 8.

One of the major differences between singular and Alexander–Spanier cohomology is this matter of tautness. In general, tautness is more likely to hold with respect to the Alexander–Spanier cohomology theory than with respect to the singular theory. In fact, the following theorem holds:

Theorem 6.1. *In each of the following four cases A is taut in X with respect to the Alexander–Spanier cohomology theory:*

(1) *A is compact and X is Hausdorff.*
(2) *A is closed and X is paracompact Hausdorff.*
(3) *A is arbitrary and every open subset of X is paracompact Hausdorff.*
(4) *A is a retract of some open subset of X.*

This theorem is due to Spanier [10]; for a proof, see Massey, [7], pp. 238–241. One case of this theorem is proved in Spanier [9], pp. 316–317.

A more precise comparison of singular and Alexander–Spanier cohomology is possible, because there is defined for any pair (X,A), any coefficient group G, and any integer q a homomorphism

$$\lambda: \bar{H}^q(X,A;G) \to H^q(X,A;G).$$

This homomorphism is natural, in the sense that it commutes with homomorphisms induced by continuous maps. There are various theorems which assert that for certain classes of nice topological spaces, λ is an isomorphism for all G and q. For a discussion of this question, see Spanier, [9], Chapter 6, Section 9, or Massey, [7], §8.8. For our purposes, the following are the two most important cases in which $\lambda: \bar{H}^q(X;G) \to H^q(X;G)$ is known to be an isomorphism for all G and q:

(a) X a paracompact n-manifold.
(b) X is a CW-complex, or a space which has the homotopy type of a CW-complex.

Using these properties of the homomorphism λ, we can easily prove the following propositions:

Proposition 6.2. Let M be a paracompact n-manifold, and let A be a closed subset of M. Then A is taut in M (with respect to singular cohomology) if and only if $\lambda: \bar{H}^q(A;G) \to H^q(A;G)$ is an isomorphism for all q and G.

Thus in this case, the question of tautness depends only on A.

Proposition 6.3. Let M be a paracompact n-manifold, and let A be a closed subset of M. Then

$$\text{dir lim } H^q(N;G) \approx \bar{H}^q(A;G),$$

where the direct limit is taken over all neighborhoods N of A in M.

The proof of both of these propositions depends on the naturality of the homomorphism λ. The open neighborhoods of A are cofinal in the family of all neighborhoods of A; and every open neighborhood N of A is also a paracompact manifold. Therefore $\lambda: \bar{H}^q(N) \to H^q(N)$ is an isomorphism. The rest of the details of the proofs may be left to the reader.

Remark: In Dold [4], the conclusion of Proposition 6.3 is taken as the definition of the Čech–Alexander–Spanier cohomology groups $\bar{H}^q(A)$.

We will now use these results to derive important relations between the homology groups of an open subset of a compact manifold and the cohomology groups of its complement. Let M be a compact, oriented n-manifold, U an open subset of M, and $A = M - U$ the closed complement.

For any compact set $K \subset U$, consider the following diagram:

$$H^q(M, M - K) \longrightarrow H^q(M) \longrightarrow H^q(M - K) \overset{\delta}{\longrightarrow} H^{q+1}(M, M - K)$$

$$\downarrow k^* \qquad\qquad\qquad \downarrow P \qquad\qquad\qquad \downarrow 2 \qquad\qquad\qquad \downarrow k^*$$

$$H^q(U, U - K) \qquad\qquad\qquad H_{n-q}(M - K, U - K) \quad H^{q+1}(U, U - K)$$

$$\downarrow 1 \qquad\qquad\qquad\qquad\qquad\qquad \downarrow l_* \qquad\qquad\qquad \downarrow 1$$

$$H_{n-q}(U) \overset{i_*}{\longrightarrow} H_{n-q}(M) \overset{j_*}{\longrightarrow} H_{n-q}(M, U) \overset{\partial}{\longrightarrow} H_{n-q-1}(U).$$

In this diagram, the top line is the cohomology sequence of the pair $(M, M - K)$, the bottom line is the homology sequence of the pair (M, U), and $k:(U, U - K) \to (M, M - K)$ and $l:(M - K, U - K) \to (M, U)$ are inclusion maps which induce isomorphisms by the excision property. The homomorphisms denoted by Arrows 1 and 2 are defined by

$$x \to x \cap (k_*^{-1} \mu_K)$$
$$y \to y \cap (l_*^{-1} \mu_A)$$

for any $x \in H^q(U, U - K)$ and $y \in H^q(M - K)$; here $\mu_K \in H_n(M, M - K)$ and $\mu_A \in H_n(M, M - A)$ have the same meaning as in the definition of the Poincaré duality isomorphism. In addition, each square of this diagram is commutative; this is a consequence of Lemma 8.1 in the appendix to this chapter.

Now pass to the direct limit as K ranges over all compact subsets of U. Note that

$$\text{dir lim } H^q(M - K) = \bar{H}^q(A)$$

since as K ranges over all compact subsets of U, $M - K$ ranges over all open neighborhoods of A (see Proposition 6.3). Hence we obtain the following commutative diagram:

$$H^q_c(U) \overset{\tau}{\longrightarrow} H^q(M) \overset{h^*}{\longrightarrow} \bar{H}^q(A) \overset{\delta}{\longrightarrow} H^{q+1}_c(U)$$

$$\downarrow P \qquad\qquad \downarrow P \qquad\qquad \downarrow P' \qquad\qquad \downarrow P \qquad\qquad (6.1)$$

$$H_{n-q}(U) \overset{i_*}{\longrightarrow} H_{n-q}(M) \overset{j_*}{\longrightarrow} H_{n-q}(M, U) \overset{\partial}{\longrightarrow} H_{n-q-1}(U).$$

Each square of this diagram is commutative, and the top line is exact, since direct limits preserve exactness. The vertical arrows labelled P are the Poincaré duality isomorphisms for M and U. It follows from the five-lemma that *the homomorphism labelled P' is also an isomorphism*. For future reference, we state this as follows:

Proposition 6.4. *Let M be a compact orientable n-manifold, A a closed subset of M, and $U = M - A$ the complementary set. Then the relative homology group $H_{n-q}(M, U; G)$ is isomorphic to the Čech–Alexander–Spanier cohomology group $\bar{H}^q(A; G)$.*

Of course the most interesting cases of Diagram (6.1) and Proposition 6.4 are those cases where the Alexander–Spanier cohomology group, $\bar{H}^q(A)$, and the singular cohomology group, $H^q(A)$, are isomorphic. In that case, it is easily verified that h^* is the homomorphism induced by the inclusion of A in M. However, the reader must not lose sight of the fact that it is absolutely necessary to use Alexander–Spanier cohomology for the correct statement of this proposition. The following example illustrates this point: Consider the 2-sphere, S^2, as the compactification of the plane $\mathbf{R}^2$, obtained by adjoining to it a point labelled ∞. Let A be the closed subset of S^2 which is the union of the graph of the equation $y = \sin(1/x)$ $(x \neq 0)$, the segment $-1 \leq y \leq +1$ of the y-axis, and the point ∞. As above, let $U = S^2 - A = \mathbf{R}^2 - A$. Then U has two components, and it may be shown that each component is homeomorphic to an open disc. Consider the following portion of the reduced homology sequence of (S^2, U):

$$H_1(S^2) \xrightarrow{j_*} H_1(S^2, U) \xrightarrow{\partial} \tilde{H}_0(U) \xrightarrow{i_*} \tilde{H}_0(S^2).$$

Since $\tilde{H}_0(U)$ is infinite cyclic, we deduce that $H_1(S^2, U)$ is also infinite cyclic. Hence by Proposition 6.4, $\bar{H}^1(A)$ is infinite cyclic. On the other hand, the singular cohomology group $H^1(A)$ is zero. The set A has the same Alexander–Spanier cohomology groups as a circle, while its singular cohomology groups are the same as a space consisting of two points. However, the complement of A in S^2 is homeomorphic to the complement of a circle imbedded in S^2.

Proposition 6.5. *Let A be a closed, proper subset of a compact, connected, orientable n-manifold. Then $\bar{H}^q(A; G) = 0$ for all $q \geq n$ and all coefficient groups G.*

This is a direct consequence of Proposition 6.4. It is of interest to note that this proposition is false in general for the singular cohomology groups $H^q(A; G)$; for a spectacular counterexample, see Barrett and Milnor, [1].

Theorem 6.6 (Alexander duality theorem). *Let M be a compact, connected, orientable n-manifold and q an integer such that $H_q(M, G) = H_{q+1}(M, G) = 0$. Then for any closed subset $A \subset M$,*

$$\bar{H}^{n-q-1}(A) \approx H_q(M - A).$$

The most important example of a manifold satisfying the hypotheses of this theorem is the n-sphere, S^n. Obviously, we must have $0 < q < n - 1$, because, $H_0(M, G)$ and $H_n(M, G)$ are always nonzero for a compact, connected, orientable n-manifold. However, there is no difficulty in stating versions of this theorem corresponding to the cases $q = 0$ and $q = n - 1$; this we will now do.

Theorem 6.6, continued. *Let M be a compact connected orientable n-manifold, and let A be a closed, proper subset of M.*

(a) *If $H_1(M;G) = 0$, then $\bar{H}^{n-1}(A;G) \approx \tilde{H}_0(M - A;G)$.*
(b) *$\bar{H}^0(A;G)$ always contains a direct summand isomorphic to G; if $H_{n-1}(M;G) = 0$, then the quotient group of $\bar{H}^0(A)$ modulo this summand is isomorphic to $H_{n-1}(M - A;G)$.*

This direct summand of $\bar{H}^0(A;G)$ can be more precisely described as follows: let P denote a space consisting of one point, and let $f: A \to P$ be the unique map. Then the subgroup in question is $f^*(\bar{H}^0(P;G))$. The corresponding quotient group is the "reduced" 0-dimensional Alexander–Spanier cohomology group.

The proof of the Alexander duality theorem follows immediately from Diagram (6.1); the details are left to the reader. The theorem can be considered a far-reaching generalization of the Jordan–Brouwer separation theorem and the other theorems which were proved in §III.6. Various applications of it are given in the exercises below. One of the main consequences is that if A is a closed subset of S^n, the homology groups of $S^n - A$ are independent of how A is imbedded in S^n. We have already seen special examples of this phenomenon in §III.6.

EXERCISES

6.1. Let A be a compact connected orientable $(n - 1)$-manifold imbedded in S^n. Prove that $S^n - A$ has exactly two components.

6.2. Prove that a nonorientable compact $(n - 1)$-manifold can not be imbedded in S^n. [*Hint*: If M is such a manifold, prove first that $H^{n-1}(M;\mathbf{Z})$ is a finite group of order 2. Then apply the Alexander duality theorem.]

6.3. Let A be a compact subset of $\mathbf{R}^n$. Derive a relation between the Alexander–Spanier cohomology groups of A and the singular homology groups of $\mathbf{R}^n - A$.

6.4. Let M be a compact, connected, orientable 2-manifold. We say a homology class $u \in H_1(M;\mathbf{Z})$ can be *represented by an inbedded circle* if there exists a subset $A \subset M$ such that A is homeomorphic to a circle, and the obvious homomorphism $H_1(A) \to H_1(M)$ sends a generator of $H_1(A;\mathbf{Z})$ onto u. Prove that if $u \neq 0$ and u can be represented by an imbedded circle, then u is not divisible (i.e., there does not exist an integer $d > 1$ and a homology class v such that $u = dv$; an equivalent condition is that the subgroup of $H_1(M)$ generated by u should be a direct summand). Prove also that if M is a torus, every nondivisible homology class can be represented by an imbedded circle.

6.5. State and prove the analogues of the theorems of this section for nonorientable manifolds, using $\mathbf{Z}_2$ coefficients for all homology and cohomology groups.

6.6. Let A be a compact subset of Euclidean 3-space $\mathbf{R}^3$ which is tautly imbedded and has finitely generated integral homology groups. Prove that the integral homology and cohomology groups of A are torsion-free.

§7. Duality Theorems for Manifolds with Boundary

We recall the definition: An n-dimensional manifold with boundary M is a Hausdorff space such that each point has an open neighborhood homomorphic to $\mathbf{R}^n$, or to $\mathbf{R}^n_+ = \{(x_1, \ldots, x_n) \in \mathbf{R}^n \mid x_n \geq 0\}$. For simple examples of manifolds with boundary, and for the classification of compact, connected 2-dimensional manifolds with boundary, the reader is referred to Massey [6], Chapter I, Sections 9–12. The set of all points of M having an open neighborhood homomorphic to $\mathbf{R}^n$ is called the *interior* of M, and the complementary set is called the *boundary* of M. Whether a point x belongs to the interior or boundary of M can be determined by means of the local homology groups of M at x (cf. the exercises to §III.2). The interior is an open, everywhere dense subset of M, which is an n-dimensional manifold; the boundary is a closed subset which is an $(n-1)$-dimensional manifold.

Our main objective is to state and prove an analog of the Poincare duality theorem for manifolds with boundary. For this purpose, it will be convenient to use the following fundamental theorem of Morton Brown:

Theorem 7.1. *Let M be a compact n-dimensional manifold with boundary B. Then there exists an open neighborhood V of B and a homeomorphism g of $B \times [0,1)$ onto V such that $g(b,0) = b$ for any $b \in B$.*

For a short proof of this theorem, see R. Connelly [3]. Connelly's proof is reproduced in the appendix to Vick [11].

This theorem has many consequences; among them are the following:

Corollary 7.2. *The inclusion map of $M - B$ into M is a homotopy equivalence.*

Corollary 7.3. *Let $V_t = g(B \times [0,t))$ for $0 < t < 1$, and $K_t = M - V_t$. Then V_t is an open neighborhood of B in M, B is a deformation retract of V_t, and the collection $\{K_t \mid 0 < t < 1\}$ is cofinal in the family of all compact subsets of $M - B$.*

Next, for $0 < t < 1$ let $i_t : (M,B) \to (M, M - K_t) = (M,V_t)$ denote the inclusion map. It follows that the induced homomorphisms

$$i_{t*} : H_q(M,B) \to H_q(M, M - K_t),$$
$$i_t^* : H^q(M, M - K_t) \to H^q(M,B)$$

are isomorphisms.

Corollary 7.4. $H_c^q(M - B; G)$ *is naturally isomorphic to* $H^q(M,B;G)$.

This corollary follows from the definition of $H_c^q(M - B)$ as a direct limit, the fact that $H^q(M - B, (M - B) - K) \approx H^q(M, M - K)$ for any compact set $K \subset M - B$, and the cofinality of the family $\{K_t\}$.

We will define a manifold M with boundary B to be *oriented* if the manifold $M - B$ is oriented in the sense defined in §2. This implies that for each compact set $K \subset M - B$, there is a unique homology class $\mu_K \in H_n(M - B, M - B - K; \mathbf{Z})$ such that $\rho_x(\mu_K)$ is the local orientation of $M - B$ at x. But as was observed above,

$$H_n(M - B, M - B - K) \approx H_n(M, M - K),$$

by the excision property. In addition, if M is compact and $K = K_t$, $H_n(M, M - K) \approx H_n(M, B)$. Thus the fact that M is oriented and compact implies the existence of a unique homology class $\mu_M \in H_n(M, B; \mathbf{Z})$ such that for any $x \in M - B$, the homomorphism $H_n(M, B) \to H_n(M, M - \{x\})$ maps μ_M onto the local orientation μ_x. μ_M is called *the fundamental homology class of M*.

Theorem 7.5. *Let M be a compact orientable n-dimensional manifold with boundary B. Then the homomorphism*

$$H^q(M, B; G) \to H_{n-q}(M; G),$$

(defined by $x \to x \cap \mu_M$ for any $x \in H^q(M, B; G)$) is an isomorphism.

PROOF. We already know that $H^q(M, B; G)$ is isomorphic to $H_{n-q}(M; G)$. For, by Corollary 7.4, $H^q(M, B) \approx H_c^q(M - B)$; then we have the Poincare duality isomorphism $P : H_c^q(M - B) \approx H_{n-q}(M - B)$. Finally, by Corollary 7.2, there is an isomorphism $H_{n-q}(M - B) \approx H_{n-q}(M)$ induced by inclusion. Thus it suffices to prove that the composition of these three isomorphisms is the same as the homomorphism $H^q(M, B) \to H_{n-q}(M)$ occurring in the statement of the theorem. In order to prove this, consider the following commutative diagram:

$$
\begin{array}{ccccc}
H^q(\mathscr{I}M, \mathscr{I}M - K) & \longleftarrow & H^q(M, M - K) & \longrightarrow & H^q(M, B) \\
\otimes & & \otimes & & \otimes \\
H_n(\mathscr{I}M, \mathscr{I}M - K) & \longrightarrow & H_n(M, M - K) & \longleftarrow & H_n(M, B) \\
\downarrow & & \downarrow & & \downarrow \\
H_{n-q}(\mathscr{I}M) & \longrightarrow & H_{n-q}(M) & \longleftarrow & H_{n-q}(M).
\end{array}
$$

In this diagram, $\mathscr{I}M = M - B$ denotes the interior of M, $K = K_t$, all three vertical arrows denote cap products, and all horizontal arrows denote isomorphisms which are induced by inclusion maps. The left-hand vertical arrow defines the Poincaré duality isomorphism $P : H_c^q(\mathscr{I}M) \to H_{n-q}(\mathscr{I}M)$, and the right-hand vertical arrow denotes the cap product occurring in the statement of the theorem. Putting all these facts together, the reader should have no difficulty deducing the theorem. Q.E.D.

The isomorphism of the theorem just proved is one-half of the Lefschetz–Poincaré duality theorem for manifolds with boundary. As a preliminary to the other half of this duality theorem, we need the following important result:

Theorem 7.6. *Let M be a compact, oriented, n-dimensional manifold with boundary B, and let $\partial_*: H_n(M,B;\mathbf{Z}) \to H_{n-1}(B,\mathbf{Z})$ denote the boundary operator of the pair (M,B). Then $\partial_*(\mu_M)$ is a fundamental homology class for some orientation of B; in particular, B is orientable.*

PROOF. In order to prove this theorem, it is necessary to show that for any $b \in B$, $j\partial_*(\mu_M)$ is a generator of the infinite cyclic group $H_{n-1}(B, B - \{b\}; \mathbf{Z})$. Here j denotes the homomorphism $H_{n-1}(B) \to H_{n-1}(B, B - \{b\})$ induced by inclusion. Note that $j\partial_* = \partial'$ is the boundary operator of the exact homology sequence of the triple $(M, B, B - \{b\})$.

By the definition of a manifold with boundary, there exists an open neighborhood U of b and a homeomorphism h of U onto $\mathbf{R}^n_+$. Since b is a boundary point of M, $h(b)$ must lie in the subspace of $\mathbf{R}^n_+$ defined by the equation $x_n = 0$. Obviously, we can assume that h is chosen so that $h(b) = (0,0,\ldots,0)$. We may as well identify each point $x \in U$ with its image $h(x) \in \mathbf{R}^n_+$; thus the coordinates $x_1, \ldots, x_n$ in $\mathbf{R}^n_+$ are actually coordinate functions in U. Then $B \cap U$ is the subset of U defined by the equation $x_n = 0$. Let $a \in U$ be the point with coordinates $(0,\ldots,0,1)$, and let N and W be the following subsets of U:

$$N = \{(x_1, \ldots, x_n) \in U \mid \sum x_i^2 \le 4\},$$
$$W = \{(x_1, \ldots, x_n) \in N \mid \sum x_i^2 < 4 \text{ and } x_n > 0\},$$
$$E = N \cap B.$$

Now consider the following commutative diagram:

In this diagram, the arrows labelled ∂', ∂_1, and ∂_2 denote the boundary operators of certain triples; all other arrows denote homomorphisms induced by inclusion maps. It is a routine matter to prove that ∂_2 and the homomorphisms numbered 1 through 6 are isomorphisms. Thus all the groups in the diagram, except possibly $H_n(M,B)$, are infinite cyclic, and are related by a unique isomorphism. We know that $\rho(\mu_M)$ is a generator of the infinite cyclic group $H_n(M, M - \{a\})$. It therefore follows that $\partial'(\mu_M)$ is a generator of the infinite cyclic group $H_{n-1}(B, B - \{b\})$, as was to be proved.

                                                                                Q.E.D.

We can now derive the remaining half of the Lefschetz–Poincaré duality theorem for manifolds with boundary. Let M be a compact, oriented n-dimensional manifold with boundary B, and let $\mu_M \in H_n(M,B;\mathbf{Z})$ denote the fundamental homology class of M. Consider the following diagram involving the exact homology and cohomology sequences of the pair (M,B):

(D)
$$\begin{array}{ccccccc}
H^q(M,B;G) & \xrightarrow{j^*} & H^q(M;G) & \xrightarrow{i^*} & H^q(B;G) & \xrightarrow{\delta} & H^{q+1}(M,B;G) \\
\downarrow{\scriptstyle 1} & & \downarrow{\scriptstyle 2} & & \downarrow{\scriptstyle 3} & & \downarrow{\scriptstyle 1} \\
H_{n-q}(M;G) & \xrightarrow{j_*} & H_{n-q}(M,B;G) & \xrightarrow{\partial} & H_{n-q-1}(B;G) & \xrightarrow{i_*} & H_{n-q-1}(M;G).
\end{array}$$

In this diagram, homomorphisms denoted by arrows 1 and 2 are cap product with the fundamental class, i.e., $x \to x \cap \mu_M$. Arrow number 3 denotes the Poincaré duality isomorphism for B, defined by $y \to y \cap (\partial\mu_M)$. On account of the basic properties of cap products, each square in this diagram is commutative up to a $\pm$ sign. We have already proved that arrows 1 and 3 are isomorphisms. It follows from the five-lemma that arrow 2 is also an isomorphism. Thus we have proved the following result:

Theorem 7.7. *Let M be a compact, oriented n-dimensional manifold with boundary B and fundamental class $\mu_M \in H_n(M, M - B; \mathbf{Z})$. Then there are Lefschetz–Poincaré duality isomorphisms*

$$H^q(M,B;G) \to H_{n-q}(M;G)$$

and

$$H^q(M;G) \to H_{n-q}(M,B;G)$$

defined by cap product with μ_M. In addition the homology sequence of (M,B) and the cohomology sequence of (M,B) are isomorphic as indicated in Diagram (D) above.

EXERCISES

7.1. Let M be a compact, connected, orientable n-manifold with nonempty boundary B (B need not be connected). Prove the following relations for any abelian group G:

$H_n(M;G) = 0$.
$H_n(M,B;G) \approx G$.
$H_{n-1}(M;\mathbf{Z})$ and $H_{n-1}(M,B;\mathbf{Z})$ are torsion-free abelian groups.

7.2. State and prove analogues of the theorems of this section for nonorientable manifolds with boundary, using $\mathbf{Z}_2$ coefficients.

7.3. Let M be a compact n-dimensional manifold with boundary B. If n is odd, prove that

$$\chi(B) = 2\chi(M) = -2\chi(M,B),$$

where χ denotes the Euler characteristic. (Note: It may be proved that the integral homology groups of a compact manifold with boundary are all finitely generated. Hence the Euler characteristic $\chi(M)$ and $\chi(M,B)$ are well defined.)

7.4. Let M be a compact, oriented n-dimensional manifold with boundary B. (a) For any field F, prove that the bilinear form

$$H^q(M,B;F) \otimes H^{n-q}(M;F) \to F,$$

defined by $u \otimes v \to \langle u \cup v, \mu_M \rangle$, is nonsingular (cf. Theorem 5.1). (b) By analogy with Theorem 5.2, prove that the bilinear form

$$B^q(M,B) \otimes B^{n-q}(M) \to \mathbf{Z},$$

defined by $u \otimes v \to \langle u \cup v, \mu_M \rangle$, is nonsingular.

7.5. Prove that the integral homology groups $H_q(M,B)$ and $H_{n-q}(M)$ have the same rank, and that the torsion subgroups of $H_q(M,B)$ and $H_{n-q-1}(M)$ are isomorphic, where (M,B) is as in the preceding exercise.

7.6. Let M be a compact, orientable $2q$-dimensional manifold with boundary B, where q is odd, and let F be a field of characteristic $\neq 2$. Prove that the homomorphism $j^*:H^q(M,B;F) \to H^q(M;F)$ has even rank. (*Hint*: See the proof of Proposition 5.3.)

7.7. Let M_i be a manifold with boundary B_i for $i = 1, 2$. Prove that $M_1 \times M_2$ is a manifold with boundary. What is the boundary of $M_1 \times M_2$?

§8. Appendix: Proof of Two Lemmas about Cap Products

For the statement of the first lemma, assume that $\{A,B\}$ is an excisive couple in the space X, and $X = A \cup B$. We then have the following diagram of homology groups and homomorphisms:

$$H_n(A, A \cap B) \xrightarrow{e_{1*}} H_n(X,B) \xleftarrow{l_*} H_n(X) \xrightarrow{j_*} H_n(X,A) \xleftarrow{e_{2*}} H_n(B, A \cap B).$$

All homomorphisms are induced by inclusion maps; e_{1*} and e_{2*} are isomorphisms because $\{A,B\}$ is excisive. Assume that $v \in H_n(X)$ is given; let

$$v_1 = (e_{1*})^{-1}l_*(v) \in H_n(A, A \cap B),$$
$$v_2 = (e_{2*})^{-1}j_*(v) \in H_n(B, A \cap B).$$

Now consider the following diagram:

The top line is the exact cohomology sequence of the pair (X,A), the bottom line is the exact homology sequence of the pair (X,B), and the vertical arrows are induced either by the inclusion maps e_1 or e_2, or else by cap product with the indicated homology class.

Lemma 8.1. *Each square in the above diagram is commutative.*

PROOF. In §VIII.3, we defined a slant product

$$C^p(Y,G_1) \otimes C_q(X \times Y; G_2) \to C_{q-p}(X; G_1 \otimes G_2)$$

by the formula

$$u\backslash v = u\backslash\backslash\xi(v)$$

for any $u \in C^p(Y)$ and $v \in C_q(X \times Y)$. This slant product satisfies the following formula:

$$\partial(u\backslash v) = (\delta u)\backslash v + (-1)^p u\backslash(\partial v).$$

On passing to homology and cohomology classes, it determines a homomorphism

$$H^p(Y) \otimes H_q(X \times Y) \to H_{q-p}(X),$$

which is also called the slant product.

For the purposes of this appendix, it is convenient to define in a similar way, a cap product on the chain-cochain level. This will be a homomorphism

$$C^p(X;G_1) \otimes C_q(X;G_2) \xrightarrow{\cap} C_{q-p}(X;G_1 \otimes G_2)$$

defined by

$$u \cap v = u\backslash d_\#(v),$$

where $d_\#:C_q(X) \to C_q(X \times X)$ is the chain map induced by the diagonal map d. It satisfies the following boundary-coboundary formula,

$$\partial(u \cap v) = (\delta u) \cap v + (-1)^p u \cap (\partial v), \tag{8.1}$$

for any $u \in C^p(X)$ and $v \in C_q(X)$. On passage to cohomology and homology classes, it gives rise to the cap product defined in §VIII.3. The naturality condition

$$f_\#((f^\# u) \cap v) = u \cap (f_\# v)$$

obviously holds for any continuous map $f:X \to X', u \in C^p(X')$ and $v \in C_q(X)$. Moreover, this definition can be generalized easily to cover the case of relative chain and cochain groups which we need below.

Since $\{A,B\}$ is excisive and $X = A \cup B$, the inclusion map

$$C(A) + C(B) \to C(X)$$

induces isomorphisms on homology groups. Therefore we can choose a representative cycle z for the homology class $v \in H_n(X)$ such that $z \in C_n(A) + C_n(B)$. In other words,

$$z = z_1 + z_2$$

where $z_1 \in C_n(A)$ and $z_2 \in C_n(B)$. Although z is a cycle, i.e., $\partial(z) = 0$, it does not follow that z_1 and z_2 are cycles. All we can conclude is that

$$\partial(z_1) = -\partial(z_2) \in C_{n-1}(A \cap B).$$

Let $z_1' \in C_n(A, A \cap B)$ and $z_2' \in C_n(B, A \cap B)$ denote the images of z_1 and z_2 respectively in these quotient groups. Then

$$\partial(z_1') = \partial(z_2') = 0$$

and

$$e_{1\#}(z_1') = l_\#(z), \qquad e_{2\#}(z_2') = j_\#(z).$$

Therefore z_1' and z_2' are representative cycles for v_1 and v_2 respectively. Now consider the following diagram of chain and cochain complexes, and chain-cochain maps:

$$
\begin{array}{ccccccccc}
0 & \longrightarrow & C^*(X,A) & \xrightarrow{\ j^*\ } & C^*(X) & \xrightarrow{\ i^*\ } & C^*(A) & \longrightarrow & 0 \\
 & & \downarrow {\scriptstyle e_2^\#} & & \downarrow & & \downarrow {\scriptstyle \cap z_1'} & & \\
 & & C^*(B, A \cap B) & & {\scriptstyle \cap z} & & C(A, A \cap B) & & \qquad (8.2) \\
 & & \downarrow {\scriptstyle \cap z_2'} & & \downarrow & & \downarrow {\scriptstyle e_{1\#}} & & \\
0 & \longrightarrow & C(B) & \xrightarrow{\ k_\#\ } & C(X) & \xrightarrow{\ l_\#\ } & C(X,B) & \longrightarrow & 0.
\end{array}
$$

Although the homomorphisms denoted by the vertical arrows do not have degree 0, they commute with the boundary and coboundary operators because z, z_1', and z_2' are cycles, and because of Formula (8.1) above. The top and bottom lines of this diagram are exact. We assert that each square of this diagram is commutative. This is a consequence of the "commutativity" of the following diagram:

$$
\begin{array}{ccccccccc}
C^*(B, A \cap B) & \xleftarrow{\ e_2^\#\ } & C^*(X,A) & \xrightarrow{\ j^*\ } & C^*(X) & = & C^*(X) & \xrightarrow{\ i^*\ } & C^*(A) \\
\otimes & & \otimes & & \otimes & & \otimes & & \otimes \\
C_n(B, A \cap B) & \xrightarrow[e_{2\#}]{} & C_n(X,A) & \xleftarrow[j_\#]{} & C_n(X) & \xrightarrow[l_\#]{} & C_n(X,B) & \xleftarrow[e_{1\#}]{} & C_n(A, A \cap B) \\
\downarrow {\scriptstyle \cap} & & \downarrow {\scriptstyle \cap} & & \downarrow {\scriptstyle \cap} & & \downarrow {\scriptstyle \cap} & & \downarrow {\scriptstyle \cap} \\
C(B) & \xrightarrow[k_\#]{} & C(X) & = & C(X) & \xrightarrow[l_\#]{} & C(X,B) & \xleftarrow[e_{1\#}]{} & C(A, A \cap B).
\end{array}
$$

The "commutativity" of each of the four squares of this diagram expresses a naturality relation for cap products.

The proof of the lemma may now be completed by passing from Diagram (8.2) to the corresponding diagram of homology and cohomology groups, induced homomorphisms, etc.   Q.E.D.

The statement and proof of the second lemma are somewhat longer. Assume that M is an oriented n-manifold and that $M = U \cup V$, where U

and V are open subsets of M. Let K and L be compact subsets of U and V respectively. Since M is oriented, by Theorem 2.1, there exist unique homology classes

$$\mu_{K \cup L} \in H_n(M, M - K \cup L),$$
$$\mu_K \in H_n(M, M - K),$$
$$\mu_L \in H_n(M, M - L),$$

and

$$\mu_{K \cap L} \in H_n(M, M - K \cap L)$$

which restrict to the chosen local orientations at each point. Consider the following diagram:

$$
\cdots \xrightarrow{\ \Delta^* \ } H^q(M, M - K \cap L) \longrightarrow H^q(M, M - K) \otimes H^q(M, M - L) \longrightarrow H^q(M - K \cup L) \xrightarrow{\ \Delta_* \ } \cdots
$$

In the top line of this diagram, we have the relative Mayer–Vietoris cohomology sequence of the triad $(M; M - K, M - L)$, while the bottom line is the usual Mayer–Vietoris homology sequence. The maps

$$p : (U \cap V, U \cap V - K \cap L) \to (M, M - K \cap L),$$
$$k_1 : (U, U - K) \to (M, M - K),$$
$$k_2 : (V, V - L) \to (M, M - L),$$

are inclusion maps which induce isomorphisms on homology and cohomology by the excision property; also,

$$v_{K \cap L} = p_*^{-1}(\mu_{K \cap L}),$$
$$v_K = k_{1*}^{-1}(\mu_K),$$
$$v_L = k_{2*}^{-1}(\mu_L).$$

Lemma 8.2. *Each square of the above diagram is commutative.*

It is understood that the diagram is extended to the right and left indefinitely, and that the lemma applies to each square of the extended diagram.

If we pass to the direct limit over all such compact sets $K \subset U$ and $L \subset V$, we obtain a commutative diagram involving two exact sequences which played a crucial role in the proof of the Poincaré duality theorem in §4.

Lemma 8.2 is a special case of a more general lemma which we will now state. Let $X_1, X_2, Y_1,$ and Y_2 be subspaces of a topological space X such that $X = (\text{Interior } X_1) \cup (\text{Interior } X_2)$ and $\{Y_1, Y_2\}$ is an excisive couple. Assume we have given homology classes

$$\mu \in H_n(X, Y_1 \cap Y_2)$$
$$v_\alpha \in H_n(X_\alpha, X_\alpha \cap Y_\alpha), \qquad \alpha = 1, 2,$$

and

$$v \in H_n(X_1 \cap X_2, X_1 \cap X_2 \cap (Y_1 \cup Y_2))$$

such that

$$i_{\alpha*}(\mu) = k_{\alpha*}(v_\alpha),$$

and

$$q_{\alpha*}(v_\alpha) = m_{\alpha*}(v)$$

for $\alpha = 1, 2$, where

$$i_\alpha : (X, Y_1 \cap Y_2) \to (X, Y_\alpha),$$
$$k_\alpha : (X_\alpha, X_\alpha \cap Y_\alpha) \to (X, Y_\alpha),$$
$$q_\alpha : (X_\alpha, X_\alpha \cap Y_\alpha) \to (X_\alpha, X_\alpha \cap (Y_1 \cup Y_2)),$$

and

$$m_\alpha : (X_1 \cap X_2, X_1 \cap X_2 \cap (Y_1 \cup Y_2)) \to (X_\alpha, X_\alpha \cap (Y_1 \cup Y_2))$$

are all inclusion maps. Consider the following diagram:

$$
\begin{array}{ccccccc}
H^q(X, Y_1 \cup Y_2) & \xrightarrow{\Phi} & H^q(X, Y_1) \otimes H^q(X, Y_2) & \xrightarrow{\Psi} & H^q(X, Y_1 \cap Y_2) & \xrightarrow{\Delta^*} & H^{q+1}(X, Y_1 \cup Y_2) \\
\downarrow{\alpha} & 1 & \downarrow & 2 & \downarrow{\gamma} & 3 & \downarrow{\alpha} \\
H_{n-q}(X_1 \cap X_2) & \xrightarrow{\Phi'} & H_{n-q}(X_1) \otimes H_{n-q}(X_2) & \xrightarrow{\Psi'} & H_{n-q}(X) & \xrightarrow{\Delta_*} & H_{n-q-1}(X_1 \cap X_2).
\end{array}
$$

The top line is the relative Mayer–Vietoris cohomology sequence of the triad $(X; Y_1, Y_2)$, while the bottom line is the usual Mayer–Vietoris homology sequence. The homomorphisms α, β, and γ are defined as follows:

$$\alpha(x) = (p^*x) \cap v, \qquad x \in H^*(X, Y_1 \cup Y_2),$$
$$\beta(u,v) = ((k_1^* u) \cap v_1, (k_2^* v) \cap v_2), \qquad u \in H^*(X, Y_1), v \in H^*(X, Y_2)$$
$$\gamma(w) = w \cap \mu, \qquad w \in H^*(X, Y_1 \cap Y_2).$$

Here $p : (X_1 \cap X_2, X_1 \cap X_2 \cap (Y_1 \cup Y_2)) \to (X, Y_1 \cup Y_2)$ is an inclusion. From the basic properties of cap products, it is easy to check that the squares 1 and 2 in the above diagram are commutative. However, square 3 need *not* be commutative. In fact, we have the following precise statement:

Lemma 8.3. *There exists a homology class $y \in H_{n+1}(X, Y_1 \cup Y_2)$ such that for any integer q and any $w \in H^q(X, Y_1 \cap Y_2)$,*

$$\Delta_* \gamma(w) - \alpha \Delta^*(w) = \Delta_*((\Delta^* w) \cap y)$$

(the homology class y is not unique, in general).

Before proving this lemma, we will indicate how it implies Lemma 8.2. Let

$$X = M, \qquad X_1 = U, \qquad X_2 = V,$$
$$Y_1 = M - K, \quad \text{and} \quad Y_2 = M - L.$$

Then $H_{n+1}(X, Y_1 \cup Y_2) = H_{n+1}(M, M - K \cap L) = 0$ since M is an n-dimensional manifold. Hence $y = 0$ in this case, and Lemma 8.2 follows.

PROOF OF LEMMA 8.3. The standard situation which leads to a commutative diagram of exact sequences is the following:

$$
\begin{array}{ccccccccc}
0 & \longrightarrow & K' & \overset{i}{\longrightarrow} & K & \overset{j}{\longrightarrow} & K'' & \longrightarrow & 0 \\
& & \downarrow{\scriptstyle \varphi'} & & \downarrow{\scriptstyle \varphi} & & \downarrow{\scriptstyle \varphi''} & & \\
0 & \longrightarrow & L' & \overset{k}{\longrightarrow} & L & \overset{l}{\longrightarrow} & L'' & \longrightarrow & 0.
\end{array}
\tag{8.3}
$$

In this diagram, the following two hypotheses are assumed:

(i) The top and bottom lines are short exact sequences of chain complexes and chain maps.
(ii) The chain maps φ', φ, and φ'' satisfy the following commutativity relations:

$$\varphi i = k\varphi', \quad \text{and} \quad \varphi'' j = l\varphi.$$

Unfortunately, this situation does not apply to the case at hand, because neither of these two hypotheses holds when we go back to chains and cochains. In order to prove Lemma 8.3, it is necessary to investigate what happens when we relax these hypotheses. The first (and more interesting) step is to relax the commutativity condition (ii), and require only commutativity up to a chain homotopy. To be precise, assume that the following chain homotopy relations hold in the above diagram:

$$\varphi i - k\varphi' = \partial D + D\partial',$$
$$\varphi'' j - l\varphi = \partial'' E + E\partial,$$

where $D: K' \to L$ and $E: K \to L''$ are homomorphisms of degree $+1$. An easy calculation then shows that

$$\partial''(Ei + lD) = -(Ei + lD)\partial',$$

i.e., the homomorphism $Ei + lD: K' \to L''$ commutes with the boundary operator (up to a minus sign). Therefore it induces homomorphisms

$$(Ei + lD)_*: H_{q-1}(K') \to H_q(L'')$$

for all q. We assert that this homomorphism gives us a measure of the lack of commutativity of the following diagram:

$$
\begin{array}{ccccccc}
\cdots \overset{j_*}{\longrightarrow} & H_q(K'') & \overset{\partial_K}{\longrightarrow} & H_{q-1}(K') & \overset{i_*}{\longrightarrow} \cdots \\
& \downarrow{\scriptstyle \varphi''_*} & & \downarrow{\scriptstyle \varphi'_*} & \\
\cdots \overset{l_*}{\longrightarrow} & H_q(L'') & \overset{\partial_L}{\longrightarrow} & H_{q-1}(L') & \overset{k_*}{\longrightarrow} \cdots .
\end{array}
$$

In fact, the following equation holds:

$$\partial_L \varphi''_* - \varphi'_* \partial_K = \partial_L(Ei + lD)_* \partial_K. \tag{8.4}$$

To prove this equation, one must prove that for any $u \in H_q(K'')$,

$$\varphi'_* \partial_K(u) = \partial_L(\varphi''_*(u) - (Ei + lD)_* \partial_K(u)).$$

Choose a representative cycle for the homology class u, and then compute representative cycles for the left- and right-hand side of this equation. We leave it to the reader to verify that the two representative cycles are homologous.

Next, we will consider relaxing Hypothesis (i), the exactness hypothesis. We will assume given a diagram

$$K' \xrightarrow{i} K \xrightarrow{j} K''$$

of chain complexes and chain maps such that i is a monomorphism, j is an epimorphism, and image i is contained in kernel j. However, we do *not* assume that image $i = $ kernel j; this is the assumption we have to avoid. We also have to consider the following two additional chain complexes:

$$\mathcal{K}(j) = \text{kernel } j,$$
$$\mathcal{C}(i) = \text{cokernel } i.$$

We then have the following commutative diagram of chain complexes and chain maps:

$$
\begin{array}{ccccccccc}
0 & \longrightarrow & K' & \xrightarrow{i} & K & \longrightarrow & \mathcal{C}(i) & \longrightarrow & 0 \\
& & \downarrow{\scriptstyle \alpha} & & \| & & \downarrow{\scriptstyle \beta} & & \\
0 & \longrightarrow & \mathcal{K}(j) & \longrightarrow & K & \xrightarrow{j} & K'' & \longrightarrow & 0.
\end{array}
$$

Each row of this diagram is exact. Using the five-lemma, it is readily seen that $\alpha_* : H_q(K') \to H_q(\mathcal{K}(j))$ is an isomorphism for all q if and only if $\beta_* : H_q(\mathcal{C}(i)) \to H_q(K'')$ is an isomorphism for all q. If that is the case, we can define a long exact homology sequence

$$\cdots \to H_q(K') \xrightarrow{i_*} H_q(K) \xrightarrow{j_*} H_q(K'') \xrightarrow{\partial_*} H_{q-1}(K') \to \cdots$$

in a natural way.

Let us agree to say that the sequence of chain complexes and chain maps

$$K' \xrightarrow{i} K \xrightarrow{j} K''$$

is *almost exact* if all the assumptions listed in the preceding paragraph (including that α_* and β_* are isomorphisms) hold. The point is that almost exact sequences are just as good as short exact sequences when it comes to defining long exact homology sequences.

EXAMPLE 8.1. Assume that $\{A,B\}$ is an excisive couple in the space X. We then have the following almost exact sequence of chain complexes,

$$C(X, A \cap B) \to C(X,A) \oplus C(X,B) \to C(X, A \cup B)$$

which gives rise to the relative Mayer–Vietoris homology sequence (cf. §VIII.6). The dual sequence of cochain complexes,

$$C^*(X, A \cup B; G) \to C^*(X,A;G) \oplus C^*(X,B;G) \to C^*(X, A \cap B)$$

is also almost exact, and gives rise to the relative Mayer–Vietoris sequence in cohomology.

We will now apply these ideas to generalize Diagram (8.3) and Equation (8.4) above. Assume we have given the following diagram of chain complexes and chain maps:

$$
\begin{array}{ccccc}
K' & \xrightarrow{\ i\ } & K & \xrightarrow{\ j\ } & K'' \\
\downarrow{\scriptstyle \varphi'} & & \downarrow{\scriptstyle \varphi} & & \downarrow{\scriptstyle \varphi''} \\
L' & \xrightarrow{\ k\ } & L & \xrightarrow{\ l\ } & L''.
\end{array}
\qquad (8.5)
$$

It is assumed that both rows of this diagram are almost exact (instead of exact), and that each square is chain-homotopy commutative; in other words, there exist chain homotopies $D: K' \to L$ and $E: K \to L''$ such that

$$\varphi i - k\varphi' = \partial D + D\partial',$$

$$\varphi'' j - l\varphi = \partial'' E + E\partial.$$

Then exactly as before, we can verify that the homomorphism

$$Ei + lD : K' \to L''$$

commutes with the boundary operators (up to a minus sign) and induces homomorphisms

$$(Ei + lD)_* : H_{q-1}(K') \to H_q(L'').$$

Then this homomorphism suffices to describe the lack of commutativity in the following diagram:

$$
\begin{array}{ccc}
H_q(K'') & \xrightarrow{\ \partial_K\ } & H_{q-1}(K') \\
\downarrow{\scriptstyle \varphi''_*} & & \downarrow{\scriptstyle \varphi'_*} \\
H_q(L'') & \xrightarrow{\ \partial_L\ } & H_{q-1}(L')
\end{array}
$$

by means of the following equation:

$$\partial_L \varphi''_* - \varphi'_* \partial_K = \partial_L (Ei + lD)_* \partial_L \qquad (8.6)$$

PROOF OF EQUATION (8.6): Consider the following diagram:

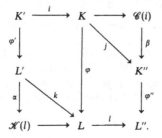

It follows that we can write down the analog of Equation (8.4) for the following diagram:

$$
\begin{array}{ccccc}
K' & \xrightarrow{\ i\ } & K & \longrightarrow & \mathscr{C}(i) \\
{\scriptstyle \alpha\varphi'}\big\downarrow & & \big\downarrow & & \big\downarrow{\scriptstyle \varphi''\beta} \\
\mathscr{K}(l) & \longrightarrow & L & \xrightarrow{\ l\ } & L''.
\end{array}
$$

Since α_* and β_* are isomorphisms, Equation (8.6) is then an easy consequence.

Q.E.D.

We are now ready to apply these ideas to prove Lemma 8.3. Choose representative cycles

$$
\begin{aligned}
&\mu' \in C_n(X, Y_1 \cup Y_2) \\
&v'_\alpha \in C_n(X_\alpha, X_\alpha \cap Y_\alpha), \qquad \alpha = 1, 2, \\
&v' \in C_n(X_1 \cap X_2, X_1 \cap X_2 \cap (Y_1 \cup Y_2))
\end{aligned}
$$

for the homology classes μ, v_α, and v respectively. Now consider the following diagram of chain and cochain complexes, and chain maps:

$$
\begin{array}{ccccc}
C^*(X, Y_1 \cup Y_2) & \xrightarrow{\ \varphi\ } & C^*(X,Y_1) \oplus C^*(X,Y_2) & \xrightarrow{\ \psi\ } & C^*(X, Y_1 \cap Y_2) \\
{\scriptstyle \alpha'}\big\downarrow & & {\scriptstyle \beta'}\big\downarrow & & {\scriptstyle \gamma'}\big\downarrow \\
C(X_1 \cap X_2) & \xrightarrow{\ \varphi'\ } & C(X_1) \oplus C(X_2) & \xrightarrow{\ \psi''\ } & C(X_1) + C(X_2).
\end{array}
$$

The homomorphisms in this diagram are defined as follows (see the diagram at the end of the proof):

$$
\begin{aligned}
\varphi(x) &= (j_1^{\#}x, j_2^{\#}x) & & x \in C^*(X, Y_1 \cup Y_2) \\
\alpha'(x) &= (p^{\#}x) \cap v' & & x \in C^*(X, Y_1 \cup Y_2) \\
\psi(u,v) &= i_1^{\#}u - i_2^{\#}v & & u \in C^*(X,Y_1),\, v \in C^*(X,Y_2) \\
\beta(u,v) &= ((k_1^{\#}u) \cap v'_1, (k_2^{\#}v) \cap v'_2) & & u \in C^*(X,Y_1),\, v \in C^*(X,Y_2) \\
\gamma'(w) &= w \cap \mu', & & w \in C^*(X, Y_1 \cap Y_2) \\
\varphi(x) &= (m_1 x, m_2 x), & & x \in C(X_1 \cap X_2) \\
\psi'(u,v) &= k'_1 u - k'_2 v, & & u \in C(X_1),\, v \in C(X_2).
\end{aligned}
$$

The top line is almost exact; on passage to cohomology, one obtains the relative Mayer–Vietoris sequence. The bottom line is exact; on passage to homology, it gives the usual Mayer–Vietoris sequence. At the right end of the bottom line, $C(X_1) + C(X_2)$ denotes the chain subcomplex of $C(X)$

generated by $C(X_1)$ and $C(X_2)$. In order that the image of γ' should lie in this subcomplex, we assume that the representative cycle μ' is a linear combination of singular cubes which are "small of order $\mathscr{U}$," where $\mathscr{U} = \{X_1, X_2\}$.

It is readily verified that α', β', and γ' are chain maps. Moreover, both squares of this diagram are chain homotopy commutative. Explicit chain homotopies may be defined as follows. The hypothesis that $i_{\alpha*}(\mu) = k_{\alpha*}(v_\alpha)$ implies the existence of chains

$$a_\alpha \in C_{n+1}(X, Y_\alpha), \qquad \alpha = 1, 2$$

such that

$$\partial a_\alpha = i_{\alpha\#}(\mu') - k_{\alpha\#}(v').$$

Similarly, the hypothesis that $q_{\alpha*}(v_\alpha) = m_{\alpha*}(v)$ implies the existence of chains

$$b_\alpha \in C_{n+1}(X_\alpha, X_\alpha \cap (Y_1 \cup Y_2))$$

such that

$$\partial b_\alpha = q_{\alpha\#}(v'_\alpha) - m_{\alpha\#}(v').$$

Then one defines chain homotopies

$$D: C^*(X, Y_1 \cup Y_2) \to C(X_1) \oplus C(X_2)$$
$$E: C^*(X, Y_1) \oplus C^*(X, Y_2) \to C(X_1) + C(X_2)$$

by the formulas

$$D(x) = (-1)^{|x|}((n_1^\# x) \cap b_1, (n_2^\# x) \cap b_2))$$
$$E(u,v) = (-1)^{|u|}(u \cap a_1 - v \cap a_2).$$

It is then easy to verify that

$$\beta'\varphi - \varphi'\alpha' = \partial D + D\delta,$$
$$\gamma'\psi - \psi'\beta' = \partial E + E\delta$$

as required. Thus we are in the situation described in Diagram (8.5) above, and Formula (8.6) is applicable. Using the definition of D and E above, and the naturality properties of the cap product, an easy computation gives the following formula:

$$(E\varphi + \psi'D)(x) = (-1)^{|x|}x \cap y'$$

for any $x \in C^*(X, Y_1 \cup Y_2)$, where

$$y' = j_{1\#}a_1 + n_{1\#}b_1 - j_{2\#}a_2 - n_{2\#}b_2.$$

In view of the way the chains a_1, a_2, b_1, and b_2 were chosen, it is easy to check that $\partial y' = 0$, i.e., y' is a cycle. Let $y \in H_{n+1}(X, Y_1 \cup Y_2)$ denote the homology class of y'; then it follows from Formula (8.6) that $\pm y$ has the properties stated in Lemma 8.3; this completes the proof. To assist the reader in following the above proof, we offer the following commutative diagram of the chain complexes and chain maps which occur in the above proof:

$$C(X, Y_1 \cap Y_2)$$

$$i_{1\#} \swarrow \qquad \searrow i_{2\#}$$

$$C(X, Y_1) \qquad\qquad\qquad C(X, Y_2)$$

$$k_{1\#} \nearrow \qquad\qquad\qquad\qquad \nwarrow k_{2\#}$$

$$C(X, X_1 \cap Y_1) \qquad\qquad j_{1\#} \quad j_{2\#} \qquad\qquad C(X_2, X_2 \cap Y_2)$$

$$q_{1\#} \downarrow \qquad\qquad C(X, Y_1 \cup Y_2) \qquad\qquad \downarrow q_{2\#}$$

$$n_{1\#} \nearrow \qquad p_{\#} \uparrow \qquad n_{2\#} \nwarrow$$

$$C(X_1, X_1 \cap (Y_1 \cup Y_2)) \qquad\qquad C(X_2, X_2 \cap (Y_1 \cup Y_2))$$

$$m_{1\#} \searrow \qquad C(X_1 \cap X_2, X_1 \cap X_2 \cap (Y_1 \cup Y_2)) \qquad \swarrow m_{2\#}$$

All chain maps in this diagram are induced by inclusion maps.

Remark: The homology class y is not unique; for, the chains a_1, a_2, b_1, and b_2 can each be changed by adding a cycle from the chain group to which it belongs. We leave it to the interested reader to investigate in more detail the indeterminancy of the homology class y.

Bibliography for Chapter IX

[1] M. Barratt and J. Milnor, An example of anomalous singular homology, *Proc. Amer. Math. Soc.*, **13** (1962), 293–297.

[2] H. Cartan, *Seminaire Henri Cartan 1948/49: Topologie Algébrique*, W. A. Benjamin, Inc., New York, 1967.

[3] R. Connelly, A new proof of Brown's collaring theorem, *Proc. Amer. Math. Soc.*, **27** (1971), 180–182.

[4] A. Dold, *Lectures on Algebraic Topology*, Springer-Verlag, New York, 1972.

[5] N. Jacobson, *Basic Algebra I*, W. H. Freeman and Co., San Francisco, 1974.

[6] W. S. Massey, *Algebraic Topology: An Introduction*, Springer-Verlag, New York, 1978.

[7] W. S. Massey, *Homology and Cohomology Theory: An Approach Based on Alexander–Spanier Cochains*, Marcel Dekker, Inc., New York, 1978.

[8] J. Milnor, *Lectures on Characteristic Classes*, Princeton University Press, Princeton, 1974.

[9] E. Spanier, *Algebraic Topology*, McGraw-Hill, New York, 1966.

[10] E. Spanier, Tautness for Alexander–Spanier cohomology, *Pac. J. Math.*, **75** (1978), 561–563.

[11] J. Vick, *Homology Theory*, Academic Press, New York, 1973.

CHAPTER X

Cup Products in Projective Spaces and Applications of Cup Products

§1. Introduction

In this chapter we will determine cup products in the cohomology of the real, complex, and quaternionic projective spaces. The cup products (mod 2) in real projective spaces will be used to prove the famous Borsuk–Ulam theorem. Then we will introduce the mapping cone of a continuous map, and use it to define the Hopf invariant of a map $f : S^{2n-1} \to S^n$. The proof of existence of maps of Hopf invariant 1 will depend on our determination of cup products in the complex and quaternionic projective plane.

§2. The Projective Spaces

We defined the n-dimensional real, complex, and quaternionic projective spaces (denoted by RP^n, CP^n, and QP^n respectively) in §IV.3. We also defined CW-complex structures on them, and then determined the homology groups of CP^n and QP^n. Now we are going to prove that they are compact, connected manifolds, and then use the Poincaré duality theorem to determine the cup products in their cohomology.

Since the universal covering space of RP^n is S^n, it is clear that RP^n is a compact, connected manifold (see Exercise 2.1 in the preceding chapter).

Next, we will prove that CP^n is a $2n$-dimensional manifold. Let $(z_0, z_1, \ldots, z_n)$ denote homogeneous coordinates in CP^n (see IV.3), and let

$$U_i = \{(z_0, \ldots, z_n) \in CP^n \,|\, z_i \neq 0\}$$

for $i = 0, 1, \ldots, n$. Then U_i is an open subset of CP^n. We may "normalize" the homogeneous coordinates of a point in U_i by requiring that $z_i = 1$. With this normalization, each point of U_i has unique homogeneous coordinates. These unique coordinates define an obvious homomorphism of U_i with $\mathbf{C}^n = \mathbf{R}^{2n}$. Since the collection of sets $\{U_i \mid i = 0,1, \ldots ,n\}$ is clearly a covering of CP^n, this suffices to prove that CP^n is a $2n$-manifold.

Remark: In the preceding paragraph, we have neglected various details of point set topology which arise because of the fact that CP^n is defined as a quotient space. The reader can either work these details out for himself, or consult some reference such as Bourbaki [3].

That CP^n is compact and connected follows from the CW-complex defined on it in §IV.3.

An analogous proof, using quaternions instead of complex numbers, shows that QP^n is a compact, connected manifold of dimension $4n$.

A method of proving that RP^n is orientable for n odd and nonorientable for n even is outlined in Exercises 2.2 to 2.5, of Chapter IX. We will not make use of this result in this chapter, except in the exercises. In §IV.4, we proved that the integral homology groups $H_{2n}(CP^n)$ and $H_{4n}(QP^n)$ are infinite cyclic. This implies that CP^n and QP^n are orientable for all n.

We will now discuss cup products in these projective spaces. For the sake of brevity, it will be convenient to write uv instead of $u \cup v$. For any integer $n \geq 1$, u^n will denote the product $uu \cdots u$ (n factors), while $u^0 = 1$.

In order to describe cup products in the cohomology of CP^n and QP^n, note that

$$H^i(CP^n;\mathbf{Z}) \approx \begin{cases} \mathbf{Z} & \text{for } i \text{ even and } 0 \leq i \leq 2n, \\ 0 & \text{otherwise.} \end{cases}$$

This follows from determination of the homology of CP^n in §IV.4 and the universal coefficient theorem. Similarly,

$$H^i(QP^n;\mathbf{Z}) \approx \begin{cases} \mathbf{Z} & \text{for } i \equiv 0 \bmod 4 \text{ and } 0 \leq i \leq 4n, \\ 0 & \text{otherwise.} \end{cases}$$

Theorem 2.1. *Let u be a generator of the infinite cyclic group $H^2(CP^n;\mathbf{Z})$. Then u^k is a generator of $H^{2k}(CP^n;\mathbf{Z})$ for $0 \leq k \leq n$.*

Theorem 2.2. *Let v be a generator of $H^4(QP^n;\mathbf{Z})$. Then v^k is a generator of the infinite cyclic group $H^{4k}(QP^n;\mathbf{Z})$ for $0 \leq k \leq n$.*

PROOF OF THEOREM 2.1. The proof is by induction on n, using Theorem 5.2 of the preceding chapter. For $n = 1$, the theorem is a triviality, while for $n = 2$, it follows directly from Theorem IX.5.2. Assume that the theorem is true for CP^n, $n \geq 2$; we will show this implies the theorem for CP^{n+1}.

In §IV.3, we defined a structure of CW-complex on CP^{n+1}, such that the skeleton of dimension $2k$ is CP^k for $0 \leq k \leq n + 1$. From this it follows that we may consider CP^n as a closed subspace of CP^{n+1}, and the relative cohomology groups of the pair (CP^{n+1}, CP^n) are given by

$$H^k(CP^{n+1}, CP^n; \mathbf{Z}) = \begin{cases} \mathbf{Z} & \text{for } k = 2n + 2, \\ 0 & \text{otherwise.} \end{cases}$$

Let $i: CP^n \to CP^{n+1}$ denote the inclusion map; from the exact cohomology sequence we deduce that

$$i^*: H^k(CP^{n+1}; \mathbf{Z}) \to H^k(CP^n; \mathbf{Z})$$

is an isomorphism for all $k \neq 2n + 2$. Let u denote a generator of $H^2(CP^{n+1}; \mathbf{Z})$; by the inductive hypothesis, $(i^*u)^k$ is a generator of $H^{2k}(CP^n; \mathbf{Z})$ for $0 \leq k \leq n$; it follows that u^k is a generator of $H^{2k}(CP^{n+1}; \mathbf{Z})$ for the same values of k. By applying Theorem IX.5.2 to the cup product

$$H^{2n}(CP^{n+1}; \mathbf{Z}) \otimes H^2(CP^{n+1}; \mathbf{Z}) \to H^{2n+2}(CP^{n+1}; \mathbf{Z})$$

we conclude that u^{n+1} is a generator of $H^{2n+2}(CP^{n+1})$, completing the inductive step. Q.E.D.

The proof of Theorem 2.2 is similar, and is left to the reader. To obtain an analogous result for real projective space, RP^n, it is necessary to use mod 2 cohomology.

Theorem 2.3. *The mod 2 cohomology group $H^k(RP^n; \mathbf{Z}_2)$ is cyclic of order 2 for $0 \leq k \leq n$. If w is a generator of $H^1(RP^n; \mathbf{Z}_2)$, then w^k is a generator of $H^k(RP^n; \mathbf{Z}_2)$ for $0 \leq k \leq n$.*

PROOF. Once again the proof is by induction on n, using the CW-complex structure on RP^n which is given in §IV.3. The theorem is true for $n = 1$, because RP^1 is homomorphic to S^1. We determined the integral homology groups of RP^2 in III.4; from this one can show that $H^k(RP^2; \mathbf{Z}_2) = \mathbf{Z}_2$ for $k = 0, 1, 2$. Determination of the cup products in $H^*(RP^2; \mathbf{Z}_2)$ then follows from the analog for nonorientable manifolds of Theorem 5.1 of the preceding chapter.

The inductive step is slightly more complicated than that in the proof of Theorem 2.1. Recall that RP^n is a CW-complex with one cell in each dimension $\leq n$, and the k-skeleton is RP^k for $0 \leq k \leq n$. It follows that

$$H^k(RP^n, RP^{n-1}; \mathbf{Z}_2) = \begin{cases} \mathbf{Z}_2 & \text{for } k = n, \\ 0 & \text{for } k \neq n. \end{cases}$$

From this it follows that

$$i^*: H^k(RP^n; \mathbf{Z}_2) \to H^k(RP^{n-1}; \mathbf{Z}_2)$$

is an isomorphism for $k < n - 1$. We will prove that it is also an isomorphism for $k = n - 1$. Consider the following portion of the mod 2 exact cohomology sequence of the pair (RP^n, RP^{n-1}):

$$0 \to H^{n-1}(RP^n) \overset{i_*}{\to} H^{n-1}(RP^{n-1}) \overset{\delta^*}{\to} H^n(RP^n, RP^{n-1}) \overset{j^*}{\to} H^n(RP^n) \overset{i^*}{\to} H^n(RP^{n-1}).$$

First of all, $H^n(RP^{n-1}; \mathbb{Z}_2) = 0$ because RP^{n-1} is only $(n-1)$-dimensional. Therefore $j^*: H^n(RP^n; RP^{n-1}) \to H^n(RP^n)$ is an epimorphism. Next, $H^n(RP^n; \mathbb{Z}_2)$ is cyclic of order 2, because RP^n is a compact, connected n-manifold. Since j^* is an epimorphism of a group of order 2 onto a group of order 2, it must be an isomorphism. It follows by exactness that $\delta^*: H^{n-1}(RP^{n-1}) \to H^n(RP^n, RP^{n-1})$ is the zero homomorphism. Hence $i^*: H^{n-1}(RP^n) \to H^{n-1}(RP^{n-1})$ is an isomorphism, as was asserted.

The remainder of the inductive step is similar to that in the proof of Theorem 2.1, and may be left to the reader. The only difference is that one uses the analog for nonorientable manifolds of Theorem 5.1 rather than Theorem 5.2 of Chapter IX.

One can express Theorem 2.1 by means of the following ring isomorphism:

$$H^*(CP^n; \mathbb{Z}) \approx \mathbb{Z}[u]/(u^{n+1});$$

in other words, the integral cohomology ring $H^*(CP^n; \mathbb{Z})$ is isomorphic to the integral polynomial ring $\mathbb{Z}[u]$ modulo the ideal generated by u^{n+1}. Similarly,

$$H^*(QP^n; \mathbb{Z}) \approx \mathbb{Z}[v]/(v^{n+1}),$$
$$H^*(RP^n; \mathbb{Z}_2) \approx \mathbb{Z}_2[w]/(w^{n+1}).$$

Rings with this type of structure are often called *truncated polynomial rings*.

We will now use this result on the structure of $H^*(RP^n; \mathbb{Z}_2)$ to prove the famous Borsuk–Ulam theorem (for a discussion of some of the interesting consequences of this theorem, the reader is referred to *Algebraic Topology: An Introduction*, Chapter 5, Section 9). Recall that a map $f: S^m \to S^n$ is called *antipode preserving* in case $f(-x) = -f(x)$ for any $x \in S^m$.

Theorem 2.4. *There does not exist any continuous antipode preserving map* $f: S^n \to S^{n-1}$.

PROOF. We will only give the proof for $n > 2$; the proof for $n \le 2$ is contained in *Algebraic Topology: An Introduction* (loc. cit.) The proof is by contradiction. Assume that $f: S^n \to S^{n-1}$ is an antipode preserving map. Hence f induces a map $g: RP^n \to RP^{n-1}$, since RP^n is the quotient space obtained by identifying antipodal points of S^n. Thus we get a commutative diagram

$$
\begin{array}{ccc}
S^n & \overset{f}{\longrightarrow} & S^{n-1} \\
\downarrow{\scriptstyle p} & & \downarrow{\scriptstyle q} \\
RP^n & \underset{g}{\longrightarrow} & RP^{n-1},
\end{array}
$$

where p and q are the projections of S^n and S^{n-1} onto their quotient spaces. Because $n > 2$, both S^n and S^{n-1} are simply connected. Thus they are the universal covering spaces of RP^n and RP^{n-1} respectively and the fundamental groups, $\pi(RP^n)$ and $\pi(RP^{n-1})$, are both cyclic of order 2. The induced homomorphism

$$g_*: \pi(RP^n) \to \pi(RP^{n-1})$$

must be an isomorphism; this may be proved by an easy argument which is given on p. 172 of *Algebraic Topology: An Introduction*. Now consider the following commutative diagram:

$$
\begin{array}{ccc}
\pi(RP^n) & \xrightarrow{\ g_*\ } & \pi(RP^{n-1}) \\
\Big\downarrow{\scriptstyle h} & & \Big\downarrow{\scriptstyle h} \\
H_1(RP^n) & \xrightarrow{\ g_*\ } & H_1(RP^{n-1}).
\end{array}
$$

The homomorphisms denoted by h are the natural homomorphisms of the fundamental group onto the first homology group which were defined in III.7. Since the fundamental groups involved are abelian, these homomorphisms are both isomorphisms (cf. Theorem III.7.1). It follows that $g_*: H_1(RP^n) \to H_1(RP^{n-1})$ is an isomorphism.

Next, consider the following commutative diagram:

$$
\begin{array}{ccc}
H^1(RP^n; Z_2) & \xrightarrow{\ \alpha\ } & \mathrm{Hom}(H_1(RP^n); Z_2) \\
\Big\uparrow{\scriptstyle g^*} & & \Big\uparrow{\scriptstyle \mathrm{Hom}(g_*,1)} \\
H^1(RP^{n-1}; Z_2) & \xrightarrow{\ \alpha\ } & \mathrm{Hom}(H_1(RP^{n-1}), Z_2).
\end{array}
$$

The homomorphisms labelled α are those which occur in the universal coefficient theorem (§VII.4); in this case they are both isomorphisms. It follows from this that

$$g^*: H^1(RP^{n-1}; Z_2) \to H^1(RP^n; Z_2)$$

is also an isomorphism. Let w be a generator of $H^1(RP^{n-1}; Z_2)$; then $g^*(w)$ is a generator of $H^1(RP^n; Z_2)$. By Theorem 2.3, $(g^*w)^n \neq 0$. However, this is a contradiction, since

$$(g^*w)^n = g^*(w^n)$$

and $w^n = 0$.   Q.E.D.

EXERCISES

2.1. For $k < n$, consider CP^k as the $2k$-skeleton of CP^n. Prove that CP^k is not a retract of CP^n. Similarly, prove that for $k < n$, QP^k is not a retract of QP^n, and RP^k is not a retract of RP^n.

2.2. Determine the integral homology groups of RP^n by induction on n. Use the fact that RP^n is a CW-complex, as described in §IV.3, and that it is orientable for n odd, and nonorientable for n even.

2.3. Use the results of the preceding exercise and the universal coefficient theorem to determine the structure of the integral cohomology groups $H^k(RP^n; Z)$. Then determine the cup products in the integral cohomology of RP^n. (*Hint*: Use the homomorphism $H^k(RP^n; \mathbf{Z}) \to H^k(RP^n; \mathbf{Z}_2)$ induced by reduction mod 2 of the integer.)

§3. The Mapping Cylinder and Mapping Cone

The techniques developed in this section will be used in the next section to define certain homotopy invariants of continuous maps.

Let $f: X \to Y$ be a continuous map. The *mapping cylinder of f*, denoted by $M(f)$, is the topological space defined as follows: Assume that $X \times I$ and Y are disjoint; if they are not, take disjoint copies. Then form the quotient space of the disjoint union of $X \times I$ and Y by identifying the points $(x,0)$ and $f(x)$ for each $x \in X$.

The mapping cylinder $M(f)$ can be visualized as a space which contains a copy of X (namely, $X \times \{1\}$), a copy of Y, and corresponding to each $x \in X$ a copy of the unit interval connecting the points x and $f(x)$. This space is topologized so that if x_1 and x_2 are points in X, that are close to each other, then the corresponding segments from x_1 to $f(x_1)$ and from x_2 to $f(x_2)$ are also close to each other.

The obvious deformation retraction of $X \times I$ onto $X \times \{0\}$ gives rise to a deformation retraction of $M(f)$ onto Y. If we denote by $i: X \to M(f)$ the inclusion map (defined by $i(x) = (x,0)$) and by $r: M(f) \to Y$ the retraction, then the following diagram is commutative:

Thus an arbitrary continuous map f is the composition of an inclusion map i and a homotopy equivalence r.

The *mapping cone* of $f: X \to Y$, denoted by $C(f)$, is the quotient space of the mapping cylinder $M(f)$ obtained by identifying the subset $X \times \{1\}$ to a single point. Alternatively, the mapping cone can be constructed as follows: let $C(X)$, called the *cone over X*, denote the quotient space of $X \times I$ obtained by identifying all of $X \times \{1\}$ to a single point. Then $C(f)$ is the quotient space of the (disjoint) union of Y and $C(X)$ obtained by identifying the point $(x,0) \in C(X)$ with the point $f(x) \in Y$ for all $x \in X$.

EXAMPLE 3.1. If $X = S^n$, the n-sphere, then it is easily seen that $C(X)$ is homeomorphic to the $(n+1)$-dimensional ball E^{n+1}. In this case, $C(f)$ is the same as the space $X^* = X \cup e^{n+1}$ obtained by adjoining an $(n+1)$-cell to the space X by means of the map f, as described in §IV.2. In particular, if K^m denotes the m-dimensional skeleton of a CW-complex, then we can regard K^{n+1} as the mapping cone of a certain map $f: X \to K^n$, where X is a disjoint union of n-spheres (assuming that the number of $(n + 1)$-cells is finite).

One of the basic facts about the spaces $M(f)$ and $C(f)$ is that they satisfy certain naturality conditions. Let

$$
\begin{array}{ccc}
X & \xrightarrow{f} & Y \\
\downarrow{\varphi_1} & & \downarrow{\varphi_2} \\
X' & \xrightarrow{f'} & Y'
\end{array}
$$

be a commutative diagram of topological spaces and continuous maps. Then it is readily seen that φ_1 and φ_2 induce continuous maps of quotient spaces, $M(f) \to M(f')$ and $C(f) \to C(f')$; let us agree to denote both of these induced maps by the symbol φ. Then it follows that the following two diagrams are commutative:

$$
\begin{array}{ccccc}
X & \xrightarrow{i} & M(f) & \xrightarrow{r} & Y \\
\downarrow{\varphi_1} & & \downarrow{\varphi} & & \downarrow{\varphi_2} \\
X' & \xrightarrow{i'} & M(f') & \xrightarrow{r'} & Y',
\end{array}
$$

$$
\begin{array}{ccc}
Y & \xrightarrow{j} & C(f) \\
\downarrow{\varphi_2} & & \downarrow{\varphi} \\
Y' & \xrightarrow{j'} & C(f').
\end{array}
$$

In the second diagram, the symbols j and j' denote obvious inclusion maps.

Lemma 3.1. *Let* $p: M(f) \to C(f)$ *denote the natural map which identifies the subset* $X = X \times \{1\}$ *of* $M(f)$ *to a single point* P *of* $C(f)$. *Then the induced homomorphism of relative cohomology groups*

$$
p^*: H^q C(f), P) \to H^q(M(f), X)
$$

is an isomorphism for all q.

PROOF. Let $\bar{X}$ denote the subset $X \times [\tfrac{1}{2}, 1]$ of $M(f)$, and let $\bar{P}$ denote the image of $\bar{X}$ under p. Consider the following commutative diagram:

$$
\begin{array}{ccccc}
H^q(M(f), X) & \xleftarrow{1} & H^q(M(f), \bar{X}) & \xrightarrow{3} & H^q(M(f) - X, \bar{X} - X) \\
\uparrow{p^*} & & \uparrow{p_1^*} & & \uparrow{p_2^*} \\
H^q(C(f), P) & \xleftarrow{2} & H^q(C(f), \bar{P}) & \xrightarrow{4} & H^q(C(f) - P, \bar{P} - P).
\end{array}
$$

In this diagram, the horizontal arrows denote homomorphisms induced by inclusion maps, and the vertical arrows denote homomorphisms induced by p. Arrows 1 and 2 are isomorphisms because X is a deformation retract of $\bar{X}$ and P is a deformation retract of $\bar{P}$. Arrows 3 and 4 are isomorphisms by the excision property; and p_2^* is an isomorphism, because p maps $M(f) - X$ and $\bar{X} - X$ homomorphically onto $C(f) - P$ and $\bar{P} - P$ respectively. It follows that p^* is an isomorphism, as desired.   Q.E.D.

Now let $k : Y \to M(f)$ denote the inclusion map; k is a homotopy equivalence because Y is a deformation retract of $M(f)$. Consider the following diagram:

$$
\begin{array}{ccc}
H^q(C(f),P) & \xrightarrow{\;j^*\;} & H^q(Y) \\
\downarrow{\scriptstyle p^*} & & {\scriptstyle k^*}\big\uparrow\downarrow{\scriptstyle r^*} \qquad \searrow{\scriptstyle f^*} \\
H^{q-1}(X) \xrightarrow{\;\delta\;} H^q(M(f),X) \longrightarrow & H^q(M(f)) & \xrightarrow[\;i^*\;]{} H^q(X).
\end{array}
$$

The bottom line is the cohomology sequence of the pair $(M(f),X)$. All the vertical arrows are isomorphisms, and k^* and r^* are inverses of each other. Finally, the diagram is readily seen to be commutative. As a consequence of these facts, we see that the following sequence of cohomology groups and homomorphisms is exact:

$$\to H^{q-1}(X) \xrightarrow{\Delta} H^q(C(f),P) \xrightarrow{j^*} H^q(Y) \xrightarrow{f^*} H^q(X) \to.$$

Here $\Delta = (p^*)^{-1}\delta$. This exact sequence will be called the *cohomology sequence of the map f*. Observe that a commutative diagram

$$
\begin{array}{ccc}
X & \xrightarrow{\;f\;} & Y \\
\downarrow{\scriptstyle \varphi_1} & & \downarrow{\scriptstyle \varphi_2} \\
X' & \xrightarrow{\;f'\;} & Y'
\end{array}
$$

gives rise to an induced map of the cohomology sequence of f into the cohomology sequence of f'; that is, we get a ladder-like diagram involving the two exact sequences, and every square in the diagram is commutative.

Now let us apply these ideas to study the cohomology sequences of two maps which are homotopic. Let $f_0, f_1 : X \to Y$ be continuous maps, and let $f : X \times I \to Y$ be a homotopy between f_0 and f_1, i.e., $f_0(x) = f(x,0)$ and $f_1(x) = f(x,1)$. This gives rise to the following commutative diagram:

$$
\begin{array}{ccc}
X & \xrightarrow{\;f_0\;} & Y \\
\downarrow{\scriptstyle h_0} & & \| \\
X \times I & \xrightarrow{\;f\;} & Y \\
\uparrow{\scriptstyle h_1} & & \| \\
X & \xrightarrow{\;f_1\;} & Y
\end{array}
$$

Here $h_i(x) = (x,i)$ for $i = 0$ or 1. Corresponding to this diagram, we get a bigger diagram involving the cohomology sequences of $f_0, f,$ and f_1 together with homomorphisms between them. By making use of the five-lemma together with the fact that h_0 and h_1 are homotopy equivalences, we easily deduce that *the cohomology sequences of the maps f_0 and f_1 are isomorphic.* To be precise, any homotopy between f_0 and f_1 gives rise to an isomorphism between the corresponding cohomology sequences. Presumably different homotopies could give rise to different isomorphisms.

We could also word this conclusion as follows: *the cohomology sequence of a map f is a homotopy invariant of f.*

EXAMPLE 3.2. Suppose $f : X \to Y$ is a constant map. Then it is clear that Y is a retract of $C(f)$. Hence there exists a homomorphism $r^* : H^q(Y) \to H^q(C(f))$ such that $j^* r^*$ is the identity map of $H^q(Y)$. Moreover, r^* preserves cup products, i.e., $r^*(x \cup y) = (r^* x) \cup (r^* y)$. Because of the invariance of the cohomology sequence of f under homotopies, we can conclude that this same result is true in case $f : X \to Y$ is only assumed to be homotopic to a constant map. As a matter of fact, it is easy to prove directly that f is homotopic to a constant map if and only if Y is a retract of $C(f)$.

EXERCISE

3.1. As in the above discussion, let $f : X \times I \to Y$ be a continuous map, and let $f_0, f_1 : X \to Y$ be defined by $f_i(x) = f(x,i)$, $i = 0, 1$. Prove that $M(f_i)$ is a deformation retract of $M(f)$, and $C(f_i)$ is a deformation retract of $C(f)$ for $i = 0, 1$. Then deduce that the pairs $(C(f_0), Y)$ and $(C(f_1), Y)$ are of the same homotopy type.

§4. The Hopf Invariant

The Hopf invariant associates with each map $f : S^{2n-1} \to S^n$ an integer that is a homotopy invariant of f. Using it, we will be able to prove that for n even and ≥ 2, there are infinitely many different homotopy classes of such maps.

In order to define the Hopf invariant, we will assume that the spheres S^{2n-1} and S^n are "oriented," in the sense that definite generators $a \in H^{2n-1}(S^{2n-1}, \mathbf{Z})$ and $b \in H^n(S^n; \mathbf{Z})$ have been chosen for these infinite cyclic groups. We will also assume that $n \geq 2$. As in the preceding section, let $C(f)$ denote the mapping cone of f. It follows from the exactness of the cohomology sequence of the map f that the following two homomorphisms

$$\Delta : H^{2n-1}(S^{2n-1}) \to H^{2n}(C(f))$$
$$j^* : H^n(C(f)) \to H^n(S^n)$$

are both isomorphisms. Let $a' = \Delta(a) \in H^{2n}(C(f); \mathbf{Z})$, and let $b' \in H^n(C(f), \mathbf{Z})$ be the unique element such that $j^*(b') = b$. Since $H^{2n}(C(f); \mathbf{Z})$ is infinite cyclic, there exists a unique integer $H(f)$ such that

$$b' \cup b' = H(f) \cdot a'.$$

In view of the homotopy invariance of the cohomology sequence of f, the integer $H(f)$ depends only on the homotopy class of f.

We will now list some of the principal properties of the Hopf invariant:

(1) If n is odd and >1 then $H(f) = 0$ for any map $f: S^{2n-1} \to S^n$. This follows from the anti-commutative law for cup products. As a consequence, the Hopf invariant is useless in this case.

(2) If $n = 2, 4,$ or 8, there exist maps $f: S^{2n-1} \to S^n$ such that $H(f) = \pm 1$. For $n = 2$ we may choose f such that $C(f) = CP^2$, the complex projective plane; while for $n = 4$, we may choose f such that $C(f) = QP^2$. The case $n = 8$ is more complicated; in essence, we must choose f so that $C(f)$ is the so-called *Cayley projective plane*. An explicit description of such a map f is given by Steenrod [5], pp. 109–110. A complete discussion of the Cayley projective plane is given by H. Freudenthal, [4].

(3) For any even integer $n \geq 2$, there exist maps f such that $H(f) = \pm 2$. To prove this, recall that S^n may be considered as a CW-complex with a single vertex, e^0, a single n-cell e^n, and no cells of any other dimension. Hence $S^n \times S^n$ may be represented as a CW-complex with one vertex, $e^0 \times e^0$, two n-cells, $e^0 \times e^n$ and $e^n \times e^0$, and one $2n$-cell, $e^n \times e^n$. The n-skeleton of this CW-complex is the subspace

$$S^n \vee S^n = (S^n \times e^0) \cup (e^0 \times S^n)$$

of $S^n \times S^n$. Let $g: S^{2n-1} \to S^n \vee S^n$ denote the attaching map for the single $2n$-cell of this CW-complex, and let $h: S^n \vee S^n \to S^n$ be defined by $h(x, e^0) = h(e^0, x) = x$ for $x \in S^n$ (h is sometimes called the *folding map*). We assert that if we define

$$f = hg: S^{2n-1} \to S^n,$$

then (for n even), $H(f) = \pm 2$. To prove this assertion, consider the following commutative diagram:

$$
\begin{array}{ccccc}
S^{2n-1} & \xrightarrow{\;g\;} & S^n \vee S^n & \xrightarrow{\;j_1\;} & C(g) \\
\| & & \downarrow{\scriptstyle h} & & \downarrow{\scriptstyle h'} \\
S^{2n-1} & \xrightarrow{\;f\;} & S^n & \xrightarrow{\;j_2\;} & C(f)
\end{array}
$$

Here h' is induced by h. By definition, $C(g) = S^n \times S^n$. Let b denote the chosen generator of $H^n(S^n; \mathbf{Z})$. Then $\{b \times 1, 1 \times b\}$ is a basis for $H^n(S^n \times S^n)$ and $(b \times 1) \cup (1 \times b) = b \times b$ is a generator of $H^{2n}(S^n \times S^n)$, (cf. §VIII.11). Now, consider the following commutative diagram:

$$
\begin{array}{ccc}
H^n(S^n \vee S^n) & \xleftarrow{\;j_1^*\;} & H^n(S^n \times S^n) \\
\uparrow{\scriptstyle h^*} & & \uparrow{\scriptstyle h'^*} \\
H^n(S^n) & \xleftarrow{\;j_2^*\;} & H^n(C(f)).
\end{array}
$$

Both j_1^* and j_2^* are isomorphisms, and $j_2^*(b') = b$. We leave it to the reader to convince himself that

$$h'^*(b') = (b \times 1) + (1 \times b).$$

We also have the following commutative diagram:

Both Δ_1 and Δ_2 are isomorphisms, hence h'^* is an isomorphism. Let us assume that the generator $a \in H^{2n-1}(S^{2n-1})$ is chosen so that $\Delta_1(a) = b \times b$; hence $h'^*(a') = b \times b$. To prove our assertion, apply the homomorphism h'^* to the equation

$$b' \cup b' = H(f) \cdot a'.$$

The result is

$$(b \times 1 + 1 \times b) \cup (b \times 1 + 1 \times b) = H(f)(b \times b),$$

hence $H(f) = 2$. If we had used the orientation of S^{2n-1} determined by the generator $-a$, we would have obtained $H(f) = -2$.

(4) Let $f: S^{2n-1} \to S^n$, be a continuous map, and $h: S^n \to S^n$ a map of degree k (i.e., $h^*(b) = kb$). Then

$$H(hf) = k^2 H(f).$$

(5) Let $h: S^{2n-1} \to S^{2n-1}$ be a map of degree k (i.e., $h^*(a) = ka$) and $f: S^{2n-1} \to S^n$ a continuous map. Then

$$H(fh) = k \cdot H(f).$$

The proof of Assertions (4) and (5) are left to the reader as exercises.

Remarks. Assume that n is even and ≥ 2. It follows from the preceding paragraphs that given any integer $2m$, there exists a map $f: S^{2n-1} \to S^n$ such that $H(f) = 2m$. It is known that $H(f)$ is of necessity an even integer, except when $n = 2, 4$, or 8. This was proved by José Adem [2] for $n \neq 2^k$, and by J. F. Adams [1] for $n = 2^k, k > 3$.

It is also known that two maps $f_0, f_1: S^3 \to S^2$ are homotopic if and only if $H(f_0) = H(f_1)$. In general, such a statement is not true for maps of S^{2n-1} into S^n, $n > 2$. However, it is known that there are only a finite number of homotopy classes of such maps having a given integer as Hopf invariant.

EXERCISES

4.1. Given any space X, define the *suspension of X*, denoted $S(X)$, to be the quotient space of $X \times I$ obtained by identifying each of the subsets $X \times 0$ and $X \times 1$ to a point; it is a sort of "double cone" over X. Similarly, if $f: X \to Y$ is a continuous map, define $S(f):S(X) \to S(Y)$ to be the map induced on quotient spaces by the map of $X \times I$ into $Y \times I$ which sends (x,t) to (fx,t).

(a) If $X = S^n$, prove that $S(X)$ is homeomorphic to S^{n+1}.

(b) What is the relation between the homology groups of X and those of $S(X)$?

(c) If $u \in H^p(S(X))$ and $v \in H^q(S(X))$, where $p > 0$ and $q > 0$, prove that $u \cup v = 0$.

(d) If $f_0, f_1 : X \to Y$, and f_0 is homotopic to f_1, prove that $S(f_0)$ is homotopic to $S(f_1)$.

(e) Let $f : X \to Y$; we would like to prove that $C(Sf) = S(Cf)$. Unfortunately, this is not quite true. Prove that there is a natural map $S(Cf) \to C(Sf)$ which induces isomorphisms of homology and cohomology groups.

(f) Let $f : S^{2n-2} \to S^{n-1}$ be a continuous map; in view of (a), the Hopf invariant $H(Sf)$ is defined. Prove that $H(Sf) = 0$. Remark: The converse of this last statement is true "up to homotopy." To be more explicit, let $g : S^{2n-1} \to S^n$ be a map such that $H(g) = 0$. Then there exists a map $f : S^{2n-2} \to S^{n-1}$ such that g is homotopic to $S(f)$; see G. W. Whitehead, [6].

Bibliography for Chapter X

[1] J. F. Adams, On the nonexistence of elements of Hopf invariant one, *Ann. Math.*, **72** (1960), 20–104.

[2] J. Adem, The iteration of Steenrod squares in algebraic topology, *Proc. Nat. Acad. Sci.*, **38** (1952), 720–726.

[3] N. Bourbaki, *Topologie Générale*, Hermann et Cie., Paris, 1947, Chapters VI and VIII.

[4] H. Freudenthal, *Oktaven, Ausnahme-gruppen, und Oktavengeometrie* (mimeographed), Utrecht, 1951, revised ed., 1960.

[5] N. E. Steenrod, *The Topology of Fibre Bundles*, Princeton University Press, Princeton, 1951.

[6] G. W. Whitehead, On the Freudenthal theorems, *Ann. of Math.*, **57** (1953), 209–228.

Appendix:
A Proof of De Rham's Theorem

§1. Introduction

In Chapter I we mentioned that some of the motivating ideas for the development of homology theory in the Nineteenth century arose in connection with such topics as Stokes's theorem, Green's theorem, Gauss's divergence theorem, and the Cauchy integral theorem. De Rham's theorem may be looked on as the modern culmination of this particular line of thought. It relates the homology and cohomology of a differentiable manifold to the exterior differential forms on the manifold. Exterior differential forms are objects which can serve as integrands of line integrals, surface integrals, etc., such as occur in the statement of the classical Green's theorem and Stokes's theorem. De Rham's theorem is of obvious importance, because it is a connecting link between analysis on manifolds and the topological properties of manifolds.

In this appendix we will assume that the reader is familiar with the basic properties of differentiable manifolds, differential forms on manifolds, and the integration of differential forms over (differentiable) singular cubes. These topics are explained in many current textbooks, and there would be little point in our repeating such an exposition here. As examples of such texts, we list the following: M. Spivak [6], Flanders [3], Warner [9], and Whitney [10].

The first part of this chapter is devoted to using differentiable singular cubes to define the homology and cohomology groups of a differentiable manifold. We prove that in studying the homology and cohomology groups of such a manifold, it suffices to consider only *differentiable* singular cubes; the nondifferentiable ones can be ignored.

Next, we introduce what may be called the *De Rham cochain complex* of a differentiable manifold. This cochain complex consists of the exterior differential forms, with the exterior derivative serving as the coboundary operator. There is a natural homomorphism from this De Rham complex to the cochain complex (with coefficient group **R**, the real numbers) based on differentiable singular cubes. This homomorphism is defined on any exterior differential form of degree p by integrating that form over differentiable singular p-cubes. The general form of Stokes's theorem is precisely the assertion that this natural homomorphism is a cochain map. De Rham's theorem asserts that this natural cochain map induces an isomorphism on cohomology.

The proof we give of De Rham's theorem is modelled on Milnor's proof of the Poincaré duality theorem in Chapter IX. The reader who has worked through that proof should have no trouble grasping the structure of our proof of De Rham's theorem. Curtis and Dugundji [11] have also given a proof of De Rham's theorem along somewhat similar lines.

§2. Differentiable Singular Chains

Let M be an n-dimensional differentiable manifold of class C^∞ (we assume the reader is familiar with this concept). In order to define a differentiable singular cube, we must make use of the fact that the standard unit p-cube,

$$I^p = \{(x_1, \ldots, x_p) \in \mathbf{R}^p \,|\, 0 \le x_i \le 1, i = 1, 2, \ldots, p\}$$

is a subset of Euclidean space $\mathbf{R}^p$. For $p > 0$, a singular p-cube $T : I^p \to M$ will be called *differentiable* if there exists an open neighborhood U of I^p in $\mathbf{R}^p$ and an extension $T' : U \to M$ of T such that T' is differentiable (of class C^∞). We complete this definition by defining any singular 0-cube to be differentiable.

Remark: If a singular p-cube $T : I^p \to M$ is differentiable, there will, in general, be many different choices for the open neighborhood U and the extension $T' : U \to M$.

We now introduce the following notation:

$$Q_p^S(M) = \text{subgroup of } Q_p(M) \text{ generated by the}$$
$$\text{differentiable singular } p\text{-cubes,}$$
$$D_p^S(M) = D_p(M) \cap Q_p^S(M),$$
$$C_p^S(M) = Q_p^S(M)/D_p^S(M).$$

The superscript S in the above notation is intended to suggest the word "smooth." We will refer to $C_p^S(M)$ as the group of differentiable or smooth p-chains of M. Note that $C_0^S(M) = Q_0^S(M) = Q_0(M) = C_0(M)$.

Next, observe that if $T : I^p \to M$ is a differentiable singular p-cube, then the faces $A_i T$ and $B_i T$, $1 \le i \le p$, are all obviously differentiable singular $(p - 1)$-cubes. It follows that $\partial_p(T) \in Q^S_{p-1}(M)$. Thus $Q^S(M) = \{Q^S_p(M), \partial_p\}$ is a subcomplex of $Q(M)$, and $C^S(M) = \{C^S_p(M)\}$ is a subcomplex of $C(M)$. We will also introduce the following notation: for any abelian group G,

$$C^S(M; G) = C^S(M) \otimes G,$$
$$C^*_S(M; G) = \mathrm{Hom}(C^S(M), G),$$
$$H^S_p(M; G) = H_p(C^S(M; G)),$$
$$H^p_S(M; G) = H^p(C^*_S(M; G)).$$

We can now state the main theorem of this section:

Theorem 2.1. *Let M be a differentiable manifold. The inclusion map of chain complexes,*

$$C^S(M) \to C(M)$$

induces an isomorphism of homology groups,

$$H^S_p(M) \approx H_p(M).$$

Corollary 2.2. *For any abelian group G, we have the following isomorphisms of homology and cohomology groups:*

$$H^S_p(M; G) \approx H_p(M; G),$$
$$H^p_S(M; G) \approx H^p(M; G).$$

The corollary follows from the theorem by use of standard techniques (cf. Theorem V.2.3). Before we can prove the theorem, it is necessary to discuss to what extent the methods and results of Chapters II and III on homology theory carry over to the homology groups $H^S_p(M; G)$ for any differentiable manifold M. We will now do this in a brief but systematic fashion.

(a) Let M_1 and M_2 be differentiable manifolds, and let $f : M_1 \to M_2$ be a *differentiable* maps of class C^∞. If $T : I^p \to M_1$ is a differentiable singular p cube, in M_1, then $fT : I^p \to M_2$ is also differentiable. Hence we get an induced chain map

$$f_\# : C^S(M_1) \to C^S(M_2)$$

with all the usual properties.

(b) Two differentiable maps $f_0, f_1 : M_1 \to M_2$ will be called *differentiably homotopic* if there exists a map $f : I \times M_1 \to M_2$ such that $f_0(x) = f(0, x)$ and $f_1(x) = f(1, x)$ for any $x \in M_1$, and in addition, there exists an open neighborhood U of $I \times M_1$ in $\mathbf{R} \times M_1$ and a map $f' : U \to M_2$ which is an extension of f, and is differentiable of class C^∞. The technique of §II.4 can now be applied *verbatim* to prove that the induced chain maps $f_{0\#}$,

$f_{1\#}: C^S(M_1) \to C^S(M_2)$ are chain homotopic. This has all the usual consequences; in particular, the induced homomorphisms on homology and cohomology groups are the same.

(c) An open, convex subset if $\mathbf{R}^n$ is *differentiably* contractible to a point; in fact, the standard formulas for proving that such a subset is contractible are differentiable homotopies in the sense of the preceding definition. From this it follows that if U is an open, convex subset of $\mathbf{R}^n$, then

$$H_p^S(U;G) = \begin{cases} G & \text{for } p = 0, \\ 0 & \text{for } p \neq 0, \end{cases}$$

with similar formulas for $H_S^p(U;G)$.

(d) Let M be a differentiable manifold, and let A be a subspace of M which is a differentiable submanifold. For example, A could be an arbitrary open subset of M, or A could be a closed submanifold of M. Then we can consider $C^S(A)$ as a subcomplex of $C^S(M)$; hence we can consider the quotient complex $C^S(M)/C^S(A) = C^S(M,A)$ and we obtain exact homology and cohomology sequences for the pair (M,A) using differentiable singular cubes.

(e) If $T: I^n \to M$ is a differentiable singular cube, the subdivision of T, $\mathrm{Sd}_n(T)$ as defined in §II.7, is readily seen to be a linear combination of differentiable singular cubes. Hence the subdivision operator defines a chain map

$$\mathrm{sd}: C^S(M) \to C^S(M)$$

just as in §II.7. Unfortunately, the chain homotopy $\varphi_n: C_n(M) \to C_{n+1}(M)$ defined in §II.7 does *not* map $C_n^S(M)$ into $C_{n+1}^S(M)$. This is because the function $\eta_1: I^2 \to [\frac{1}{2},1]$ is not differentiable (the function $\eta_0: I^2 \to I$ is differentiable). However, it is not difficult to get around this obstacle. Consider the real-valued function η_1' defined by

$$\eta_1'(x_1,x_2) = \frac{1 + x_1 - x_1 x_2}{2 - x_2}.$$

It is readily verified that η_1' maps I^2 into the interval $[\frac{1}{2},1]$, and that η_1 and η_1' are equal along the boundary of the square I^2. Obviously, η_1' is differentiable in a neighborhood of I^2. Thus if we substitute η_1' for η_1 in the formula for $G_e(T)$ in §II.7, then $G_e(T)$ will be a linear combination of differentiable singular cubes whenever T is a differentiable singular cube. Moreover, the operator G_e will continue to satisfy identities (f.1) to (f.4) of §II.7. Thus we can define a chain homotopy $\varphi_n: C_n^S(M) \to C_{n+1}^S(M)$ using the modified definition of G_e. From this point on, everything proceeds exactly as in §II.7. The net result is that we can prove an analog of Theorem II.6.3 for singular homology based on differentiable singular cubes, and the excision property (Theorem II.6.2) holds for this kind of homology theory.

(f) Suppose that the differentiable manifold M is the union of two open subsets,

$$M = U \cup V.$$

Then we can obtain an exact Mayer–Vietoris sequence for this situation by the method described in §III.5.

(g) Finally, we note that an analog of Proposition III.6.1 must hold for homology groups based on differentiable singular cubes; this is practically obvious.

With these preparations out of the way, we can now prove Theorem 2.1. The pattern of proof is similar to Milnor's proof of the Poincaré duality theorem in §4 of Chapter IX, only this proof is much easier. We prove the theorem for the easiest cases first, and then proceed to successively more general cases.

Case 1: M is a single point. This case is completely trivial.

Case 2: M is an open convex subset of Euclidean n-space, $\mathbf{R}^n$. This follows easily from Case 1, since M is differentiably contractible to a point in this case.

Case 3: $M = U \cup V$, where U and V are open subsets of M, and the theorem is assumed to be true for U, V, and $U \cap V$. This case is proved by use of the Mayer–Vietoris sequence and the five-lemma.

Case 4: M is the union of a nested family of open sets, and the theorem is assumed to be true for each set of the family. Then the theorem is true for M. The proof is by an easy argument using direct limits, and Proposition III.6.1.

Case 5: M is an open subset of $\mathbf{R}^n$. Every open subset of $\mathbf{R}^n$ is a countable union of convex open subsets,

$$M = \bigcup_{i=1}^{\infty} U_i.$$

For each U_i the theorem is true by Case 2. For any finite union, $\bigcup_{i=1}^{n} U_i$ the theorem is true by induction on n, using Case 3 and the basic properties of convex sets. Then one uses Case 4 to prove the theorem for M.

Case 6: The general case. Any differentiable manifold can be covered by coordinate neighborhoods, each of which is diffeomorphic to an open subset of Euclidean space. Using Case 4, Case 5, and Zorn's lemma, we see that there must exist a nonempty open subset $U \subset M$ such that the theorem is true for U, and U is maximal among all open sets for which the theorem is true. If $U \neq M$, then we can find a coordinate neighborhood V such that V is not contained in U. By Case 3, the theorem is true for $U \cup V$, contradicting the maximality of U. Hence $U = M$, and the proof is complete.

§3. Statement and Proof of De Rham's Theorem

For any differentiable manifold M, we will denote by $D^q(M)$ the set of C^∞ differential forms on M of degree q. $D^q(M)$ is a vector space over the field of real numbers. As usual, $d: D^q(M) \to D^{q+1}(M)$ will denote the exterior differential. Since $d^2 = 0$,

$$D^*(M) = \{D^q(M), d\}$$

is a cochain complex, which will be referred to as the *De Rham complex* of M. If $f: M_1 \to M_2$ is a differentiable map (or class C^∞), then there is defined in a well-known way a homomorphism $f^*: D^q(M_2) \to D^q(M_1)$. The homomorphism f^* commutes with the exterior differential d, and hence it is a cochain map of $D^*(M_2)$ into $D^*(M_1)$.

Given any differentiable singular n-cube $T: I^n \to M$, and any differential form $\omega \in D^n(M)$, there is defined the integral of ω over T, denoted by

$$\int_T \omega$$

(cf. Spivak, [6], p. 100ff.). The basic idea of the definition is quite simple: $T^*(\omega)$ is a differential form of degree n on the cube I^n, hence it can be written

$$T^*(\omega) = f \, dx_1 \, dx_2 \cdots dx_n$$

in terms of the usual coordinate system $(x_1, x_2, \ldots, x_n)$ in I^n. Then $\int_T \omega$ is defined to be the n-fold integral of the C^∞ real-valued function f over the cube I^n. Actually, the preceding definition only makes sense if $n > 0$; in case $n = 0$, ω is a real-valued function, and $I^n = I^0$ is a point. In this case $\int_T \omega$ is defined to be the value of the function ω at the point $T(I^0) \in M$.

More generally, if

$$u = \sum a_i T_i$$

is a linear combination of differentiable singular n-cubes, then we define

$$\int_u \omega = \sum a_i \int_{T_i} \omega.$$

With this notation, we can write the generalized Stokes's theorem as follows: For any $u \in Q_n^S(M)$ and any $\omega \in D^{n-1}(M)$,

$$\int_u d\omega = \int_{\partial u} \omega.$$

For the proof, see Spivak [6], p. 102–104.

At this stage, we should mention three formal properties of the integral of a differential form over a singular chain. The proofs are more or less obvious.

(a) The integral $\int_u \omega$ is a bilinear function

$$Q_n^S(M) \times D^n(M) \to \mathbf{R}.$$

In other words, for each u it is a linear function of ω, and for each ω it is a linear function of u.

(b) Let $f : M_1 \to M_2$ be a differentiable map, $u \in Q_n^S(M_1)$, and $\omega \in D^n(M_2)$. Then

$$\int_u f^*(\omega) = \int_{f_\#(u)} \omega.$$

(c) If u is a *degenerate* singular n-chain, i.e., $u \in D_n^S(M)$, then

$$\int_u \omega = 0$$

for any differential form ω of degree n.

In view of Property (a), we can define a homomorphism

$$\varphi : D^n(M) \to \mathrm{Hom}(Q_n^S(M), \mathbf{R})$$

by the formula

$$\langle \varphi\omega, u \rangle = \int_u \omega$$

for any $\omega \in D^n(M)$ and any $u \in Q_n^S(M)$. The generalized Stokes's theorem now translates into the assertion that φ is a cochain map

$$D^*(M) \to \mathrm{Hom}(Q^S(M), \mathbf{R})$$

and Property (c) translates into the assertion that the image of φ is contained in the subcomplex $\mathrm{Hom}(C^S(M), \mathbf{R}) = C_S^*(M; \mathbf{R})$; thus we can (and will) look on φ as a cochain map

$$\varphi : D^*(M) \to C_S^*(M; \mathbf{R}).$$

Finally, Property (b) is equivalent to the assertion that the cochain map φ is natural *vis-a-vis* differentiable maps of manifolds.

Theorem 3.1 (De Rham's theorem). *For any paracompact differentiable manifold M, the cochain map φ induces a natural isomorphism $\varphi^* : H^n(D^*(M)) \approx H_S^n(M; \mathbf{R})$ of cohomology groups.*

If we combine this result with Corollary 2.2, we see that $H^n(D^*(M))$ is naturally isomorphic to $H^n(M; \mathbf{R})$ for any paracompact differentiable manifold M.

PROOF OF DE RHAM'S THEOREM. The proof proceeds according to the same basic pattern as Milnor's proof of the Poincaré duality theorem in Chapter IX.

Case 1: M is an open, convex subset of Euclidean n-space, $\mathbf{R}^n$. In this case, we know from the results of §2 that

$$H_n^S(M; \mathbf{R}) = \begin{cases} \mathbf{R} & \text{if } n = 0, \\ 0 & \text{if } n \neq 0. \end{cases}$$

Similarly,

$$H^n(D^*(M)) = \begin{cases} \mathbf{R} & \text{if } n = 0, \\ 0 & \text{if } n \neq 0. \end{cases}$$

This is essentially the content of the so-called Poincare lemma (see Spivak, [6], p. 94). Thus to prove the theorem in this case, we only have to worry about what happens in degree 0. This is made easier by the fact that in degree 0, every cohomology class contains exactly one cocycle. The details of the proof are simple, and may be left to the reader.

Case 2: M is the union of two open subsets, U and V, and De Rham's theorem is assumed to hold for U, V, and $U \cap V$. Then De Rham's theorem holds for M.

To prove the theorem in this case we use Mayer–Vietoris sequences. We already have a Mayer–Vietoris sequence for cohomology based on differentiable singular cubes; we will now derive such a sequence for the De Rham cohomology. Let $i: U \cap V \to U$, $j: U \cap V \to V$, $k: U \to M$, and $l: V \to M$ denote inclusion maps. Define cochain maps

$$\alpha: D^*(M) \to D^*(U) \oplus D^*(V),$$
$$\beta: D^*(U) \oplus D^*(V) \to D^*(U \cup V)$$

by

$$\alpha(\omega) = (k^*\omega, l^*\omega),$$
$$\beta(\omega_1, \omega_2) = i^*(\omega_1) - j^*(\omega_2).$$

We assert that the following sequence

$$0 \to D^*(M) \xrightarrow{\alpha} D^*(U) \oplus D^*(V) \xrightarrow{\beta} D^*(U \cap V) \to 0 \qquad (3.1)$$

is exact. The only part of this assertion which is not easy to prove is the fact that β is an epimorphism. This may be proved as follows. Let $\{g, h\}$ be a C^∞ partition of unity subordinate to the open covering $\{U, V\}$ of M. This means that g and h are C^∞ real-valued functions defined on M such that the following conditions hold: $g + h = 1, 0 \le g(x) \le 1$ and $0 \le h(x) \le 1$ for any $x \in M$, the closure of the set $\{x \in M \,|\, g(x) \ne 0\}$ is contained in U, and the closure of the set $\{x \in M \,|\, h(x) \ne 0\}$ is contained in V. The hypothesis that M is paracompact implies the existence of such a partition of unity. The proof is given in many textbooks, e.g., De Rham [2], p. 4, Sternberg, [8], Chapter II, §4, Auslander and MacKenzie, [1], §5–6. Now let ω be a differential form on $U \cap V$. Then $g\omega$ can be extended to C^∞ differential form ω_V on V by defining $\omega_V(x) = 0$ at any point $x \in V - U$. Similarly, $h\omega$ can be extended to a C^∞ differential form ω_U on U by defining $\omega_U(y) = 0$ at any point $y \in U - V$. Then it is easily verified that

$$\beta(\omega_U - \omega_V) = \omega$$

as desired.

On passage to cohomology, the short exact sequence (1) gives rise to a Mayer–Vietoris sequence for De Rham cohomology.

Similarly, the Mayer–Vietoris sequence for cohomology based on differentiable singular cubes is a consequence of the following short exact sequence of cochain complexes (cf. §III.5):

$$0 \to C_{\mathcal{S}}^*(M,\mathcal{U}) \xrightarrow{\alpha'} C_{\mathcal{S}}^*(U) \oplus C_{\mathcal{S}}^*(V) \xrightarrow{\beta'} C_{\mathcal{S}}^*(U \cap V) \to 0. \qquad (3.2)$$

Here $\mathcal{U} = \{U,V\}$ is an open covering of M, and the definition of the cochain maps α' and β' is similar to that of α and β above.

Finally, we may put these two short exact sequences together in a commutative diagram as follows:

The cochain map labelled a is induced by the inclusion of the subcomplex $C^S(M,\mathcal{U})$ in $C^S(M)$; it induces an isomorphism on cohomology. Clearly, each square of this diagram is commutative. On passage to cohomology we obtain the diagram we need to prove this case of De Rham's theorem.

Case 3: $M = \bigcup_{i=1}^{\infty} U_i$, where $U_1 \subset U_2 \subset \cdots \subset U_i \subset U_{i+1} \subset \cdots$ is a nested sequence of open sets, and for each i, $\overline{U}_i$ is compact. It is assumed that De Rham's theorem holds for each U_i; we will show that it holds for M. To carry out the proof in this case, we need to make use of inverse limits. The reader can find all the required material on inverse limits in the appendix, pp. 381–410 of Massey [5].

First of all, for each index i there is a cochain map $D^*(M) \to D^*(U_i)$ induced by inclusion of U_i in M. This is a compatible family of maps, and $D^*(M)$ is the inverse limit of the inverse system of cochain complexes $\{D^*(U_i)\}$ (this is practically obvious from the definitions of inverse limit and differential form). Moreover, for each q, the inverse sequence or tower $\{D^q(U_i)\}$ satisfies the Mittag-Leffler condition; this is an easy consequence of the assumption that each $\overline{U}_i$ is compact. It follows that the first derived functor

$$\lim^1 D^q(U_i) = 0$$

for all q. Hence we can apply Theorem A.19 on pp. 407–408 of Massey [5] to conclude that there exists a natural short exact sequence

$$0 \to \lim^1 H^{q-1}(D^*(U_i)) \to H^q(D^*(M)) \to \lim \text{inv } H^q(D^*(U_i)) \to 0. \quad (3.3)$$

Next, we will prove similar facts about the cochain complexes $C_{\mathcal{S}}^*(U_i;\mathbf{R})$ and $C^*(M;\mathbf{R})$. We know that the chain complex $C^S(M)$ is the direct limit of the chain complexes $C^S(U_i)$,

$$C^S(M) = \text{dir lim } C^S(U_i).$$

Applying the functor $\text{Hom}(\ ,\mathbf{R})$, we see that

$$
\begin{aligned}
C_{\mathcal{S}}^*(M;\mathbf{R}) &= \text{Hom}(C^S(M);\mathbf{R}) \\
&= \text{inv lim } \text{Hom}(C^S(U_i);\mathbf{R}) \\
&= \text{inv lim } C_{\mathcal{S}}^*(U_i;\mathbf{R});
\end{aligned}
$$

(compare Exercise 2 on p. 397 of Massey [5]). Moreover, for each index i, the homomorphism

$$
C_{\mathcal{S}}^*(U_{i+1};\mathbf{R}) \rightarrow C_{\mathcal{S}}^*(U_i;\mathbf{R})
$$

is obviously an epimorphism. Therefore the Mittag-Leffler condition holds for the inverse sequence of cochain complexes $\{C_{\mathcal{S}}^*(U_i;\mathbf{R})\}$. Applying Theorem A.19 of Massey [5] to this situation, we obtain the following natural short exact sequence:

$$
0 \rightarrow \lim{}^1 H_{\mathcal{S}}^{q-1}(U_i;\mathbf{R}) \rightarrow H_{\mathcal{S}}^q(M;\mathbf{R}) \rightarrow \lim \text{ inv } H_{\mathcal{S}}^q(U_i;\mathbf{R}) \rightarrow 0. \quad (3.4)
$$

We may now apply the cochain map φ to obtain a homomorphism from Sequence (3.3) into the Sequence (3.4). This homomorphism enables one to easily complete the proof in this case.

Case 4: M is an open subset of Euclidean space. Every such M is obviously the union of a countable family of convex open subsets $\{U_i\}$ having the property that each $\bar{U}_i$ is compact and $\bar{U}_i \subset M$. Then one proves that De Rham's theorem holds true for finite unions

$$
\bigcup_{i=1}^n U_i
$$

by an induction on n, using Case 2 and the basic properties of convex sets. Next one passes to the limit as $n \rightarrow \infty$, using Case 3.

Case 5: M is a connected paracompact manifold. It is known that any connected paracompact manifold has a countable basis of open sets (for a thorough discussion of the topology of paracompact manifolds, see the appendix to Volume I of Spivak [7]). It follows that M is the union of a countable family of open sets $\{U_i\}$ such that each U_i is a coordinate neighborhood (and hence diffeomorphic to an open subset of Euclidean space) and $\bar{U}_i$ is compact. Let $V_n = U_1 \cup U_2 \cup \cdots \cup U_n$. Using Cases 2 and 4, we can prove by induction on n that De Rham's theorem is true for each V_n. Note that $\bar{V}_n$ is compact, and $M = \bigcup_{n=1}^\infty V_n$. Hence it follows from Case 3 that De Rham's theorem holds for M.

Case 6: The general case. By Case 5, De Rham's theorem is true for each component of M. It follows easily that it is true for M.

This completes the proof of De Rham's theorem. We conclude by pointing out two directions in which De Rham's theorem can be extended:

(a) One of the basic operations on differential forms is the product: if ω and θ are differential forms of degree p and q respectively, then their product, $\omega \wedge \theta$, is a differential form of degree $p + q$. Moreover, the differential of such a product is given by the standard formula:

$$d(\omega \wedge \theta) = (d\omega) \wedge \theta + (-1)^p \omega \wedge (d\theta).$$

It follows that this product in the De Rham complex $D^*(M)$ gives rise to a product in $H^*(D^*(M))$, just as the cup product in the cochain complex $C_S^*(M;\mathbf{R})$ gives rise to cup products in $H_S^*(M,\mathbf{R})$. It can then be proved that the De Rham isomorphism,

$$\varphi^*: H^*(D^*(M)) \to H_S^*(M;\mathbf{R})$$

preserves products. However, the proof is of necessity rather roundabout, since the cochain map $\varphi: D^*(M) \to C_S^*(M;R)$ definitely is *not* a ring homomorphism. For a discussion and proof of these matters in a context somewhat similar to that of this appendix, see V. Gugenheim [4]. Gugenheim's paper makes heavy use of the technique of acyclic models.

(b) Given any differential form ω on M, we define the *support* of ω to be the closure of the set $\{x \in M \mid \omega(x) \neq 0\}$. With this definition, it is readily seen that the set of all differential forms of degree p which have compact support is a vector subspace of $D^p(M)$, which we will denote by $D_c^p(M)$. Moreover, if the support of ω is compact, then so is the support of $d(\omega)$. Hence $D_c^*(M) = \{D_c^p(M), d\}$ is a cochain subcomplex of $D^*(M)$.

Now consider the cochain map $\varphi: D^*(M) \to C_S^*(M;\mathbf{R})$. It is clear that if ω is a differential form with compact support, then $\varphi(\omega)$ is a cochain with compact support in accordance with the definition in §IX.3 (to be precise, that definition has to be modified slightly because we are using cochains which are defined only on *differentiable* singular cubes). It can now be proved that φ induces an isomorphism of $H^q(D_c^*(D_c^*(M))$ onto the q-dimensional cohomology group of M with compact supports and real coefficients. The details are too lengthy to include in this appendix. Such a theorem is usually proven in books on sheaf theory.

Bibliography for the Appendix

[1] L. Auslander and R. MacKenzie, *Introduction to Differentiable Manifolds*, McGraw-Hill, New York, 1963.

[2] G. De Rham, *Variétés Différentiables: Formes, Courants, Formes Harmoniques.* Hermann et Cie., Paris, 1955.

[3] H. Flanders, *Differential Forms with Applications to the Physical Sciences*, Academic Press, New York, 1963 (especially Chapters I, II, III, and V).

[4] V. Gugenheim, On the multiplicative structure of De Rham theory, *J. Differential Geometry*, **11** (1976), 309–314.

[5] W. S. Massey, *Homology and Cohomology Theory: An Approach Based on Alexander–Spanier Cochains*, Marcel Dekker, Inc., New York, 1978.

[6] M. Spivak, *Calculus on Manifolds*, W. A. Benjamin, Inc., Menlo Park, 1965 (especially Chapters 4 and 5).

[7] M. Spivak, *A Comprehensive Introduction to Differential Geometry*, Vol. I, Publish or Perish, Inc., Boston, 1970.

[8] S. Sternberg, *Lectures on Differential Geometry*, Prentice Hall, Englewood Cliffs, 1964.

[9] F. W. Warner, *Foundations of Differentiable Manifolds and Lie Groups*, Scott and Foresman, Glenview, 1971.

[10] H. Whitney, *Geometric Integration Theory*, Princeton University Press, Princeton, 1957 (especially Chapters II and III).

[11] M. L. Curtis and J. Dugundji, A proof of De Rham's theorem, *Fund. Math.*, **48** (1970), 265–268.

Index

Acyclic models, method of 138, 148
Alexander duality theorem 222−223
Algebraic homotopy 20
Algebraic mapping cone 109
Almost exact sequence 234
Almost simplicial complex 99−100
Anticommutative product 192
Augmentation 14, 138

Ball, n-dimensional 40, 77
Betti, E. 10
Betti group 215
Bockstein operator 119, 158
Borsuk−Ulam theorem 242
Boundaries, group of 13
Boundary of a cell 90
Boundary operator 12−13
  of a pair 24
Brouwer fixed point theorem 40
Brouwer, L. E. J. 38, 62, 67

Cap product 177, 181
  in a product space 191
Cayley projective plane 248
Čech−Alexander−Spanier
  cohomology 219
Cell, n-dimensional 77
Cellular map 83
  induced homomorphism 88
Chain complex 105
  acyclic 138

augmented 116, 138
  positive 138
Chain groups of a CW-complex 84
Chain homotopy 20, 107
Chain map 106
Characteristic map 77
Coboundaries 156
Coboundary operator 155
Cochain
  complex 155
  homotopy 155
  map 155
  with compact support 206, 261
Cocycles 156
Coefficient homomorphism 117−118, 158
Coherent orientations 101
Cohomology group 156
Cohomology sequence of a map 246
Compact pair 63
Complex, CW 79−83
  regular 94
Connecting homomorphism 108
Contractible space 21
Cross product 174, 179
Cup product 174, 180−181, 218
  in a product space 191
  in projective spaces 240−241
CW-complex 79−83
Cycles 13

Defect of singular homology theory 184
Deformation retract 21

Degenerate singular cube   12
Degree of a map   41−43
De Rham complex   256
DeRham's theorem   257
Diagonal map   148, 174
Differentiable singular chains   252−253
Differentiable singular cube   252
Dimension of a CW-complex   80
Direct limit   207
Disc, n-dimensional   40, 77
Divisible group   160

Edge of a graph   44
Eilenberg−Zilber theorem   134
Euler characteristic
    of a CW-complex   86−87
    of a graph   46
Exact homology sequence of a chain
        map   110
Exact sequence of chain complexes   108
Excision property   28
Excisive couple   151
Ext functor   160−162

Face of a cell   94
Faces of a singular cube   12
Five-lemma   37
Fundamental group   69

Graded module   192
Graded ring   192
Graph   44

Homology sequence of a pair   24
Homology group
    relative   21−22, 26−31
    singular   13
Homomorphism induced by a continuous
        map   16
Homotopy
    classes   19
    differentiable   253
    equivalence   21
    type   21
Hopf invariant   247−249

Incidence numbers   91, 96, 98−99
Injective group   160−161

Injective resolution   162
Inner product   159, 173
Invariance of domain theorem   67
Inverse limit   259

Jordan−Brouwer separation theorem   66
Jordan curve theorem   62

Künneth theorem   135

Lebesgue number   33
Lefschetz−Poincaré duality
        theorem   225, 227
Local homology group   43
Local orientation   200

Magic formula   137
Manifold
    n-dimensional   199
    non-orientable   201
    orientable   201
    with boundary   224
Map of pairs   26
    homotopic   26−27
Mapping cone   244
Mapping cylinder   244
Mayer−Vietoris sequence   58−62, 117,
        169−170
    relative   186−187
    with compact supports   209−210
Mysterious facts of life   183

Orientable manifold   201
Orientation of a cell   90, 98−99
Orientation of a manifold   200
Orientation of a product complex   131
Oriented edges   47

Poincaré duality theorem   199, 208, 213
Poincaré, H.   10
Poincaré series   153
Projective
    group   160
    plane, Cayley   248
    resolution   162
    spaces   81−83
        complex   82−83

cup products in   240—241
homology groups of   87—88
quaternionic   83
real   81—82
Proper map   207
Pseudomanifold   100
orientable   101

Rank of an abelian group   87
Reduced cohomology group   157
Reduced homology group   14
Regular CW-complex   94
Relative homology group   22—23
Retract   18
Riemann, G. F. B.   10

Simplex, singular   148
Simplicial complex (see Almost simplicial
complex)
Simplicial singular chains   148
Singular cube   11
Singular cycle   13
Singular homology group   13
Skeleton of a CW-complex   80
Skew commutative product   192
Slant product   176, 179—180
Small of order $\mathcal{U}$ 29
Sphere, n-dimensional   38, 77
Split exact sequence   114
Splitting homomorphism   109, 114
Subcomplex of a CW-complex   83

Subdivision operator   32, 254
Support
compact   206
of a chain   165—166
of a cochain   167—168
of a differential form   261
Suspension of a map   41, 250
Suspension of a space   249

Taut subspace   219
Tensor product
of chain complexes   132
of graded rings   196
Tor functor   121
Triad   185
exact sequence of   185—186
Triple   185
exact sequence of   185

Unit n-cube   11
Universal coefficient theorem
for cohomology   163
for homology   121—122

Vector field, tangent   42
Vertices of a cube   32
Vertices of a graph   44

Weak topology   77, 80

Graduate Texts in Mathematics

Soft and hard cover editions are available for each volume up to vol. 14, hard cover only from Vol. 15

1   TAKEUTI/ZARING. Introduction to Axiomatic Set Theory.

2   OXTOBY. Measure and Category.

3   SCHAEFFER. Topological Vector Spaces.

4   HILTON/STAMMBACH. A Course in Homological Algebra.

5   MACLANE. Categories for the Working Mathematician.

6   HUGHES/PIPER. Projective Planes.

7   SERRE. A course in Arithmetic.

8   TAKEUTI/ZARING. Axiomatic Set Theory.

9   HUMPHREYS. Introduction to Lie Algebras and Representation Theory. 2nd printing, revised.

10 COHEN. A Course in Simple Homotopy Theory.

11 CONWAY. Functions of One Complex Variable. 2nd ed.

12 BEALS. Advanced Mathematical Analysis.

13 ANDERSON/FULLER. Rings and Categories of Modules.

14 GOLUBITSKY/GUILLEMIN. Stable Mappings and Their Singularities.

15 BERBERIAN. Lectures in Functional Analysis and Operator Theory.

16 WINTER. The Structure of Fields.

17 ROSENBLATT. Random Processes. 2nd ed.

18 HALMOS. Measure Theory.

19 HALMOS. A Hilbert Space Problem Book.

20 HUSEMOLLER. Fibre Bundles. 2nd ed.

21 HUMPHREYS. Linear Algebraic Groups.

22 BARNES/MACK. An Algebraic Introduction to Mathematical Logic.

23 GREUB. Linear Algebra. 4th ed.

24 HOLMES. Geometric Functional Analysis and Its Applications.

25 HEWITT/STROMBERG. Real and Abstract Analysis. 4th printing.

26 MANES. Algebraic Theories.

27 KELLEY. General Topology.

28 ZARISKI/SAMUEL. Commutative Algebra I.

29 ZARISKI/SAMUEL. Commutative Algebra II.

30 JACOBSON. Lectures in Abstract Algebra I: Basic Concepts.

31 JACOBSON. Lectures in Abstract Algebra II: Linear Algebra.

32 JACOBSON. Lectures in Abstract Algebra III: Theory of Fields and Galois Theory.

33 HIRSCH. Differential Topology.

34 SPITZER. Principles of Random Walk. 2nd ed.

35 WERMER. Banach Algebras and Several Complex Variables. 2nd ed.

36 KELLEY/NAMIOKA. Linear Topological Spaces.

37 MONK. Mathematical Logic.

38 GRAUERT/FRITZSCHE. Several Complex Variables.

39 ARVESON. An Invitation to C^*-Algebras.

40 KEMENY/SNELL/KNAPP. Denumerable Markov Chains. 2nd ed.

41 APOSTOL. Modular Functions and Dirichlet Series in Number Theory.

42 SEERE. Linear Representations of Finite Groups.

43 GILLMAN/JERISON. Rings of Continuous Functions.

44 KENDIG. Elementary Algebraic Geometry.

45 LOEVE. Probability Theory. 4th ed. Vol. 1.

46 LOEVE. Probability Theory. 4th ed. Vol. 2.

47 MOISE. Geometric Topology in Dimensions 2 and 3.

48 SACHS/WU. General Relativity for Mathematicians.

49 GRUENBERG/WEIR. Linear Geometry. 2nd ed.

50 EDWARDS. Fermat's Last Theorem.

51 KLINGENBERG. A Course in Differential Geometry.

52 HARTSHORNE. Algebraic Geometry.

53 MANIN. A Course in Mathematical Logic.

54 GRAVER/WATKINS. Combinatorics with Emphasis on the Theory of Graphs.

55 BROWN/PEARCY. Introduction to Operator Theory. Vol. 1: Elements of Functional Analysis.

56 MASSEY. Algebraic Topology: An Introduction.

57 CROWELL/FOX. Introduction to Knot Theory.

58 KOBLITZ. p-adic Numbers, p-adic Analysis, and Zeta-Functions.

59 LANG. Cyclotomic Fields.

60 ARNOLD. Mathematical Methods in Classical Mechanics.

61 WHITEHEAD. Elements of Homotopy Theory.

62 KARGAPOLOV/MERZLJAKOV. Fundamentals of the Theory of Groups.

63 BOLLOBAS. Graph Theory—An Introductory Course.

64 EDWARDS. Fourier Series, Volume 1. second edition.

65 WELLS. Differential Analysis on Complex Manifolds.

66 WATERHOUSE. Introduction to Affine Group Schemes.

67 SERRE. Local Fields.

68 WEIDMANN. Linear Operators in Hilbert Spaces.

69 LANG. Cyclotomic Fields II.

70 MASSEY. Singular Homology Theory.

71 FARKAS/KRA. Riemann Surfaces.

Printed in the United States
By Bookmasters